Stochastic Methods in Scientific Computing

Stochastic Methods in Scientific Computing: From Foundations to Advanced Techniques introduces the reader to advanced concepts in stochastic modelling, rooted in an intuitive yet rigorous presentation of the underlying mathematical concepts. A particular emphasis is placed on illuminating the underpinning Mathematics, and yet have the practical applications in mind. The reader will find valuable insights into topics ranging from Social Sciences and Particle Physics to modern-day Computer Science with Machine Learning and AI in focus. The book also covers recent specialised techniques for notorious issues in the field of stochastic simulations, providing a valuable reference for advanced readers with an active interest in the field.

Features

- Self-contained, starting from the theoretical foundations and advancing to the most recent developments in the field
- Suitable as a reference for post-graduates and researchers or as supplementary reading for courses in numerical methods, scientific computing and beyond
- Interdisciplinary, laying a solid ground for field-specific applications in finance, physics and biosciences on common theoretical foundations
- Replete with practical examples of applications to classic and current research problems in various fields.

Numerical Analysis and Scientific Computing Series

Series Editors:
Frederic Magoules, Choi-Hong Lai

About the Series
This series, comprising of a diverse collection of textbooks, references, and handbooks, brings together a wide range of topics across numerical analysis and scientific computing. The books contained in this series will appeal to an academic audience, both in mathematics and computer science, and naturally find applications in engineering and the physical sciences.

Modelling with Ordinary Differential Equations
A Comprehensive Approach
Alfio Borzì

Numerical Methods for Unsteady Compressible Flow Problems
Philipp Birken

A Gentle Introduction to Scientific Computing
Dan Stanescu, Long Lee

Introduction to Computational Engineering with MATLAB
Timothy Bower

An Introduction to Numerical Methods
A MATLAB® Approach, Fifth Edition
Abdelwahab Kharab, Ronald Guenther

The Sequential Quadratic Hamiltonian Method
Solving Optimal Control Problems
Alfio Borzì

Advances in Theoretical and Computational Fluid Mechanics
Existence, Blow-up, and Discrete Exterior Calculus Algorithms
Terry Moschandreou, Keith Afas, Khoa Nguyen

Stochastic Methods in Scientific Computing
From Foundations to Advanced Techniques
Massimo D'Elia, Kurt Langfeld, Biagio Lucini

For more information about this series please visit: https://www.crcpress.com/Chapman--HallCRC-Numerical-Analysis-and-Scientific-Computing-Series/book-series/CHNUANSCCOM

Stochastic Methods in Scientific Computing

From Foundations to Advanced Techniques

Massimo D'Elia
University of Pisa, Italy

Kurt Langfeld
University of Leeds, United Kingdom

Biagio Lucini
Swansea University, United Kingdom

CRC Press
Taylor & Francis Group
Boca Raton London New York

CRC Press is an imprint of the
Taylor & Francis Group, an **informa** business
A CHAPMAN & HALL BOOK

Designed cover image: ShutterStock Images

First edition published 2024
by CRC Press
2385 NW Executive Center Drive, Suite 320, Boca Raton FL 33431

and by CRC Press
4 Park Square, Milton Park, Abingdon, Oxon, OX14 4RN

CRC Press is an imprint of Taylor & Francis Group, LLC

ISBN: 978-1-498-79633-0 (hbk)
ISBN: 978-1-032-77559-3 (pbk)
ISBN: 978-1-315-15615-6 (ebk)

DOI: 10.1201/ 9781315156156

Typeset in CMR10
by KnowledgeWorks Global Ltd.

Publisher's note: This book has been prepared from camera-ready copy provided by the authors.

*To our families, many friends
and many cores*

Contents

Preface

Alongside the advances in computer hardware and the steady increase of available computer resources, computer simulations for advanced scientific computing have enjoyed an upsurge in interest. Practitioners and scientists realised that many applications have stochastic elements for a number of reasons: the lack of data is compensated by a stochastic model replacing those data; a deterministic description of a system is neither possible nor desirable. Examples of the latter category are the description of gases with molecules as protagonists of the financial market, where it is impossible to be deterministic because the stock brokers are sometimes forced to make ad hoc micro-decisions with random outcomes.

The book takes the vantage point of a practitioner but also addresses fundamental questions such as: How do we mathematically grasp randomness and yet come to definite model predictions? We pay tribute to the "understanding" of the main concepts by providing justifications and a sketch of proof in some cases. We refrain, however, from devoting much book space to the rigour of proof, which the reader can pick up from the texts in Pure Mathematics if interested. Proof of, say, convergence of a stochastic approach does not necessarily guarantee that the method is feasible in practice. For example, there is proof of convergence for the Metropolis-Hasting algorithm, but there exists a class of models, namely, those with a so-called "overlap" problem (see Chapter 7), for which an acceptable answer cannot be produced with reasonable computational resources. These exemptions are usually a focus of active research, and advanced techniques that alleviate the problem make a chapter in his book.

How our approach to this book works is best illustrated in chapter 1: to satisfy the reader with an interest in fundamental questions, we introduce fundamental concepts such as the " probability space" and give thoughts to the axioms of the " probability measure". After the definition of "probability density", we focus on practical questions, e.g. how can we generate in a computer experiment a sequence of random variables for a *given* probability density? We then draw the reader's attention to the uncertainty in average values, which are at the heart of the usefulness of statistical analysis in everyday life. We offer an "unusual" proof of the Central Limit Theorem, which emphasises the importance of the conditions when the theorem applies. Our approach stipulates curiosity: what happens to the probability distribution if a pre-condition is *not* met? The answer then sends us on a journey in Chapter 2 to discover phenomena such as *anomalous diffusion*.

In Chapter 3, we introduce Monte Carlo simulations, which are the Swiss Army Knife of stochastic simulations. We emphasise the importance of the "error bar" of a numerical result from such simulations. Even an Artificial Intelligences or a Gaussian Process (see Chapter 8) produces an estimate for the value of a particular stock next week. Nobody would make a significant investment based upon the prediction if the *reliability* of the result is unknown. We introduce the important concept of autocorrelations, which can lead to a systematic underestimation of error bars and equip the reader with trusted techniques (Bootstrap and Jackknife) to achieve reliable error estimates.

Chapter 4 then uses the developed concepts and illustrates their power by applying them to statistical models, which are, in their own right, an active field of scientific computing. The chapter showcases *thermodynamics* and explains how laws of nature can be derived (such as the gas laws) without the need for a deterministic description of the system. The chapter also introduces the phenomenon of *phase transitions*. While this topic sounds deeply anchored in academic literature, concepts are transferrable and are. e.g. used in the description of the spread of infectious diseases [1] or even of the formation of traffic jams at a high density of cars [2].

"There is no free lunch" or in mathematical terms, the objectives for some classes of stochastic methods fall into the category of so-called NP hard problems: it is proven that an answer for all generic problems in the class cannot be produce in polynomial time. In practical terms, it usually means that a generic, satisfactory answer cannot be produced at all. It does not mean that for specific problems, a target and efficient simulation strategy *can* be found. This observation brings us to specific and advanced simulation methods in Chapter 5. Some of the simulation strategies are for targeted models and evading the complexity issue altogether (see 5.2, cluster algorithms). Others offer vast improvements for the generic type of models but ultimately would be prone to exponential simulation costs for large systems. Among the latter class are certainly the Hybrid Monte Carlo techniques (section 5.1).

Chapter 6 is the only "stared" chapter of the book, the star indicating that it has a particular focus that makes it specialised reading. Over half a century, quantum physics has benefitted from a correspondence between the Quantum Field Theory (QFT) and a Statistical Model. This correspondence has opened up QFTs for computer simulations that deliver the first principle insights into these complex theories with controllable error margins. The reader is invited to combine their understanding of phase transitions and advanced simulation techniques to get a first glimpse of the fascinating world of quantum physics with the computer experiment.

As already mentioned, Chapter 7 serves as an eye opener that some classes of statistical questions cannot be answered at a practical level, leads the reader to the forefront of research and triggers thoughts by example of how advanced simulation strategies can be designed.

The Data Science revolution over the last decade has delivered many practical solutions using algoritms called Machine Learning (ML) or Artificial Intelligence (AI). It has triggered a proliferation of tools for everyday life with tremendous impacts, such as speech recognition, voice, image and video generation and medical and other expert systems to name just a few. In this day in time, Universities see a surge of interest from professionals for mathematically oriented Data Sciences courses. They recognise that mathematics has always underpinned the AI and ML systems and that a mathematical understanding is needed to leave the application layer and research the next level of those systems. With the successes came a proliferation of articles and books in this area, and navigating this literature can be a challenge in itself. Chapter 8 draws a close connection between Data Analytics and statistical systems and delivers the reader an understanding of the basic mathematical building blocks that will help the reader explore and research these exciting systems.

Kurt Langfeld

Massimo D'Elia

Biagio Lucini

Author

Massimo D'Elia is Professor of Theoretical Physics at the Physics Department of the University of Pisa. Alumnus of the Scuola Normale Superiore in Pisa, he got his PhD in Physics from Pisa University in 1998. He has been Postdoctoral Fellow at the University of Cyprus and at ETH, Zurich, and Assistant Professor at the Physics Department of the University of Genoa. He is an expert in Lattice Gauge Theories and their implementation on HPC infrastructures, with interests also in Quantum Computing and numerical approaches to Quantum Gravity. He has obtained various achievements, especially in the study of the Phase Diagram of strong interactions. He has served as a reviewer for major scientific journals and funding agencies. He is author of two textbooks ("Lezioni di Meccanica Classica", Pisa University Press, 2020, 2022; "Introduction to the Basic Concepts of Modern Physics", with Carlo Maria Becchi, Springer, 2007, 2010, 2016) and of more than 200 papers.

Kurt Langfeld is Professor of Theoretical Physics and Head of School of Mathematics at the University of Leeds, England. His work in numerical methods for simulating Quantum Field Theories and Particle Physics is widely respected, with over 180 articles published in international journals.

In 1991, he was awarded a PhD in Theoretical Physics from Technical University of Munich. He went on to serve as Researcher and Lecturer at the University of Tübingen, Germany from 1991 to 2006; during this time he also enjoyed research visits to CEA, Saclay (Paris) and KIAS (Seoul). In 1999, he achieved the highly esteemed Venia Legendi award at the University of Tübingen. 2005 saw him become Professor for Theoretical Physics at the University of Tübingen, before moving on to Plymouth Univeristy as part of their Particle Physics group in 2006. April 2012 saw him assume full Professorship for Theoretical Physics at Plymouth and 2016 saw him become Professor and Head of Department of Mathematical Sciences at the University of Liverpool. In 2020, he took up his current role as Head of School of Mathematics at the University of Leeds.

He has dedicated himself to giving back to the community; he has been an active reviewer for the Engineering and Physical Sciences Research Council (EPSRC), Austrian council FWF, Swiss National Supercomputer Center (CSCS). He is a member of the Parliament Scientific Committee, devoted trustee of the University of Liverpool Maths School (ULMaS) and part of the

Steering Group opening an University sponsored Mathematics specialist 6th form in Leeds.

Biagio Lucini is a Fellow of the Learned Society of Wales and a Professor of Mathematics at Swansea University. He was awarded a Ph.D. in Theoretical Particle Physics by Scuola Normale (Pisa, Italy) in 2000. Before joining Swansea University in 2005, he was a Postdoctoral Fellow at the Rudolf Peierls Centre for Theoretical Physics (Oxford University, UK) from 2000 to 2003 and at the Theoretical Physics Institute of ETH (Zurich, Switzerland) from 2003 to 2005. Fellowship and awards he has received include a Royal Society Wolfson Merit Award (2017-2022) and a Leverhulme Research Fellowship (2020-2022). His research activity is centred around Monte Carlo calculations in Statistical Mechanics and Particle Physics. In particular, his interests are in Phase Transitions and Critical Phenomena, including machine learning approaches and efficient algorithms for simulations near criticality using High Performance Computing architectures. To date, his scientific contributions have resulted in over 180 research papers.

Chapter 1

Random Numbers

Randomness and probability distributions are basic concepts in statistics. In this chapter, we introduce those topics, laying the foundations for the rest of the book. We shall start presenting random numbers and probability distributions, providing some examples of the most common distributions, among which is the normal distribution (section 1.1). Then, in section 1.2, we shall discuss an important result in probability, the central limit theorem, which highlights the central role played by the normal distribution. Finally, we conclude the chapter analysing distributions that emerge in particular contexts and do not fall into the hypotheses of the central limit theorem 1.3.

1.1 Random numbers and probability distribution

Random numbers can be drawn from a finite set of whole numbers, and a relevant question is with which probability each of the numbers of the set show up. This is, e.g. the case for looking at he outcome of throwing a fair dice. The set in this case is $\{1, 2, 3, 4, 5, 6\}$. Other applications use continuous random numbers such as the real numbers. Examples are the outcome of a Physics experiment order Stock values. Even if the admissible range for the outcome

DOI: 10.1201/9781315156156-1

of the experiment is finite, we might still be dealing with infinitely many of them (which are perhaps not even countable). In this case, randomness can be quantified using *probability densities.*

1.1.1   Quantifying randomness

Probability theory is the axiomatic branch of Mathematics that underpins many areas of science such as Statistics, Finance, Game theory, Complex System, Artificial Intelligence and Clinical Trials. At the heart of this impact perhaps lies an interpretation of probability itself.

Probability is a measure for the likelihood that a random event occurs.

With the help of a number of theorems of probability theory such as the Theorem of Large Numbers, the above interpretation implies that probability serves as an estimate for abundance. This is widely used in applications for which we want to quantify risk such as financial, actuarial or health. Quantifying risk is the first step in controlling it. The same mathematical principles are put to use in vastly different fields: In statistical physics, a large number of particles confined to a box randomly hit the wall. The momentum transfer is observed as pressure, which has lost its random nature due to the large number of particles. Car crashes occur at random creating a fluctuating payout from insurance companies. Due to the large volume of insurance policies, the overall average financial burden is less fluctuating and can be reliably covered with insurance premiums.

A popular formalisation of probability was given by *Andrey Kolmogorov* [3]. Let us discuss some essentials.

Experiment E and state ω: Each time an "experiment" is performed, the world comes out in some state ω. The definition of the experiment includes the objects of interest.

Set of all states Ω: The set of all possible outcomes ω is denoted Ω and is called the universe of possible states. Note that it is obviously intricately tied to the experiment E.

Measurement $X(\omega)$: If we are interested in measuring some features of our states ω, we need to map each individual state ω to a one number X, in which we are interested. Hence, we can view X a function of ω and call it $X(\omega)$. In Mathematics, $X(\omega)$ is called *random variable*. After having performed the experiment E once, it is the result of the measurement of X on the state ω that $\mathcal{E}$ produced.

Set of outcomes A: If we input all possible states $\omega \in \Omega$ to the function $X(\omega)$ and collate the outcome in a set, we generate the set A of a possible outcomes. Mathematically, we can write:

$$X : \Omega \to A \,.$$

The set A can be finite, infinite and countable, or infinite and continuous.

Set of events F: An *event* is any subset of the set of all outcomes A. According to our interest, we can group these subsets together to form the `set of events` $\mathcal{F}$.

Probability P: Every *event* has a probability P of occurring. Mathematically, we define the *probability measure* function P as

$$P : \mathcal{F} \to [0,1] .$$

The probability measure function is constrained by three axioms:

(A1) The probability of an event E is a non-negative number: $P(E) \in \mathbb{R}, \; P(E) \geq 0, \qquad \forall E \in F$.

(A2) The probability of the outcome set A must be equal to 1: $P(A) = 1$. This simply means that the experiment E must produce an outcome. This makes sense since we would simply say that the experiment did not take place if no output was produced.

(A3) The probability of a countable union of mutually exclusive events, A and B, must be equal to the countable sum of the probabilities of each of these events: $P(A \cup B) = P(A) + P(B)$.

The triple $(A, \mathcal{F}, P)$ is called `probability space`. Let us study an example to illustrate the definitions. The experiment is "Throwing a dice". The set of all outcomes is given by

$$A \in \{1, 2, 3, 4, 5, 6\} .$$

In case we are interested whether or not the throw of the dice produces an even number, we would consider the events E_1, the number is even, $E_1 \subset A$, and the event E_2, the number is odd, $E_2 \subset A$:

$$E_1 = \{2, 4, 6\}, \qquad E_2 = \{1, 3, 5\}, \qquad F = \{E_1, E_2\} .$$

Assigning probabilities to events is called *modelling* . There is a lot to play for.

Model 1: Symmetries are a powerful way to inform this choice. For a so-called "fair" die, we expect that every side of the side shows up top with equal probability:

$$P(1) = P(2) = P(3) = P(4) = P(5) = P(6) =: p .$$

All the events of set of outcomes A are mutually exclusive: only one number shows up at the top of the dice. With the second axiom, we find:

$$P(A) = 1 \qquad \Rightarrow P(1 \cup 2 \cup 3 \cup 4 \cup 5 \cup 6) = 1 .$$

The second axiom then completely fixes the probabilities:

$$P(1 \cup 2 \cup 3 \cup 4 \cup 5 \cup 6) \;=\; \sum_{n=1}^{6} P(n) \;=\; 6p \;=\; 1 \,, \qquad p = 1/6.$$

We now can calculate the probability that we throw an even number:

$$P(E_1) \;=\; P(2 \cup 4 \cup 6) \;=\; P(2) + P(4) + P(6) \;=\; 3/6 \;=\; 1/2 \,.$$

Model 2: Let us study now a strongly skewed dice. Say the probability $P(1) = p$, the probabilities of the outcome n is modelled as $P(n) = n\,P(1)$. Again, all the events of set of outcomes A are mutually exclusive leaving us with

$$P(1 \cup 2 \cup 3 \cup 4 \cup 5 \cup 6) \;=\; p \sum_{n=1}^{6} n \;=\; 21\,p \;=\; 1 \,, \qquad p = 1/21.$$

We now can calculate the probability that we throw an even number:

$$P(E_1) \;=\; P(2 \cup 4 \cup 6) \;=\; P(2) + P(4) + P(6) \;=\; \frac{2}{21} + \frac{4}{21} + \frac{6}{21} \;=\; \frac{12}{21} \;=\; \frac{4}{7}.$$

Two important quantities that help to interpret the outcome of experiments are mean μ and standard deviation σ.

The mean and the standard deviation for a probability space (A, F, P) are defined by

$$\mu \;=\; \langle X \rangle \;=\; \sum_{X \in A} X\,P(X) \,, \qquad (1.1)$$

$$\sigma^2 \;=\; \langle (X - \mu)^2 \rangle \;=\; \sum_{X \in A} (X - \mu)^2\,P(X) \,.$$

The beauty of probability theory is that once we have defined a probability spaces, we can combine them and find new ones. Here, we discuss the important example of *repeating an experiment*. Let us assume that we have a probability space (A, F, P) with a finite outcome space and model for the probabilities P.

$$A \;=\; \{E_1, \dots, E_n\} \,, \qquad P(E_i) \text{ given.}$$

We assume that the events are mutually exclusive. We now repeat the experiment just once. The outcome of both experiments is considered as a new experiment. Its outcome space is given by

$$A_{\text{twice}} \;=\; \{E_i, E_k \,|\, i, k = 1 \dots n\} \,.$$

We now have some modelling to do: what should the associated probabilities $P(E_i, E_k)$ be? It is entirely our choice. An important one is that we say that both experiments are *independent*. The outcome from either of them does not influence the other. How should we model this?

The notation $P(E_i, E_k)$ implies that E_i was the outcome of the first and E_k of the second experiment. If the experiments are *independent*, then the order of the experiments does not play role leading us to

$$P(E_i, E_k) = P(E_k, E_i) .$$

Assume that the first experiment produces E_1 and consider the subset $\{E_2, E_3, E_4\}$ as outcome for the second experiment. We have a subset of mutually exclusive events

$$\{E_1 E_2, E_1 E_3, E_1 E_4\}$$

and the second axiom implies

$$P(E_1 E_2 \cup E_1 E_3 \cup E_1 E_4) = P(E_1 E_2) + P(E_1 E_3) + P(E_1 E_4) .$$

The second experiment is independent from the first one. It does not matter whether the outcome was E_1 or something else. Hence, we expect that

$$P(E_1 E_2 \cup E_1 E_3 \cup E_1 E_4) \propto P(E_2 \cup E_3 \cup E_4) = P(E_2) + P(E_3) + P(E_4) .$$

We conclude

$$P(E_1 E_2) + P(E_1 E_3) + P(E_1 E_4) = c \left[P(E_2) + P(E_3) + P(E_4) \right]$$

with c a normalisation constant. Rather than considering just three outcomes for the experiment, we now consider the full outcome space A for experiment two (while experiment one still has delivered E_1):

$$P(E_1 E_1 \cup E_1 E_2 \cup \ldots \cup E_1 E_n) = c \sum_{n=1}^{n} P(E_n) = c .$$

On the other hand, considering any outcome, the experiment two and insisting that experiment has given us E_1 has the probability $P(E_1)$. This fixes the constant $c = P(1)$, and we find

$$P(E_1 E_2) + P(E_1 E_3) + P(E_1 E_4) = P(1) \left[P(E_2) + P(E_3) + P(E_4) \right] .$$

We can repeat this analysis for any numbers of events for experiment two, which leads us to the important model:

Consider an experiment with probability space (A, F, P). If the experiment is carried out a second time and *independently* from the first experiment, the new outcome space and probability model is given by

$$A_2 = A \times A ,$$
$$P(E_1, E_2) = P(E_1) \, P(E_2) , \qquad E_1, E_2 \in A .$$

1.1.2 Pseudo randomness

At the heart of stochastic simulations is the ability to generate random numbers. For now, we consider the task to choose a random number $u \in [0, 1]$ in such a way that any number with the interval is generated with equal probability. Those numbers are called *uniformly distributed*. To be more precise, we firstly agree on a given number of decimal places. Each "real" number is then represented by a potentially large whole number (integer). The task is then to generate each of these integers with equal probability.

Moreover, stochastic simulation usually uses a very large amount of these random numbers. The second requirement is that these random numbers are *independent*.

Using computer hardware such as the clock cycle to generate random numbers seems to be an obvious method. Although it seems random which number you are going to get, a string of numerical operations between extracting numbers could induce a regular pattern and, depending on the calculation, introduce a dependence on the random numbers. This violates the second design principle above. It is worthwhile to point out, however, that the development of so-called True Random Number Generators (TRNG) is still an area of much debate. Hardware might serve as a source of entropy to, e.g. inform a seed of an deterministic algorithm (see the National Institute of Standards and Technology (NIST) SP 800-90 series, update 2022).

It was recognised in the early days of computer science that good control over the randomness of those numbers is only achievable if they are generated by a mathematical algorithm. This seems paradoxical at first sight since an algorithm always creates the same string of numbers, and all randomness seems lost.

On a second thought, stochastic simulations *only* need uniformly distributed and independent random numbers. True randomness of the numbers does not play a role. Usually, the algorithm uses a *seed* number and then generates a unique set of numbers satisfying (more or less) both design principles. These numbers are *deterministically* generated but random and uniformly distributed. They are called *pseudo-random* numbers. A stochastic simulation using pseudo-random numbers also have the advantage that the simulations are reproducible (as long the pseudo-random number generator and seed are known).

All of a sudden, we face the mathematical problem to algorithmically generate large integer numbers such that they occur with equal probability and are independent from each other. To give an idea how this can be achieved, we briefly address *linear congruential generators* (LCG) but stress that those are only of historical interest and cannot be used in modern stochastic simulations. The generator is defined by a recurrence relation:

$$X_{n+1} = \left(a\, X_n + c \right) \bmod m \,,$$

where $m > 0$ is the modulus, $0 < a < m$ is the multiplier, $0 \leq c < m$ is the increment and $0 \leq X_0 < m$ is the seed.

A potential weakness of pseudo random number generators is the following: assume that after $n = N < m$ iterations, the recurrence produces a number X_N that did occur before, say $X_N = X_k$, $k < N$. The recurrence relation above then implies that also $X_{N+1} = X_{k+1}$ and so on. This means the string of numbers has a period length of N. If the stochastic simulations needs more than N random numbers, this specific algorithm cannot be used since it breaks the requirement for independence. The mathematics of LCGs is simple enough that the period can be calculated and made long for a good choice of parameters. However, the quality of LCGs was found to be very sensitive to the choice of a and c. For example, the choice $a = 1$ and $m = 1$, with $X_0 = 0$ yields

$$X_n = n \bmod m .$$

These numbers have the maximal period length of m, but the sequence has a distinctive non-random pattern.

In 1997, Matsumoto and Nishimura proposed the so-called *Mersenne twister* algorithm [4] that overcame many of the problems of earlier generators. It has a period of

$$2^{10937} - 1$$

and is equidistributed in 623 dimensions by design.

A widely used and freely available[1] pseudo-random number generator was introduced by Martin Lüscher in 1994 using the phenomenon of mixing in classical dynamical systems [5]. The majority of stochastic simulations of the Particle Physics community since the nineties have been based upon Lüscher's random number generator. All results from stochastic calculations in this book are also using those numbers.

Finally, we stress that the generation of pseudo-random numbers is still an active research area world wide. The reason is modern cryptographic applications. Pseudo-random number generators applied in this fields are required to pass *all* statistical tests that are restricted to polynomial time in the size of the seed. Those generators are then called *cryptographically secure.*

1.1.3 Designing probability distributions

Throughout the book, we will mainly address *continuous random variables*, say $X \in \mathbb{R}$. The probability of finding X in the interval $[a, b]$ can be obtained by means of the *probability density P*:

$$\Pr[a \leq X \leq b] = \int_a^b P(x) \, dx . \tag{1.2}$$

---

[1] https://luscher.web.cern.ch/luscher/ranlux/

We will assume that $P(x)$ is continuous. The so-called *cumulative distribution function* $F(a)$ is defined by the probability that x is smaller or equal than a:

$$F(a) = \Pr[X \le a] = \int_{-\infty}^{a} P(x)\, dx\,. \qquad (1.3)$$

Choosing $b = a + \Delta a$ in (1.2) (with Δa being infinitesimally small), we find:

$$\Pr[a \le X \le a + \Delta a] = \int_{a}^{a+\Delta a} P(x)\, dx \doteq P(a)\, \Delta a$$

leading to

$$P(a) = \lim_{\Delta a \to 0} \frac{\Pr[a \le x \le a + \Delta a]}{\Delta a}\,. \qquad (1.4)$$

The later equation delivers an important interpretation: $P(a)\, \Delta a$ is the probability of finding the random variable x in a small interval of size Δa around a. This probability is proportional to the size of the interval meaning that $P(a)$ is of order one. Indeed, we obtain using (1.3):

$$P(a) = \lim_{\Delta a \to 0} \frac{F(a + \Delta a) - F(a)}{\Delta a} = \frac{dF}{da}(a)\,. \qquad (1.5)$$

If we are studying more than one random variable, say X_1 and X_2, the probability of finding them in intervals of size dx_1 and dx_2 around x_1 and x_2 is given by

$$P(x_1, x_2)\, dx_1\, dx_2\,.$$

A widespread application in Science and Engineering is the following: assume that we measure an observable with an exact value $\bar{x}$. Due the imperfectness of the measuring process, an actual measurement yields a random value X, which is distributed with the probability distribution P. Ideally, $P(x)$ peaks around the true experimental value, but the only connection to the experiment is via the expectation value:

$$\bar{x} = \int_{-\infty}^{\infty} x\, P(x)\, dx. \qquad (1.6)$$

Assume that we carry out to *independent measurements* yielding two random variables X_1 and X_2: due to the independent nature of the approach, the joint probability is then given by

$$P(x_1, x_2) = P(x_1)\, P(x_2)\,.$$

If more than one random numbers are involved, we are frequently not interested in full set information provided by the joint distribution $P(x_1, x_2)$, but in the distribution of a new variable Y that is constructed from the variables X_n. In our measurement example, we could be interested in the average of our two measurements:

$$Y = \frac{X_1 + X_2}{2}\,.$$

In this case, we need sum all probabilities

$$P(x_1, x_2)\, dx_1\, dx_2 \qquad \text{with the constraint} \qquad y = (x_1 + x_2)/2 \, . \qquad (1.7)$$

The *Dirac* δ-function is a convenient tool for these tasks. For a continuous function $f(x)$ that admits a Fourier transform, it is defined by

$$\int_{-\infty}^{\infty} f(x)\, \delta(x - a)\, dx \;=\; f(a) \, . \qquad (1.8)$$

There are many representations for $\delta(x)$. One intuitive is the following:

$$\delta_\epsilon(x) \;=\; \frac{1}{\epsilon\sqrt{2\pi}} \exp\left\{ -\frac{x^2}{2\epsilon^2} \right\} , \qquad \epsilon \to 0 \, . \qquad (1.9)$$

We observe that the area under this bell-shaped curve does not depend on ϵ:

$$\int_{-\infty}^{\infty} \delta_\epsilon(x)\, dx \;=\; 1 \, .$$

On the other hand, the support of this function vanishes rapidly for $|x| > \epsilon$. We find with the substitution $x = u\,\epsilon + a$

$$\int_{-\infty}^{\infty} f(x)\, \frac{1}{\epsilon\sqrt{2\pi}} \exp\left\{ -\frac{(x-a)^2}{2\epsilon^2} \right\} dx = \int_{-\infty}^{\infty} f(u\epsilon + a)\, \frac{1}{\sqrt{2\pi}} \exp\left\{ -\frac{u^2}{2} \right\} du.$$

Since $f(x)$ admits a Taylor expansion around a, we find in leading order for small ϵ:

$$\int_{-\infty}^{\infty} f(x)\, \delta_\epsilon(x - a) \doteq f(a) \int_{-\infty}^{\infty} \frac{1}{\sqrt{2\pi}} \exp\left\{ -\frac{u^2}{2} \right\} du \;=\; f(a).$$

Another representation for the Dirac δ-function, which is of great help for actual calculations, arises from the Fourier transform:

$$\delta(x) \;=\; \int \frac{dp}{2\pi}\, \mathrm{e}^{-ipx} \, . \qquad (1.10)$$

We find

$$\int_{-\infty}^{\infty} dx\, f(x)\, \delta(x - a) \;=\; \int_{-\infty}^{\infty} \frac{dp}{2\pi}\, dx\, f(x)\, \mathrm{e}^{-ip(x-a)} \, .$$

Performing first the x-integration, we obtain

$$\int_{-\infty}^{\infty} \frac{dp}{2\pi}\, \bar{f}(p)\, \mathrm{e}^{ipa} \, , \qquad \bar{f}(p) \;=\; \int_{-\infty}^{\infty} dx\, f(x)\, \mathrm{e}^{-ipx} \, ,$$

where we have introduced the Fourier transform $\bar{f}(p)$ of f. The integral at the left-hand side of the above equation is just the inverse Fourier transform for argument a. Hence, we indeed find

$$\int_{-\infty}^{\infty} dx\, f(x)\, \delta(x - a) \;=\; \int_{-\infty}^{\infty} \frac{dp}{2\pi}\, \bar{f}(p)\, \mathrm{e}^{ipa} \;=\; f(a) \, .$$

Let us now come back to the task (1.7) of finding the average distribution. We implement the constraint with the help of the Dirac δ-function (the sum over all probabilities becomes a double integral);

$$Q_2(y) = \int_{-\infty}^{\infty} dx_1 \, dx_2 \, \delta \left(y - \frac{x_1 + x_2}{2} \right) P(x_1, x_2) .$$

If we consider the representation (1.9), we note that the integrand vanishes if $|y - (x_1 + x_2)/2| > \epsilon$, and ϵ approaches zero. This indeed tells us that we only admit y values that are constrained to the average $(x_1 + x_2)/2$. Let us check the normalisation of $Q_2(y)$ as a consistency check:

$$\int_{-\infty}^{\infty} dy \, Q_2(y) = \int_{-\infty}^{\infty} dx_1 \, dx_2 \, dy \, \delta \left(y - \frac{x_1 + x_2}{2} \right) P(x_1, x_2) .$$

We firstly carry out the y-integration:

$$\int_{-\infty}^{\infty} dy \, \delta \left(y - \frac{x_1 + x_2}{2} \right) = 1 ,$$

and indeed find:

$$\int_{-\infty}^{\infty} dy \, Q_2(y) = \int_{-\infty}^{\infty} dx_1 \, dx_2 \, P(x_1, x_2) = 1 .$$

In the remainder of this sub-chapter, we would like to address the question: how do we generate one or more a random variables X in a computer experiment that are distributed according to a given probability distribution $P(x)$? Here, we study the case of $x \in [a, b]$ but results can be generalised to real axis as well. There are sophisticated algorithms that generate a uniformly distributed random number $U \in [0, 1]$. Our goal will be to find a mapping f such that $X = f(U)$. To this aim, we equate the probabilities

$$du \, 1 = c \, P(x) \, dx$$

where $c > 0$ is constant, which we determine later. Since $P(x)$ is positive, we find a monotonic mapping between u and x by:

$$\int_0^u du' = c \int_a^x P(x') \, dx'.$$

If $u = 1$, we would like to obtain $x = b$. This provides the normalisation c:

$$\int_0^1 du' = c \int_a^b P(x') \, dx' \qquad \Rightarrow \qquad c^{-1} = \int_a^b P(x') \, dx',$$

or $c = 1$ if P has been properly normalised. We, hence, find the mapping:

$$f : x \to u : \qquad \int_0^u du' = \int_a^x P(x') \, dx', \qquad du = P(x) \, dx . \qquad (1.11)$$

Finally, we illustrate that we could use either random numbers, X or U, to estimates moments:

$$M_n := \int dx \, x^n \, P(x) \, .$$

Using (1.11) as substitution, we indeed find:

$$M_n = \int_a^b dx \, x^n \, P(x) \; = \; \int_0^1 x^n(u) \, 1 \, du \, .$$

In this chapter, we denoted random numbers by capital letters (such as X) and arguments of probability distributions with small case letters (e.g. x). In the following, we will also denote random numbers with small case letters.

1.1.4 Applications: Poisson and exponential distributions

The Poisson process is a stochastic model that has widespread applications ranging from radioactive decay to insurance Mathematics. Assume that events occur

- randomly

- independently from each other

- not at the same time

- at a given "rate".

The rate μ is defined as the average number of events per unit time.

For example, a call centre receives on average 100 calls a day during opening hours. What is the probability to receive 3 calls in just 15 minutes?

Let us divide the time interval between 0 and t into N equal portions. Later, we will take the limit $N \to \infty$. Because of the third assumption above, an event occurs in one of the tiny interaval with probability p or not. The probability is proportional to the width t/N of the interval: $p = \mu T/N$, which we take as the defintion of the rate μ. What is the probability p_k that exactly k of these events fall in ot the time intervall $[0, t]$?

We need to distribute k events into N slots. The probability for a particular configuration of having k events and $N - k$ "blanks" is $p^k (1 - p)^{N-k}$. Hence we find:

$$p_k \; = \; \lim_{N \to \infty} \left(\begin{array}{c} N \\ k \end{array} \right) p^k \, (1 - p)^{N-k} \, .$$

We now take the limit $n \to \infty$ where μ, t and k are fixed and finite. Substituting p, we note

$$\left(\begin{array}{c} N \\ k \end{array} \right) p^k \, (1 - p)^{N-k} \; = \; \frac{N!}{(N - k)! N^k} \frac{(\mu t)^k}{k!} \left(1 - \frac{\mu t}{N} \right)^{N-k} \, .$$

We now prepared to take the limt. Using $k \ll N$ (k is finite while n grows over all boundaries), we obtain

$$\lim_{N \to \infty} \frac{N!}{(N-k)!N^k} = \lim_{N \to \infty} \frac{N}{N} \frac{N-1}{N} \cdots \frac{N-k+1}{N} = 1$$

and

$$\lim_{N \to \infty} \left(1 - \frac{\mu t}{N}\right)^{N-k} = \lim_{N \to \infty} \left(1 - \frac{\mu t}{N}\right)^{N} = \exp\{-\mu t\} \,.$$

Our final result is thus given by

$$p_k = \frac{(\mu t)^k}{k!} \exp\{-\mu t\} \,, \qquad (\text{ Poisson distribution}) \qquad (1.12)$$

The probability to find *any* number of events in the time interval $[0, t]$ needs to be 1. Let's check this:

$$\sum_{k=0}^{\infty} p_k = e^{-\mu t} \sum_{k=0}^{\infty} \frac{(\mu t)^k}{k!} = e^{-\mu t} e^{\mu t} = 1,$$

where we have used the Taylor expansion of the exponential function. Finally, we verify that μ is indeed the *average number* A of events per unit time $(t = 1)$:

$$A = \sum_{k=0}^{\infty} p_k \, k = e^{-\mu} \sum_{k=0}^{\infty} \frac{(\mu)^k}{k!} k = e^{-\mu} \sum_{k=0}^{\infty} \frac{1}{k!} \mu \frac{d}{d\mu} (\mu)^k$$

$$= e^{-\mu} \, \mu \frac{d}{d\mu} \sum_{k=0}^{\infty} \frac{(\mu)^k}{k!} = e^{-\mu} \, \mu \frac{d}{d\mu} e^{\mu} = \mu \,.$$

For the fun of it, let us answer the above question of the call centre. We assume a working day shift has $10h$. A 100 calls per working day, gives us a rate of $\mu = 100/10h$. We are now looking for the probability of $k = 3$ calls in $t = 0.25\, h$:

$$\mu t = 2.5 \,, \qquad p_3 = \frac{2.5^{10}}{10!} e^{-2.5} \approx 21.4\% \,.$$

Closely related to the Poisson distribution and of equal importance is the *Exponential distribution*. Let us go back to the Poisson process above and ask a slightly different question. Given that the events independently occur at rate μ, what is the probability P that in the duration t at least *one* event occurs? Clearly, we now ask for a probability as a function of t rather than a discrete set of probabilities p_k.

We can answer this question by firstly looking at the probability that "no event occurs". This is Poisson distribution with $k = 0$. Hence, we find:

$$P(t) = = 1 - p_0 = 1 - \exp\{-\mu t\} \ .$$

This is the answer to the above question. We now asking a slightly diffrent questions: what is the *probability density* $p(t)$ for an event to occur at time t? In details, we are looking for the probability $p(t) \, dt$ that an event falls into the infinitesimal intervall $[t, t + dt]$. This can be readily obtained from the result above:

$$p(t) = \frac{d}{dt} P(t) = \mu \exp\{-\mu t\} \ , \qquad \text{(Exponential distribution) . (1.13)}$$

Let us do the cross check: the probability that one event occurs at *any time* needs to be 1, and, indeed, it is:

$$\int_0^\infty p(t) \, dt = \mu \int_0^\infty \exp\{-\mu t\} \, dt = 1 \ .$$

What is the *average time* after which an event occurs? We might have the hunch that this is related to the rate μ. We calculate:

$$\langle t \rangle = \int_0^\infty t \, p(t) \, dt = \mu \int_0^\infty t \, \exp\{-\mu t\} \, dt$$

$$= \mu \left(-\frac{d}{d\mu} \right) \int_0^\infty \exp\{-\mu t\} \, dt = \mu \left(-\frac{d}{d\mu} \right) \frac{1}{\mu} = \frac{1}{\mu} \ .$$

In the remainder of this section, we ask how we can generate, for a computer experiment, random variables that follow an *Exponential distribution*.

$$p(x) = \exp\{-\mu x\} \ , \qquad x \geq 0 \ .$$

We have uniformly distributed variables w at our finger tips, and we are seeking a mapping from w to x. We want to have $x = 0$ for $w = 0$ and $x \to \infty$ for $w \to 1$. We therefore demand (c is a constant):

$$dw = dx \, c \, e^{-\mu x} \ , \qquad \int_0^w dw' = c \int_0^x dx' \, \exp\{-\mu x'\}$$

$$\int_0^1 dw' = c \int_0^\infty dx' \, \exp\{-\mu x'\}$$

leading us to $c = \mu$ and the definition:

$$w = 1 - e^{-\mu x} \quad \Rightarrow \quad x = -\frac{1}{\mu} \ln(1 - w) \ .$$

The algorithm to generate Poisson distributed variables is therefore as follows:

Exponential distribution:

1. Choose a uniformly distributed random number $w \in [0,1]$.

2. Calculate $x = -\frac{1}{\mu}\ln(1-w)$.

We are now putting the methods of the previous subsection to the test. We assume that we can generate uniformly distrubuted random numbers $w \in [0,1]$. For all parctical applications in this book, we use the random number generator from Martin Lüscher [6]. For an approximation of the probability distribution of a sequence of random numbers x_i, $i = 1 \ldots n$, we are using (1.4) with a finite, but small value Δa. We need to estimate the probability

$$\Pr[a \leq x \leq a + \Delta a] \, .$$

Since the range of x-values is unbounded, we firstly introduce a cut c and only conider values of the sequnce with $|x_i| \leq c$. We then choose the number N of intervals and devide the x-axis in segments of equal spacing:

$$a_k \; = \; -c + 2c\,\frac{k-1}{N} \, , \qquad k = 1 \ldots N \, .$$

The interval size is given by

$$\Delta a \; = \; a_{k+1} - a_k \; = \; \frac{2c}{N} \, .$$

Let $\mathcal{N}_k$ by the number of random values x that fall into the interval $[a_k, a_k + \Delta a[$. The total number of random variables within the range of the cut can be the obtained by

$$\mathcal{N} \; = \; \sum_{k=1}^{N} \mathcal{N}_k \, .$$

We can than estimate the probability by

$$\Pr[a_k \leq x \leq a_k + \Delta a] \; \approx \; \frac{\mathcal{N}_k}{\mathcal{N}} \, , \tag{1.14}$$

and thus obtain an estimate of the probability distribution:

$$P(a_k) \; \approx \; \frac{\Pr[a_k \leq x \leq a_k + \Delta a]}{\Delta a} \; \approx \; \frac{1}{\Delta a}\,\frac{\mathcal{N}_k}{\mathcal{N}} \, .$$

A better approximation is achieved by a smaller intervall size Δa. Note, however, that in practice the number N of intervals (and hence the size of Δa) is limited by the requirement that we need a good number of random numbers in each interval in order to achieve a reliable estimate of the probability (1.14). We have tested the above approach for the Exponential distribution ($\mu = 1$). We have chosen a cut $c = 3$ (note that we only have positive random variables in the present case). The number of intervals has been $N = 50$, and we made sure that $10{,}000$ random variables were generated in the relevant region $|x| \leq c$. The results are shown in Figure 1.1.

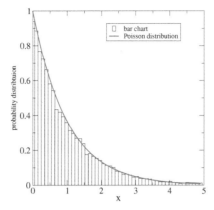

FIGURE 1.1: Numerical estimates of the probability distribution of an Exponential distribution.

1.2 Central limit theorem

The Central Limit Theorem is ubiquitous in the Sciences, Actuarial Science and Finance and Engineering. Assume that we have a random variable X with an almost[2] arbitrary probability distribution. Let us assume that we repeat the experiment (or measurement) a number of times and consider the marginal distribution of the *average* of these measurements. With very little assumptions, this distribution is known, which lets us control errors and risk.

1.2.1 Conditions and theorem

We start with the classical version of the *Central limit theorem* due to Lindeberg and Levy: Assume that the sequence of numbers x_1, x_2, ... x_n is randomly and independently chosen with probability distribution $P(x)$. We assume that this distribution possesses

- a finite expectation value $\bar{x}$ and

- a finite variance σ^2.

We are interested in the probability distribution sample average

$$y = \frac{1}{n}\left[x_1 + \ldots + x_n\right].$$ (1.15)

The theorem states that, as n approaches infinity, the distribution of the random variables $u = \sqrt{n}\,(y - \bar{x})$ converge to a Normal (or also called Gaussian)

---

[2]We will be precise what this means later in the book.

distribution:

$$\frac{1}{\sqrt{2\pi}\sigma} \exp\left\{-\frac{u^2}{2\sigma^2}\right\} .$$

What is the practical relevance of this theorem?

The impact of this theorem is enormous. We will elucidate this in the scientific context of measuring an observable in experiments. Assume that we would like to determine a physical observable $\bar{x}$. In particle physics, this could be a mass of particle produced in Heavy Ion Collisions carried out with installations such as CERN (Switzerland), GSI and DESY (Germany) or the BNL (United States of America).

A perfect experimental device would just produce the answer $\bar{x}$ with a single measurement. In practice, such a device does not exist. A realistic device produces a value x in the interval $[x, x+dx]$ with probability $P(x)\,dx$, where the probability distribution $P(x)$ characterises the apparatus. We will not assume that our experimental device is hampered by systematic errors, but we will assume that the device produces the exact value $\bar{x}$ by an average over many measurements, i.e.

$$\int dx\, x\, P(x) = \bar{x} , \tag{1.16}$$

but, depending on $P(x)$, a single measurement can be far from the true value.

As an example, we consider an observable $\bar{x} = 3$ and a crude experiment which can produce any value for between 0 and 6 with equal probability:

$$P(x) = \begin{cases} 1/6 & \text{for } x \in [0,6] \\ 0 & \text{otherwise.} \end{cases} \tag{1.17}$$

Obviously, a single measurement for x is not sufficient to reveal the true observable. The only thing we can do is to repeat the measurement n times and to consider the average (1.15), i.e.

$$y = \frac{1}{n}\left[x_1 + \ldots + x_n\right] ,$$

where x_1 to x_n are now the values obtained from each of the measurements. We assume that the measurements are *independent*, i.e. that the probability for finding a set $\{x_1 \ldots x_n\}$ of data is given by (as seen in the previous section) by the product of conditional probabilities:

$$P(x_1)\,dx_1 \ \ldots \ P(x_n)\,dx_n .$$

The crucial question is to which accuracy does y estimate the true observable $\bar{x}$?

To answer this question, we need to find the probability distribution $Q_n(y)$ for the value y:

$$Q_n(y) = \int \prod_{i=1}^{n} dx_i \, \delta\left(y - \frac{1}{n}\,[x_1 + \ldots + x_n]\right) P(x_1) \ldots P(x_n) .$$

(1.18)

Given the proper normalisation of the single event distributions, i.e.

$$\int dx_i \, P(x_i) = 1 ,$$

using (1.16), we can easily show that the average of y coincides with the observable:

$$
\begin{aligned}
\bar{y} &= \int dy \, y \, Q_n(y) = \int dy \int \prod_{i=1}^{n} dx_i \\
&\quad \frac{1}{n}\,[x_1 + \ldots + x_n]\,\delta\left(y - \frac{1}{n}\,[x_1 + \ldots + x_n]\right) P(x_1) \ldots P(x_n) . \\
&= \int \prod_{i=1}^{n} dx_i \, \frac{1}{n}\,[x_1 + \ldots + x_n]\, P(x_1) \ldots P(x_n) \\
&= \frac{1}{n}\, n \int dx \, x \, P(x) = \bar{x} .
\end{aligned}
$$

A natural measure for the error σ of our estimate is provided by the second moment:

$$\sigma^2(n) = \int dy \, \left(y - \bar{y}\right)^2 Q_n(y) , \qquad \text{(the variance)} . \qquad (1.19)$$

If the distribution $Q_n(y)$ peaks around the true value for our observable $\bar{x}$ and $\sigma(n)$ is tiny, it would mean that a single estimator y has high probability to fall close to $\bar{x}$ with high probability implying that it yields a good approximation to $\bar{x}$.

Let us study the moments of the distribution $Q(y)$:

$$q_m = \int dy \, Q_n(y) \, y^m . \qquad (1.20)$$

In order to draw further conclusions, we need to restrict the classes of single event probability distributions $P(x)$: we will assume that its Fourier transform

$$\bar{P}(\beta) = \int dx \, P(x) \, \exp\{-i\,\beta\,x\} \qquad (1.21)$$

is an analytic function of β at $\beta = 0$. As a consequence, the moments of $P(x)$ exist and are given by:

$$p_m = \int dx \, P(x) \, x^m = i^m \, \frac{d^m}{d\beta^m} \bar{P}(\beta) \,|_{\beta=0} . \qquad (1.22)$$

We will further assume that $\bar{P}(\beta)$ vanishes for $|\beta| \to \pm\infty$. This seems to be quite a weak constraint. We point out, however, that systems with rare but large fluctuations generically fail to possess higher moments. We will illustrate this below in the mundane context of stock market indices.

Our aim is to express the moments of $Q_n(y)$ in terms of the moments of $P(x)$. For this purpose, we rewrite the δ-function in (1.18) as

$$\delta\left(y - \frac{1}{n}[x_1 + \ldots + x_n]\right) = \int \frac{d\alpha}{2\pi} \exp[i\,\alpha\,y] \prod_{i=1}^{n} \exp\left\{-i\frac{\alpha}{n}x_i\right\}, \tag{1.23}$$

and find

$$Q_n(y) = \int \frac{d\alpha}{2\pi} \exp(i\,\alpha\,y) \left[\int dx\, P(x) \exp\left\{-i\frac{\alpha}{n}x\right\}\right]^n, \tag{1.24}$$

$$= \int \frac{d\alpha}{2\pi} \exp(i\,\alpha\,y) \left[\bar{P}\left(\frac{\alpha}{n}\right)\right]^n. \tag{1.25}$$

The moments of $Q_n(y)$ are then obtained from

$$q_m = \int dy \int \frac{d\alpha}{2\pi} (-i)^m \frac{d^m}{d\alpha^m}\left[\exp(i\,\alpha\,y)\right] \bar{P}^n\left(\frac{\alpha}{n}\right). \tag{1.26}$$

After a series of partial integrations with respect to α (note that boundary terms vanish by virtue of our assumptions), the latter equation is given by

$$q_m = \int dy \int \frac{d\alpha}{2\pi} \exp(i\,\alpha\,y) (i)^m \frac{d^m}{d\alpha^m}\left[\bar{P}^n\left(\frac{\alpha}{n}\right)\right] \tag{1.27}$$

$$= \int \frac{d\alpha}{2\pi} \frac{1}{n^m} \int dy\, \exp(i\,\alpha\,y) (i)^m \frac{d^m}{d\beta^m}\left[\bar{P}^n(\beta)\right]$$

$$= \frac{i^m}{n^m} \frac{d^m}{d\beta^m}\left[\bar{P}^n(\beta)\right]\Big|_{\beta=0}.$$

Of particular interest are the so-called cumulants $c_k[Q_n]$ of the distribution $Q_n(y)$. These are defined via the generating function

$$T_Q(x) = \sum_{m=0}^{\infty} \frac{1}{m!} q_m\, x^m, \qquad c_k[Q_n] := \frac{d^k}{dx^k} \ln T_Q(x)\Big|_{x=0}. \tag{1.28}$$

Note that in particular we find for the 'error' σ in (1.19)

$$\sigma^2 = q_2 - q_1^2 = c_2[Q_n]. \tag{1.29}$$

Using Taylor's theorem and the explicit expression (1.27), we find

$$T_Q(x) = \bar{P}^n\left(\frac{i\,x}{n}\right), \qquad c_k[Q_n] = \frac{i^k}{n^{k-1}}\left[\ln \bar{P}(0)\right]^{(k)}, \tag{1.30}$$

where (k) denotes the k-th derivative. Introducing the cumulants $c_k[P]$ of the single event distribution as well, i.e.

$$c_k[P] = i^k \left[\ln \bar{P}(0)\right]^{(k)}, \tag{1.31}$$

we arrive at a very important result:

$$c_k[Q_n] = \frac{1}{n^{k-1}} c_k[P]. \tag{1.32}$$

Note that the cumulants $c_k[P]$ are finite numbers which characterise the single event probability distribution. Equation (1.32) then implies that for increasing number of measurements n, the higher $(k > 1)$ cumulants of $Q_n(y)$ vanish. In particular, we find that

$$\sigma(n) = \sqrt{c_2[Q_n]} = \frac{\sqrt{c_2[P]}}{\sqrt{n}} \propto 1/\sqrt{n}. \tag{1.33}$$

For the above example (1.17), we find

$$p_1 = \frac{1}{6} \int_0^6 dx\, x = 3, \qquad p_2 = \frac{1}{6} \int_0^6 dx\, x^2 = 12, \quad (1.34)$$
$$c_2[P] = 12 - 3^2 = 3,$$

and therefore

$$\sigma(n) = \sqrt{3/n}.$$

It is well known that if $c_k[G] = 0$ for $k > 2$, the probability distribution G is a Gaussian. We therefore expect that if n is chosen sufficiently large so that we can neglect $c_k[Q_n]$ with $k > 2$, we should be able to approximate $Q_n(y)$ by a Gaussian. To illustrate this point, we start from (1.25):

$$Q_n(y) = \int \frac{d\alpha}{2\pi} \exp(i\,\alpha\,y) \exp\left\{ n \ln\left[\bar{P}\left(\frac{\alpha}{n}\right)\right] \right\},$$

and expand the logarithm with respect to α:

$$Q_n(y) = \int \frac{d\alpha}{2\pi} \exp(i\,\alpha\,y) \exp\left\{ n \sum_{k=0}^{\infty} \frac{1}{k!} \left[\ln \bar{P}(0)\right]^{(k)} \left(\frac{\alpha}{n}\right)^k \right\}$$

$$= \int \frac{d\alpha}{2\pi} \exp(i\,\alpha\,y) \exp\left\{ n \sum_{k=1}^{\infty} \frac{1}{k!} c_k[P] \left(-i\frac{\alpha}{n}\right)^k \right\},$$

where we have used $\bar{P}(0) = \int dx\, P(x) = 1$ and the definition of the cumulants of P in (1.31). Using $c_1[P] = p_1 = \bar{x} = \bar{y}$, we find:

$$Q_n(y) = \int \frac{d\alpha}{2\pi} \exp[i\,\alpha\,(y - \bar{y})] \exp\left\{ -\frac{1}{2} c_2[P] \left(\frac{\alpha^2}{n}\right) + \mathcal{O}(\alpha^3/n^2) \right\} \tag{1.35}$$

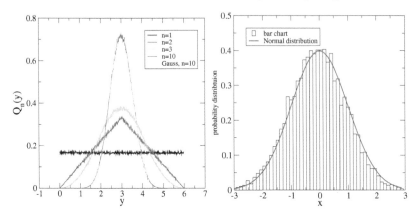

FIGURE 1.2: Illustration of the central limit theorem: probability distributions of the average y after n measurements (left). Numerical estimates of the probability distribution of a Normal distribution (right).

Note that the dominant contributions from the α integration arises from the regime where $\alpha < \sqrt{n}$. In this regime, the correction term is of order

$$\mathcal{O}(\alpha^3/n^2) \approx \mathcal{O}(1/\sqrt{n})$$

and will be neglected for sufficiently large n. The remaining integral can be easily performed:

$$Q_n(y) \approx \frac{1}{\sqrt{2\pi}\,\sigma} \exp\left\{-\frac{(y - \bar{y})^2}{2\sigma^2}\right\}, \qquad \sigma^2 = \frac{c_2[P]}{n}. \qquad (1.36)$$

which is the celebrated Gaussian or Normal distribution. We have recovered the *central limit theorem:* if the moments of probability distribution $P(x)$ exist, the probability distribution for the average y can be approximated by a Gaussian for sufficiently large n given that the standard deviation σ is properly scaled with n.

1.2.2   The normal distribution

Let us discuss the rise of the Normal distribution using the above example (1.17) for illustrative purposes. This aim, we use computer experiments (as those detailed in the previous subsection): for a given choice of n, we generate x_1 to x_n random numbers according to the desired distribution (1.17). From each sequence, we obtain its average y (1.15). We repeat this many times and produce a histogram to estimate the distribution of the y-values for the given n. Figure 1.2 shows the final result $Q_n(y)$ for $n = 1, 2, 3$ and $n = 10$. Note that $n = 1$ is original flat distribution in (1.17). We find that for already $n = 10$, the distribution not only looks as a typical bell-shaped Gaussian, but is also well fitted by a Normal distribution. This is quite remarkable: the original distribution is nothing like bell-shaped, but already $n = 10$ measurements are sufficient to quite accurately describe the distribution by a Normal distribution.

Let us study the n dependence of the distribution $Q_n(y)$ of the average y if the single event distribution $P(x)$ is normal. We find:

$$P(x, \sigma) = \frac{1}{\sigma\sqrt{2\pi}} \exp\left\{-\frac{x^2}{2\sigma^2}\right\} \Rightarrow$$

$$\bar{P}(\beta) = \int dx\, P(x)\, e^{-i\beta x} = \exp\left\{-\frac{\sigma^2}{2}\beta^2\right\}.$$

According to (1.25), the distribution for the average is then given by:

$$Q_n(y) = \int \frac{d\alpha}{2\pi} \exp(i\,\alpha\,y) \left[\bar{P}\left(\frac{\alpha}{n}\right)\right]^n$$

$$= \frac{d\alpha}{2\pi} \exp(i\,\alpha\,y) \exp\left\{-\frac{\sigma^2}{2n}\alpha^2\right\} = P\left(y, \frac{\sigma}{\sqrt{n}}\right). \quad (1.37)$$

We observe that the distribution for the average $Q_n(y)$ is *shape invariant*, i.e. the distributions for different n can be mapped onto each other by a re-scaling of the only parameter σ.

How do we create a sequence of random numbers x_n that are normal distributed numerically? We will here present an efficient algorithm that generates pairs of these numbers. Our starting point are uniformly distributed random numbers $u \in]0, 1[$. For reasons that become clear soon, we are seeking two random numbers x and y which are drawn from the joint distribution

$$dx\, dy\, \frac{1}{2\pi\sigma_1\sigma_2} \exp\left\{-\frac{(x - \mu_1)^2}{2\sigma_1^2}\right\} \exp\left\{-\frac{(y - \mu_2)^2}{2\sigma_2^2}\right\}.$$

After the scaling,

$$x = \mu_1 + \sigma_1\tilde{x}, \qquad y = \mu_2 + \sigma_2\tilde{y},$$

the new variable are distributed according to

$$\frac{d\tilde{x}\, d\tilde{y}}{2\pi} \exp\left\{-\frac{\tilde{x}^2}{2}\right\} \exp\left\{-\frac{\tilde{y}^2}{2}\right\} = \frac{d\tilde{x}\, d\tilde{y}}{2\pi} \exp\left\{-\frac{\tilde{x}^2 + \tilde{y}^2}{2}\right\}.$$

Using the substitution to polar type coordinates,

$$\tilde{x} = \sqrt{u}\,\cos\phi, \qquad \tilde{y} = \sqrt{u}\,\sin\phi, \qquad \phi \in [0, \pi[, \quad u \geq 0,$$

we realise that ϕ is uniformly distributed and u exponentially:

$$\frac{d\phi\, du}{4\pi} \exp\{-u/2\}.$$

We have already explained above how to generate a Poisson distribution (with $\mu = 1/2$ here). All these substitutions lead to the so-called Box-Müller algorithm [7]:

Normal distribution:

1. Choose two uniformly distributed random numbers $v, w \in [0, 1[$.

2. Calculate: $\phi = 2\pi v$ and $u = -2 \ln(1 - w)$.

3. calculate:

$$x = \mu_1 + \sigma_1 \sqrt{u} \cos \phi, \qquad y = \mu_2 + \sigma_2 \sqrt{u} \sin \phi.$$

The random variables x and y are normal distributed with mean μ_1 and μ_2 and standard deviations σ_1 and σ_2.

We have tested the above approach for the Normal distribution ($\sigma = 1, \mu = 0$). For a cut of $c = 3$, the number of intervals has been $N = 50$, and we made sure that $10,000$ random variables were generated in the relevant region $|x| \leq c$. The results are shown in Figure 1.2, where we compare the histogram of the estimated probability distribution with the exact analytic function.

1.2.3   Independent measurements and error propagation

Let us return to the example in (1.17), and let us assume a group of experimentalists has carried out $n = 12$ independent measurements with the result:

$$
\begin{array}{cccc}
2.813 & 2.021 & 0.331 & 0.865 \\
5.394 & 5.937 & 5.027 & 1.556 \\
0.325 & 2.750 & 1.283 & 3.890
\end{array}
$$

The average y over these values and an estimate for the variance $c_2[P] = \langle x^2 \rangle - y^2$ are given by

$$y = \frac{1}{n} \sum_{k=1}^{n} x_k \approx 2.683\,, \qquad c_2[P] \approx \frac{1}{n} \sum_{k=1}^{n} x_k^2 - y^2 \approx 3.581\,. \quad (1.38)$$

We point out that 3.581 is a poor estimate of the true value (1.34) of $c_2[P] = 3$, but it reflects the correct order of magnitude. With this estimate for $c_2[P]$, we find for the error (1.33)

$$\sigma(n = 12) \approx \sqrt{\frac{3.581}{12}} \approx 0.546\,.$$

Hence, the final 'experimental' result for the observable would be

$$\bar{x} \approx 2.683 \pm 0.546 = 2.7(5) \,. \tag{1.39}$$

Note that the true result $\bar{x} = 3$ lies well with the reach of the error bars.

The above experiment was repeated by several research labs. Depending on the budget and the focus of research, different labs carried out measurements with different numbers n:

	CERN	GSI	DESY	BNL
n	120	50	78	150
y	3.112 ± 0.163	2.764 ± 0.255	3.110 ± 0.207	3.083 ± 0.143

The smallest error was produced by the largest experiment (BNL). We could just quote their result, but it would be a pity to disregard a total of 248 measurements which were carried out by the other groups. How can we obtain a 'world average' for the observable $\bar{x}$ and how can we quantify its (statistical) error?

To answer these questions, we assume that the number n of each measurement was large enough to approximate the distribution of an individual result y_k, $k = 1 \ldots N$ (where $N = 4$ for the above example) by a Gaussian (1.36):

$$Q(y_l) \approx \frac{1}{\sqrt{2\pi}\,\sigma_l} \exp\left\{ -\frac{(y_l - \bar{x})^2}{2\sigma_l^2} \right\} \,. \tag{1.40}$$

For the world average y we make the ansatz

$$y = \sum_{l=1}^{N} a_l \, y_l \,, \qquad \sum_{l=1}^{N} a_l = 1 \tag{1.41}$$

and choose the weights a_l in an optimal way. This choice will depend on the errors σ_l of the individual experiments. In particular, the experiment with the smallest error should contribute the most to the world average. Assuming that the experiments at the different labs were carried out independently, the probability distribution of the world average is now given by

$$W(y) = \int \prod_{k=1}^{N} dy_k \; \delta\left(y - \sum_{l=1}^{N} a_l \, y_l \right) Q(y_1) \ldots Q(y_N) \,. \tag{1.42}$$

Representing the δ-function in terms of a Fourier integral over α (see (1.23)), the integrations over $y_1 \ldots y_N$ can be easily performed:

$$W(y) = \int \frac{d\alpha}{2\pi} \; \exp\{i(y - \bar{x})\} \exp\left\{ -\frac{\alpha^2}{2} \sum_l a_l^2 \sigma_l^2 \right\} \,.$$

Performing the α integration finally fields:

$$Q(y) \approx \frac{1}{\sqrt{2\pi}\,\sigma} \exp\left\{-\frac{(y-\bar{x})^2}{2\sigma^2}\right\}, \qquad \sigma^2 = \sum_l a_l^2 \sigma_l^2. \qquad (1.43)$$

The optimal result is achieved if the variance, i.e. σ^2, is as small as possible. Here, we must take into account the normalisation condition in (1.41). Adding the constraint with a Lagrange multiplier, we therefore minimise

$$\sum_l \left[a_l^2 \sigma_l^2 - \lambda a_l\right] \longrightarrow \min. \qquad (1.44)$$

The global minimum is easily obtained:

$$a_l = \frac{\lambda}{2\sigma_l^2}, \qquad \frac{2}{\lambda} = \sum_l \frac{1}{\sigma_l^2}. \qquad (1.45)$$

The minimal value for σ^2 then satisfies

$$\sigma^2 = \frac{\lambda}{2} \qquad \Rightarrow \qquad \frac{1}{\sigma^2} = \sum_l \frac{1}{\sigma_l^2}. \qquad (1.46)$$

The optimal choice for the weights can therefore also be written as

$$a_l = \frac{\sigma^2}{\sigma_l^2}. \qquad (1.47)$$

Let us return to the above example. We find

$$\sigma \approx 0.089, \qquad\qquad\qquad\qquad\qquad\qquad\qquad\qquad (1.48)$$
$$a_1 \approx 0.30, \qquad a_2 \approx 0.12, \qquad a_3 \approx 0.19, \qquad a_3 \approx 0.39.$$

With the weights at our disposal, we easily find the optimal value for the world average $y \approx 3.059$. Together with the error in (1.48), the final result is

$$\bar{x} = 3.059 \pm 0.089 = 3.06(9). \qquad (1.49)$$

Note that the true result $\bar{x} = 3$ is again covered within error bars and that the error became significantly smaller than that of the best result provided by the BNL group.

1.3   Beyond the normal distribution

The normal distribution has the property that the probability density decreases exponentially fast for the extrems $|x| \to \infty$. Certain applications, most notably in Financial and Biological Mathematics, show a different behavior with a power-law decrease for large $|x|$.

1.3.1 Cauchy-Lorentz distribution and the failure of error reduction

The Central Limit Theorem (CLT) is key for the error reduction by means of averaging the result from repeated measurements. For a further illustrations of this important fact, we are going to study an example for which error reduction fails. Let us assume that the probability distribution for obtaining a measurement x is given by the Cauchy-Lorentz distribution:

$$L(x) = \frac{1}{\pi b} \frac{1}{1 + (x/b)^2} . \tag{1.50}$$

Indeed, the second moment does not exist, i.e.

$$\int_{-\infty}^{\infty} x^2 L(x) \, dx \rightarrow \infty ,$$

so that one of the assumptions for the CLT is violated. The above distribution is bell-shaped and similar to that of the Normal distribution

$$N(x) = \frac{1}{\sigma\sqrt{2\pi}} \exp\left\{ -\frac{x^2}{2\sigma^2} \right\} . \tag{1.51}$$

Both distributions are normalised to one. In order to compare both, we want to have the same width for the bell shape. For a parameter matching, we define the width w by

$$P(w) = \frac{1}{2} P(0) , \qquad P \in \{L, N\} . \tag{1.52}$$

Thus, choosing

$$b = \sqrt{2 \ln(2)} \, \sigma$$

yields the same width for the Cauchy-Lorentz and Normal distribution. Both distributions are compared in Figure 1.3, left panel. The main difference is that $L(x)$ possess so-called "fat tails", i.e. $L(x)$ decreases according to a power-law for large x, as compared with the Normal distribution, which decreases exponentially fast. This difference has dramatic consequences for the error reduction.

Let us assume that a distribution for finding a particular value x for a single measurement is of Cauchy-Lorentz type (1.50). We repeat the measurement n-times, obtain values x_n and construct a new variable y, the average:

$$y_n = \frac{1}{n} \sum_{n=1}^{N} x_n . \tag{1.53}$$

What is the probability distribution of y?

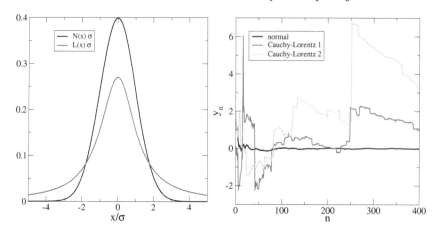

FIGURE 1.3: Left:Cauchy-Lorentz and Normal distributions with the same width. Right: Time series for the average y_n (1.53) for the normal and two Cauchy-Lorentz distributions.

We have answered this question in subsection 1.2.1 for the general context. Firstly, we need to find the Fourier transform:

$$\bar{L}(\beta) = \int dx\, L(x)\, e^{-i\beta x} = \exp\left\{ -b\,|\beta| \right\}. \tag{1.54}$$

The distribution of the average y is then given, according to (1.25), by

$$Q_n(y) = \int \frac{d\alpha}{2\pi}\, \exp(i\,\alpha\,y)\, \left[\exp\left(-b\,\frac{\alpha}{n}\right)\right]^n = L(y).$$

Apparently, the probability distribution for the average y does not change at all even if we repeat the measurements many times. This has far reaching consequences: it impossible to experimentally gain a reliable value for the observable $\bar{x}$ by repeating the experiment many times: the fluctuations of the measured value around the true answer are the same for a single measurement and for the sequence average. This is illustrated in Figure 1.3, right panel, for one particular time series with an underlying normal distribution and two generated from a Cauchy-Lorentz distribution. If the events x_n our normal distributed, the average y_n can be viewed as one event drawn from a sample with distribution $Q_n(y_n)$. Since for a normal distribution, the width shrinks to zero with increasing n, we find:

$$y_n \propto 1/\sqrt{n} \to 0 \qquad \text{(normal distribution)}.$$

On the other hand, if the "measurements" x_n are generated with a Cauchy-Lorentz distribution, the distribution Q_n does not shrink to zero width. In this case, we can obtain any value even in the limit $n \to \infty$:

$$\lim_{n\to\infty} y_n = \text{constant} \qquad \text{(Cauchy-Lorentz distribution)}.$$

This explains our observation of the time series result in Figure 1.3, right panel.

1.3.2 Pareto distribution and applications

Power law distributions are widely used to describe social, engineering, geophysical, financial and many other types phenomena. For instance, the Italian sociologist *Vilfredo Pareto* used such distributions to describe the allocation of wealth among individuals. At the beginning of the last century, Pareto showed that approximately 80% of the land in Italy was owned by 20% of the population. This so-called 80-20 rule found applications in other areas of sociology and economics. A more modern example is that, in 1989, the 20% richest of the population earned 83% of the world GDP. A computing example is the report by Paula Rooney [8] that 80% of the errors in Windows and Office are caused by 20% of the entire pool of bugs detected. The rule that "roughly 80% of the effects come from 20% of the causes" became known as the *Pareto principle.*

There might be different reasons for an 80-20 rule to emerge, but self-similarity might be a wide spread one. Let us analyse the income distribution on the basis of an 80-20 rule in greater detail. The individuals of the population been arranged along the x-axis between 0 and 1 and ordered according to their income from low to high income. $P(x)\,dx$ then quantifies the accumulated income of all individuals from the interval $[x, x+dx]$. Rather than saying that 20% of the richest earn 80% of the wealth, we recast the data to state that the fraction p earns as much as the other, high earner part, i.e. $1 - p$, of the population:

$$\int_0^p P(x)\,dx = \int_p^{1-p} P(x)\,dx . \tag{1.55}$$

The last equation just equates two numbers, i.e. the values of two integrals, and we can find many functions $P(x)$ that would satisfy (1.55). Clearly, more input is needed to specify $P(x)$. We argue that a certain skill set is needed to be part of the high income group, but there is chance as well: you also need to be "at the right time at the right place", and not *all* with the right skill set make it into the high earner class. Due to this random element, the distribution of skills of the low-earner's population $0 \le x \le p$ matches that of all of the population. Within the class of low-earners history repeats itself: the fraction p (of members of the class, i.e. $1 - p$) earns as much as the rest. This leads to:

$$\int_0^{p(1-p)} P(x)\,dx = \int_{p(1-p)}^{1-p} P(x)\,dx. \tag{1.56}$$

We found a new group of size $p(1 - p)$ of low earners. Our central assumption is again that randomness ensures that the skill distribution in this group is the same as in the whole population. This is the *self-similarity* argument. If we repeat this n-times, we obtain:

$$\int_0^{p^n(1-p)} P(x)\, dx = \int_{p^n(1-p)}^{p^{n-1}(1-p)} P(x)\, dx. \tag{1.57}$$

Note that $P(x)$ is one and only distribution and that we have now n equations at our finger tips to constrain $P(x)$. With the substitution

$$x = p^{n-1}(1-p)\, u\,,$$

we obtain:

$$\int_0^p P\Big(p^{n-1}(1-p)\, u\Big)\, du = \int_p^1 P\Big(p^{n-1}(1-p)\, u\Big)\, du. \tag{1.58}$$

This equation has the same structure as (1.55) but with a different argument for P. These n equations are not fully sufficient to determine the continuous function $P(x)$, but they make the solution plausible: choosing

$$P(x) = c\, x^\gamma\,, \tag{1.59}$$

all the n-equations (1.58) collapse into a single equation:

$$\int_0^p u^\gamma\, du = \int_p^1 u^\gamma\, du\,. \tag{1.60}$$

This leads to an equation for γ:

$$\gamma = \frac{\ln(1/2)}{\ln p} - 1\,. \tag{1.61}$$

If we say that $p = 90\%$ of the poorer part of the population earns as much as the 10% most rich, we find $\gamma \approx 5.579$:

$$P(x) = c\, x^{5.579}\,,$$

where c is highest income of the population (i.e. for $x = 1$). This distribution is illustrated in Figure 1.4). Self-similarity is a powerful tool: with the assumption of self-similarity on one data input only, i.e. that the fraction p of the populations earn as much as the rest, we were able to "derive" the income distribution (1.59). We can now answer questions like: "How much is earned by the 10% poorest?" A couple of caveats are in order: by repeatedly sorting the population in class of low and high-earners, the number of individuals in one class rapidly decreases, and the self-similarity assumption, i.e. that the same skill distribution is inherent to each class, breaks down. Secondly, we have not proven that $P(x)$ (1.59) is indeed the solution of our problem at hand. We could generate more than n-equations be refining more classes and not only the class of the lowest earners, but whether this is good enough to derive the continuous function $P(x)$ is beyond the scope of this book. Finally, there is multitude of mechanism in Economics and Finace that could lead to a powerlaw distribution. If a powerlaw in income data is observed emperically, it therefore cannot be concluded that the above mechanism of self-similarity is the cause. Excelent further readings are [9, 10].

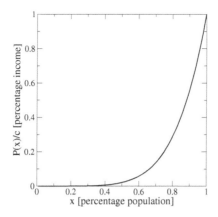

FIGURE 1.4: Income distribution (1.59) under the assumptions of self-similarity and that 90% of the poorer population earns as much as the 10% most rich.

1.4 Exercises

Exercise 1.1 *Consider two non-empty sets A and B. Use the Kolmogorov axioms on page 3 to show monotonicity:*

$$\text{If} \quad B \subseteq A \quad \text{then} \quad P(B) \leq P(A)\,.$$

Hint: Use that the sets A and A/B are disjoint and consider $P(B \cup A/B) = P(A)$.

Exercise 1.2 *Consider the probability space $(A, \mathcal{F}, P)$ and prove the complement rule*

$$P(\mathcal{F}^c) = P(A/\mathcal{F}) = 1 - P(\mathcal{F})\,.$$

Hint: Use $P(\mathcal{F} \cup A/\mathcal{F}) = P(A) = 1$.

Exercise 1.3 *Use the complement rule to show that $P(\mathcal{F}) \leq 1$.*

Exercise 1.4 *If u denote random numbers that are uniformly distributed in the interval* $]-1, 1[$, *use (1.11) and show that the random variable*

$$x = b \tan\left(\frac{\pi}{2} u\right)$$

has a Lorentz distribution

$$P(x) = \frac{1}{\pi b} \frac{1}{1 + (x/b)^2}$$

with $x \in]-\infty, \infty[$.

Exercise 1.5 *Consider* $x_i \in (-\infty, \infty)$, $i = 1\ldots n$ *random variables that follow a Gaussian distribution with mean* $\mu = 0$ *and standrad deviation* $\sigma = 1$. *Generate* $n = 1,000,000$ *random variables and create a histogram with a 100 bins covering the range from minimal to maximal value of the set* $\{x_i\}$. *Make sure that the histogram is properly normalised. If* N_i *are the number of entries in bin i and* Δx *is the uniform bin width, we have:*

$$\sum_i N_i \Delta x = 1.$$

Compare your histogram with the corresponding Normal Distribution.

Exercise 1.6 *If u denote random numbers that are uniformly distributed in the interval* $]-1, 1[$, *use (1.11) and show that the random variable*

$$x = b \tan\left(\frac{\pi}{2} u\right)$$

has a Lorentz distribution

$$P(x) = \frac{1}{\pi b} \frac{1}{1 + (x/b)^2}$$

with $x \in]-\infty, \infty[$.

Exercise 1.7 *Use the Box-Müller algorithm on page 22 to generate* $N = 1,000,000$ *random numbers that are normal distributed with mean* $\mu = 0$ *and standard deviation* $\sigma = 1$. *Create a histogram using this random numbers. Repeat this computer experiment with* $\sigma = 4$. *Has the "width" of the histogram increased by a factor 2?*

Use (1.50) from exercise 1.1 to generate $N = 1,000,000$ *random numbers that have Lorentz distribution with* $b = \sqrt{2\ln(2)}$. *Create a histogram using this random numbers and compare with the result from the normal distribution.*

Exercise 1.8 *Use the Box-Müller algorithm on page 22 to generate* $N = 1,000,000$ *random numbers that are normal distributed with mean* $\mu = 0$ *and standard deviation* $\sigma = 4$. *Create a histogram using this random numbers.*

Create $N = 1,000,000$ *sets of four Gaussian random numbers* $x_1^{(i)}$, $x_2^{(i)}$, $x_3^{(i)}$, $x_4^{(i)}$, $i = 1 \ldots N$. *Define*

$$y^{(i)} = \frac{1}{4}\left(x_1^{(i)} + x_2^{(i)} + x_3^{(i)} + x_4^{(i)}\right)$$

and create a histogram of those random numbers. Compare the widths of the y-distribution with that of the x-distribution.

Exercise 1.9 *Use (1.50) from exercise 1.1 to generate* $N = 1,000,000$ *random numbers that have Lorentz distribution with* $b = \sqrt{2\ln(2)}$. *Create a histogram using this random numbers.*

Create $N = 1,000,000$ *sets of four* **Lorentz** *random numbers* $x_1^{(i)}$, $x_2^{(i)}$, $x_3^{(i)}$, $x_4^{(i)}$, $i = 1 \ldots N$. *Define*

$$y^{(i)} = \frac{1}{4}\left(x_1^{(i)} + x_2^{(i)} + x_3^{(i)} + x_4^{(i)}\right)$$

and create a histogram of those random numbers. Compare the widths of the y-distribution with that of the x-distribution.

Exercise 1.10 *Create* $N = 1,000,000$ *sets of* $n = 10$ *random numbers* $x_1^{(i)}$, $x_2^{(i)}$, $\ldots$, $x_n^{(i)}$, $i = 1 \ldots N$, *where each of them is uniformly distributed between* $[0,1]$. *Define*

$$y^{(i)} = \frac{1}{n}\sum_{k=1}^{n} x_k^{(i)}$$

and estimate the probability distribution with the help of a histogram. Calculate mean and standard deviation of the variables $x_k^{(i)}$. *Compare the estimate from the histogram with the Gaussian approximation (1.51).*

Exercise 1.11 *Consider random numbers $X_i \in [0,4]$ that are uniformly distributed. The mean is given by $\bar{x} = 2$. Generate in a Computer experiment n random numbers and estimate the mean by*

$$\langle x \rangle = \frac{1}{n} \sum_{i=1}^{n} X_i$$

and its error. Complete the table

n	20	40	80	100
$\langle x \rangle$				
σ				

Use the data of the above table to find the "world average" for $\bar{x}$ and the corresponding error (see (1.45) and (1.46)). Generate an estimate and its

error for $\bar{x}$ with $n = 240$ and compare your result with your "world average".

Chapter 2

Random walks

If a large number of degrees is evolving in a deterministic manner, it is often impossible to describe the collective state by carrying out microscopic simulations. A stochastic approach, introducing a random element in how fundamental degrees of freedom interact, is powerful and often reveals a universal collective behavior. A famous example are ideal, classical gases with about 10^{24} interacting particles. The stochastic approach yields the fundamental gas law, which accurately defines the state of the gas and leads to the introduction of concepts like temperature and pressure. This leads us to the concept of "Random Walks".

2.1   Random walk as a Markov process

A *random walk* describes a path that consists of a sequence of random steps. A widespread example is the case for which the direction of the next step is chosen from a uniform distribution (*isotropic random walk*). They play an important role for many scientific fields ranging from ecology, psychology, computer science, physics, chemistry and biology, and also to economics. For

DOI: 10.1201/9781315156156-2

instance in chemistry, the paths of specific molecules in a solution and the related spread of the substance are called Brownian motion and lead to a process called diffusion [11]. Another example is the spreading of diseases in particular nowadays, when we observe a high degree of connectivity due to air-traffic [12]. In biology, details of the random walk significantly influence the encounter rate between organisms [13].

A large class of random walks is characterised by the *lack of memory*. For example, in Brownian motion, the next position of a particular molecule only depends on the current state of the solution and not on historic paths of the molecule. In statistics, this is called a *Markov process*, named after the Russian mathematician Andrey Markov. The path of the molecule can viewed as a so-called Markov chain. A discrete-time Markov chain is a countable set of random numbers $x_1, x_2, x_3, \ldots$, where x_k characterises the state of the Markov chain at time k (e.g. the position of the molecule). The Markov property is such that the probability of moving to the next state depends only on the present state and not on the previous states:

$$\Pr\left(x_{k+1}|x_k, x_{k-1}, \ldots x_1\right) = \Pr\left(x_{k+1}|x_k\right) .$$

There is an extensive mathematical literature on Markov processes. We can here only present a basic introduction and refer to the literature (e.g. [14]) for further readings.

2.2 Random walks in 1 and 2 dimensions

Important concepts such as "diffusion" that emerge from random walks can already understood in 1 and 2 dimensions.

2.2.1 Random walk on Z

Let us study the most simple case of an isotropic random walk: the walk in one dimension with a fixed step length. This random walk is also called a walk on a lattice or on $\mathbb{Z}$ since the positions are (or can be mapped to) integers. In the present example, we only need to decide whether we would like to take a step to the right or the left. We choose the step to right with probability p. The probability for choosing left is then $q = 1 - p$. The random walk is only isotropic for $p = q = 1/2$. A paths is characterised by a sequence of L and R each indicating the direction that we took, e.g.

$$LRRLLLRRLRRRRLLR .$$

The state of the walker is fully characterised by his or her position $x \in Z$ to the right of the origin. Let us assume that the walk has N steps. We are seeking the probability that the walker is at position x. To find the walker in x, the only fact that matters is that they must have x steps more to the right than to the left:

$$x = n_r - n_l .$$

In which order these steps have occurred is irrelevant for the position. Hence, the number of possibilities to choose n_r right steps out of N possibilities is

$$\binom{N}{n_r}.$$

We find

$$x = n_r - (N - n_r) = 2n_r - N ,$$

which, e.g. implies that n is even if the total number N of steps is even. For example, if $N = 2$, you cannot find yourself in position $n = 1$ since, for two steps, you have to move on if you are in position $n = 1$ after the first step. In the following, we will restrict ourselves to an even number N of steps, leading us to the definition

$$x = 2n , \qquad (N \text{ even}) . \tag{2.1}$$

With

$$n_r = \frac{x + N}{2} = n + \frac{N}{2} ,$$

we find for probability to be at position $x = 2n$ after N steps:

$$P(n) = \binom{N}{n_r} p^{n_r} q^{N-n_r} = \binom{N}{n + N/2} p^{N/2+n} q^{N/2-n} . \tag{2.2}$$

It is easy to check that the probabilities sum to one as they must:

$$\sum_{n=-N/2}^{N/2} P(n) = (p + q)^N = 1^N = 1 .$$

If we undertake N steps where are we on average? To answer this question, we calculate the expectation value

$$\langle n \rangle = \sum_{n=-N/2}^{N/2} n\, P(n) = \sum_{n=-N/2}^{N/2} \left(n + \frac{N}{2}\right) P(n) - \frac{N}{2} . \tag{2.3}$$

Using (2.2), we notice that

$$\left(n + \frac{N}{2}\right) P(n) = \binom{N}{n + N/2} p \frac{\partial}{\partial p} p^{N/2+n} q^{N/2-n} .$$

yielding for (2.3):

$$\langle n \rangle = p \frac{\partial}{\partial p} \sum_{n=-N/2}^{N/2} P(n) - \frac{N}{2} = p \frac{\partial}{\partial p} (p + q)^N - \frac{N}{2}$$

$$= pN (p + q)^{N-1} - \frac{N}{2} = N \left(p - \frac{1}{2}\right) . \tag{2.4}$$

For the isotropic random walk, i.e. $p = 1/2$, we remain at the origin on average. For $p > 1/2$, we find a *linear drift term* to the right: the distance from the origin linearly increases with the number of steps.

Let us also study the width of the probability $P(n)$ as a function of the number N of steps. To this aim, we study the variance

$$
\begin{aligned}
\left\langle (n - \langle n \rangle)^2 \right\rangle &= \left\langle (n + N/2 - Np)^2 \right\rangle = \left\langle (n + N/2)^2 \right\rangle \\
&- 2Np \langle n + N/2 \rangle + N^2 p^2 = \left\langle (n + N/2)^2 \right\rangle - N^2 p^2 ,
\end{aligned}
$$

where we have used (2.4). We then obtain:

$$
\begin{aligned}
\left\langle (n + N/2)^2 \right\rangle &= p \frac{\partial}{\partial p} p \frac{\partial}{\partial p} \sum_{n=-N/2}^{N/2} P(n) = p \frac{\partial}{\partial p} p \frac{\partial}{\partial p} (p + q)^N \\
&= N^2 p^2 - Np^2 + Np ,
\end{aligned}
$$

leaving us with the important result for the width w of the distribution:

$$
w^2 = \left\langle (n - \langle n \rangle)^2 \right\rangle = N p(1 - p) = N pq . \tag{2.5}
$$

The width grows with the square root of the number of steps. The factor pq in (2.5) is related to the so-called *diffusion constant* D (as will become clear below). Note that this diffusion constant is maximal (and equals $D = 1/4$) for the isotropic walk $p = q = 1/2$.

We can quite easily generalise the above results to a random walk in two (or even more dimensions). Here, $x = 2n, y = 2m$ specifies our positions on a two-dimensional lattice. At each point, we make a step in positive direction with probabilities p_x and p_y for the x and the y-direction. The choices for the steps in x and y directions are made independently, but note that we do step in each direction exactly once. This means that we move along the diagonals of the lattice. We can now group the expectation values to a vector and write:

$$
\begin{pmatrix} \langle x \rangle \\ \langle y \rangle \end{pmatrix} = 2 \begin{pmatrix} \langle n \rangle \\ \langle m \rangle \end{pmatrix} = 2 \begin{pmatrix} p_x - 1/2 \\ p_y - 1/2 \end{pmatrix} N . \tag{2.6}
$$

Hence, the vector $(p_x, p_y)^T$ can be interpreted as *drift velocity*. For further discussion, we specialise to the isotropic case $p_x = p_y = 1/2$ for which the drift velocities vanishes. What is our distance from the origin on average? We are seeking:

$$
r^2 := \langle x^2 + y^2 \rangle = 4 \left(\langle n^2 \rangle + \langle m^2 \rangle \right) = 2N , \tag{2.7}
$$

where we have used (2.5). We observe that (Euclidean) distance grows like $\sqrt{N}$ with N being the number of steps.

2.2.2 Stability

So far, we have considered a random walk on an equally spaced lattice. This is not very realistic if we would like to consider the continuous motion of a particle in a solution. In order to adapt our results for more realistic applications, we consider the lattice as *regularisation* only and try to recover the continuous random walk in the so-called *continuum limit* of infinitely many steps N within a fixed amount of time. To succeed, we need *stability*: the limit $N \to \infty$ needs to exist, and there needs to be a continuous limiting probability distribution.

To this aim, it is convenient to study the (discrete) Fourier transform of the probabilities $P(n)$ (2.2):

$$P(n) = \int_{-\pi}^{\pi} \frac{dk}{2\pi} e^{-ikn} \bar{P}(k), \qquad \bar{P}(k) = \sum_n e^{ikn} P(n). \qquad (2.8)$$

The latter equation can be easily verified by using the following representation of the Kronecker δ:

$$\delta_{nm} = \int_{-\pi}^{\pi} \frac{dk}{2\pi} e^{-ik(n-m)}, \qquad n, m \in \mathbb{Z}.$$

The transform $\bar{P}(k)$ can be calculated in closed form:

$$\bar{P}(k) = \sum_{n=-N/2}^{N/2} \binom{N}{n+N/2} p^{N/2+n} q^{N/2-n} e^{ikn}$$

$$= e^{-ikN/2} \sum_{n_r=0}^{N} \binom{N}{n_r} p^{n_r} q^{N-n_r} e^{ikn_r} = e^{-ikN/2} \left[p e^{ik} + q \right]^N.$$

By virtue of the inverse Fourier transform, we find the identity:

$$P(n) = \int_{-\pi}^{\pi} \frac{dk}{2\pi} e^{-ik(n+N/2)} \left[1 - p(1 - e^{ik}) \right]^N, \qquad (2.9)$$

where we have used $q = 1 - p$. For finding a sensible limit $N \to \infty$, we expand in powers of k:

$$\ln \left[1 - p(1 - e^{ik}) \right] = ipk - \frac{1}{2} p(1-p) k^2 + \mathcal{O}(k^3).$$

Part of the integrand in (2.9) then becomes:

$$\left[1 - p(1 - e^{ik}) \right]^N = \exp \left\{ i Npk - \frac{1}{2} Np(1-p) k^2 + \mathcal{O}(Nk^3) \right\}.$$

Note that the dominant contribution to the integral (2.9) arises from the region $|k| < 1/\sqrt{N}$. For large N, this has two consequences: (i) We might extend the integration from $-\pi \ldots \pi$ to $-\infty \ldots \infty$ since the added parts are strongly suppressed by the integrand. (ii) In the relevant region, we find

$$|Nk^3| < \frac{N}{N^{3/2}} = \frac{1}{\sqrt{N}}.$$

Thus, the higher order terms in k can be neglected. We therefore arrive at for $N \gg 1$:

$$
\begin{aligned}
P(n) &= \int_{-\infty}^{\infty} \frac{dk}{2\pi}\, e^{-ik(n+N/2)}\, \exp\left\{i\,Np\,k - \frac{1}{2}Np(1-p)\,k^2\right\} \\
&= \int_{-\infty}^{\infty} \frac{dk}{2\pi}\, e^{-ik(n-\langle n\rangle)}\, \exp\left\{-\frac{1}{2}w^2\,k^2\right\} \\
&= \frac{1}{w\sqrt{2\pi}}\, \exp\left\{-\frac{(n-\langle n\rangle)^2}{2w^2}\right\}, \quad\quad (2.10)
\end{aligned}
$$

where we have used (2.4) and (2.5). The limiting distribution is a Normal distribution with the N dependence only appearing in the parameters $\langle n\rangle$ and w.

2.2.3   Limits: Wiener process and diffusion equation

In the previous subsections, we were studying the random walk on $\mathbb{Z}$ using an underlying lattice to direct the movements. We now would like to adapt this to the more realistic scenario of a particle undergoing a continuous movement. We first introduce "macroscopic" variables for time and space. Let Δt be the time that passes during each step from n to $n+1$. After time t, we have made N steps:

$$t_k = k\,\Delta t\,, \quad k = 1\ldots N\,, \quad\quad t = N\,\Delta t\,. \quad\quad (2.11)$$

Furthermore, we associate a distance Δx as the spacing of the underlying lattice:

$$x_n = \Delta x\,n\,. \quad\quad (2.12)$$

We are exploring the limit of a large number N of steps while we keep the final time t fixed. Increasing N thus implies that we make the time step Δt smaller until it shrinks to zero in the limit $N \to \infty$.

We introduce a continuous probability distribution $\mathcal{P}(x)$ and demand:

$$\int_{x_n}^{x_n+\Delta x} \mathcal{P}(x)\, dx = P(n)\,. \quad\quad (2.13)$$

We calculate the moments in leading order of small Δx (we will justify this below):

$$
\begin{aligned}
\langle x^k \rangle &= \int x^k \, \mathcal{P}(x) \, dx = \sum_n \int_{x_n}^{x_n + \Delta x} x^k \, \mathcal{P}(x) \, dx \\
&\doteq \sum_n x_n^k \int_{x_n}^{x_n + \Delta x} \mathcal{P}(x) \, dx = \sum_n x_n^k P(n) = (\Delta x)^k \, \langle n^k \rangle .
\end{aligned}
$$

In particular, we find for the standard deviation of the distribution $\mathcal{P}(x)$:

$$
\langle x^2 \rangle - \langle x \rangle^2 = (\Delta x)^2 \left(\langle n^2 \rangle - \langle n \rangle^2 \right) = (\Delta x)^2 \, N \, p \, q = p q \frac{(\Delta x)^2}{\Delta t} \, t . \quad (2.14)
$$

The last equation crucially informs the continuum limit. To obtain a finite variance, we need to choose:

$$
\Delta x \to 0, \qquad \Delta t \to 0, \qquad \frac{1}{2} p q \frac{\Delta x^2}{\Delta t} = D = \text{finite}, \qquad (2.15)
$$

where the factor $1/2$ in the definition of D is convention. With latter definition of the continuum limit, we see that Δx vanishes in the limit $N \to \infty$, and the position variable x (2.12) becomes continuous. For large N, we also have (see (2.10)):

$$
\begin{aligned}
\mathcal{P}(x) \, \Delta x \doteq P(n) &= \frac{1}{w \sqrt{2\pi}} \exp\left\{ -\frac{(n - \langle n \rangle)^2}{2 w^2} \right\} \\
&= \frac{1}{2 \sqrt{\pi t}} \exp\left\{ -\frac{(x - \langle x \rangle)^2}{4 D t} \right\} \Delta x , \qquad (2.16)
\end{aligned}
$$

with

$$
w^2 = p q N = p q \frac{t}{\Delta t} = 2 D \frac{t}{(\Delta x)^2} .
$$

Hence, we found the continuous probability distribution that emerges in the continuum limit (2.15) of a random walk on $\mathbb{Z}$. The continuum limit of the random walk falls into the class of the Wiener processes, which are characterised by the properties:

(a) $W(t)$ is a continuous random path.

(b) Independence: $W(t) - W(s)$ is independent of $\{W(\tau)\}_{\tau \leq s}$ for all $0 \leq s \leq t$.

(c) Stationarity: The probability distribution of $W(t + s) - W(s)$ is independent of s.

(d) The probability distribution for $W(t + u) - W(u)$ is normal.

Using elementary calculus, it is straightforward to show that the probability distribution

$$\mathcal{P}(x,t) = \frac{1}{2\sqrt{\pi t}} \exp\left\{-\frac{(x - \langle x \rangle)^2}{4Dt}\right\} \Delta x$$

satisfies the diffusion equation:

$$\frac{\partial}{\partial t} \mathcal{P}(x,t) = D \frac{\partial^2}{\partial x^2} \mathcal{P}(x,t) . \tag{2.17}$$

This partial differential equation has been proposed on empiric grounds to describe the diffusion of a substance in a solution. Here, we were able to relate this diffusion to the underlying microscopic picture of a random walk.

2.2.4 Gaussian random walk and correlations

In the previous sections, we considered continuous random paths (on a lattice), i.e. the so-called Wiener process, and the limit for which the underlying lattice becomes irrelevant. In this section, we will consider an isotropic random walk consisting of discrete steps and flexible step length. If the probability distribution of the step length is Normal, we are dealing with the *Gaussian Random Walk*.

Let $P_n(x)\, dx$ be the probability that the Gaussian random is located in the interval $[x, x + dx]$ at after the nth step. The walker then picks a random direction and performs the next step with length ℓ. The probability distribution for the length ℓ is normal:

$$Q(\ell) = \frac{1}{w\sqrt{2\pi}} \exp\left\{-\frac{\ell^2}{2w^2}\right\} . \tag{2.18}$$

With the help of the Dirac δ-function (see (1.8)), we then can implement the constraint that we arrive at y starting at x with a step length ℓ, and the probability distribution for the position after step $n + 1$ is then given by:

$$P_{n+1}(y) = \int_{-\infty}^{\infty} d\ell\, dx\, P_n(x)\, Q(\ell)\, \delta\left(y - x - \ell\right) . \tag{2.19}$$

The integration sums over all possible prior positions x and all possible step lengths ℓ. Inserting (2.18) into (2.19) and performing the ℓ-integration by means of the δ-function yields:

$$P_{n+1}(y) = \int_{-\infty}^{\infty} dx\, P_n(x)\, \frac{1}{w\sqrt{2\pi}} \exp\left\{-\frac{(x - y)^2}{2w^2}\right\} . \tag{2.20}$$

To solve the above iteration, we introduce the Fourier transforms

$$\bar{P}_m(k) = \int_{-\infty}^{\infty} dx\, P_m(x)\, e^{-ikx} ,$$

$$\bar{Q}(p) = \int_{-\infty}^{\infty} d\ell\, Q(\ell)\, e^{-ip\ell} = \exp\left\{-\frac{1}{2}w^2 p^2\right\} ,$$

which turns the folding (2.20) in position space into a product in Fourier space:

$$\bar{P}_{n+1}(k) \;=\; \bar{P}_n(k)\,\bar{Q}(-k)\,. \tag{2.21}$$

We need a state distribution $P_{n=0}(x)$ for the random walker. We would like to consider the case for which the random walker is surely located at the origin $x = 0$ at the start. We can model this by using Dirac δ-function again:

$$P_{n=0}(x) \;=\; \delta(x)\,. \tag{2.22}$$

It implies that the probability distribution vanished as long as $x \neq 0$ and that its is properly normalised:

$$\int_{-\infty}^{\infty} dx\,\delta(x) \;=\; 1\,.$$

The Fourier transform of this initial distribution is then

$$\bar{P}_{n=0}(k) \;=\; \int_{-\infty}^{\infty} dx\,\delta(x)\,e^{-ikx} \;=\; 1\,, \tag{2.23}$$

where we have used the property (1.8). The solution of the iteration (2.12) is then easily obtained:

$$\bar{P}_n(k) \;=\; \exp\!\left\{-\frac{1}{2}w^2\,k^2\,n\right\} \tag{2.24}$$

or in position space:

$$P_n(x) \;=\; \int_{-\infty}^{\infty} \frac{dk}{2\pi}\,\bar{P}_n(k)\,e^{ikx} \;=\; \frac{1}{w\sqrt{2\pi n}}\,\exp\!\left\{-\frac{x^2}{2w^2 n}\right\}\,. \tag{2.25}$$

We can now introduce time by associating a time Δt passing during each step, see (2.11) and make the walk continuous in time by considering the limit $\Delta t \to 0$. In order to obtain a finite distribution, we need to enforce:

$$w^2 \;=\; 2\,D\,\Delta t\,, \qquad D:\ \text{finite}\,. \tag{2.26}$$

It means that the width for the step size distribution vanishes in the limit $\Delta t \to 0$. This does make sense for the iteration (2.21):

$$\bar{P}_{n+1}(k) \;=\; \bar{P}_n(k)\,\exp\!\left\{-D\,\Delta t\,k^2\right\}\,.$$

If a very small time Δt passes between the step from n to $n+1$, the probability distribution $\bar{P}_{n+1}$ must be very close to that of $\bar{P}_n$ since otherwise the iteration would grow overall bounds. We find for the continuous time Gaussian random walk:

$$P(x,t) \;=\; \lim_{\Delta t \to 0} P_n(x) \;=\; \frac{1}{2\sqrt{\pi D\,t}}\,\exp\!\left\{-\frac{x^2}{4Dt}\right\}\,. \tag{2.27}$$

As detailed above, this quantity has the interpretation: if the random walker starts at the origin, $P(x,t)\,dx$ is the probability that the walker can be found in the interval $[x, x + dx]$ after time t.

If we do not mind *when* the walker arrives at x, we introduce the so-called *spatial correlation* by:

$$C(x) = \int_0^\infty dt\, P(x,t) \,. \tag{2.28}$$

This integral can be performed by means of the Fourier transform:

$$P(x,t) = \int_{-\infty}^\infty \frac{dk}{2\pi}\, e^{ikx}\, \bar{P}(k,t) \,, \qquad \bar{P}(k,t) = \exp\left\{-D\,k^2\,t\right\}, \tag{2.29}$$

leading to:

$$C(x) = \frac{1}{D} \int_{-\infty}^\infty \frac{dk}{2\pi}\, e^{ikx}\, \frac{1}{k^2} \,. \tag{2.30}$$

This result can be easily generalised to d dimensions:

$$C(\vec{x}) = \frac{1}{D} \int_{-\infty}^\infty \frac{d^d k}{(2\pi)^d}\, e^{i\vec{k}\vec{x}}\, \frac{1}{\vec{k}^2} \,, \qquad \vec{x}, \vec{k} \in \mathbb{R}^d \,. \tag{2.31}$$

Note that the Fourier integral (2.31) diverges for small $\vec{k}$ for dimensions $d = 1$ and $d = 2$. We interpret this as the ability to reach any point of $\mathbb{R}$ and $\mathbb{R}^2$ but just "walking" long enough. The picture in dimensions $d \geq 3$ is very different: here, the integral is finite. For instance, in three dimensions we find using polar coordinates

$$d^3 k = k^2\, d\phi\, \sin\theta\, d\theta\, dk, \qquad \phi \in [-\pi, \pi[,\ \theta \in [0, \pi]$$

and $r = |\vec{x}|$ after the trivial ϕ-integration:

$$
\begin{aligned}
C(r) &= \frac{1}{4\pi^2 D} \int_0^\infty dk \int_0^\pi d\theta\, \sin\theta\, \exp\{ikr\cos\theta\} \\
&= \frac{1}{4\pi^2 D} \int_0^\infty dk \int_{-1}^1 d\eta\, e^{ikr\,\eta} = \frac{1}{2\pi^2 D\, r} \int_0^\infty dx\, \frac{\sin x}{x} = \frac{1}{4\pi D}\, \frac{1}{r} \,.
\end{aligned}
$$

This result is finite: it is no longer guaranteed that the vicinity of a point in distance r will be visited even after an infinite amount of time.

2.3   Levy flight

So far we have studied random walks for which the probability distribution $P(x)$ for one step x does not possess so-called "heavy tails" implying that the variance $\langle x^2 \rangle$ exists. We shall now study the arising phenomena if the latter condition does not constrain us.

2.3.1 Definition and simulations

In subsection 2.2, we started with a Wiener process on a regular lattice, which features a fixed step size for the random walker and finished with Gaussian random walk in the continuum. Here, the step size distribution was normal, which implies that all moments of the step size distribution do exist. In either case, we studied the so-called continuum limit, for which the time difference and the mean square step length vanish and found standard diffusion.

Levy flights also belong to the class of Markov processes (see subsection 2.1): the random walker has no memory of the history of the paths, but decides the direction randomly and isotropicly and the step size "on the spot". In contrast to the Gaussian random walk, however, the step size distribution has power-law behaviour for large step sizes and not all moments of its distribution do exist. Qualitatively, large step sizes are suppressed for the Gaussian walker while bigger jumps can be expected for the Levy process. Let us illustrate this in a numerical experiment in two dimensions. The step change $(\delta x, \delta y)$ of the position (x, y) for Gaussian random walker is normal distributed:

$$P(\delta x, \delta y) = \frac{1}{2\pi\sigma^2} \exp\left\{-\frac{\delta x^2 + \delta y^2}{2\sigma^2}\right\}, \qquad \sigma^2 = \frac{1}{2\ln 2}. \qquad (2.32)$$

The scale of the step is chosen such that a unit-length step has half of the maximum probability distribution:

$$P(1, 0) = \frac{1}{2} P(0, 0). \qquad (2.33)$$

For the Levy flight, we choose a step size that is Cauchy-Lorentz distributed (see (1.50)). Here, we have $(\delta r \geq 0)$

$$x \to x + \delta r \cos(\phi), \qquad y \to y + \delta r \sin(\phi), \qquad L(\delta r) = \frac{2}{\pi} \frac{1}{1 + \delta r^2},$$

where we have chosen the scale of the step in accordance with (2.33):

$$L(1) = \frac{1}{2} L(0).$$

The direction specified by $\phi \in [-\pi, \pi[$ is chosen from a uniform distribution in order to guarantee the isotropy of the walk. Our numerical result for a 1000 steps in each case are shown in Figure 2.1. Indeed, the Levy flight occasional bridges a large distance, which coined the name "flight". The Cauchy-Lorentz distribution gives rise to just one example of a Levy flight. We can consider any "heavy-tail" distribution characterised by

$$H(x) \propto \frac{1}{|x|^\mu} \qquad |x| \text{ large}, \qquad 1 < \mu < 3. \qquad (2.34)$$

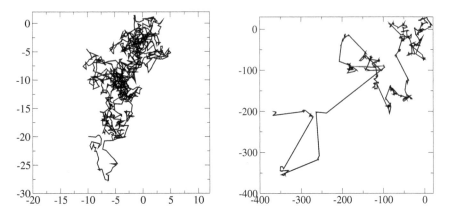

FIGURE 2.1: Gaussian Random walk (left) and Levy flight (right) for a 1000 steps each. Note the difference in scales on the axis.

The Cauchy-Lorentz random flight corresponds to the case $\mu = 2$. To characterise the spread of the walker in space, we define the diffusion length l_θ for a given $\theta < 2$ by the fractional moment

$$l_\theta = \left\langle |x|^\theta \right\rangle^{1/\theta} . \tag{2.35}$$

For small enough θ, the moment does exist:

$$\left\langle |x|^\theta \right\rangle = \int dx \, x^\theta \, H(x) \rightarrow \text{finite for } \theta < \mu - 1 .$$

For $\mu > 3$, the standard deviation exists implying that the distribution for many compounded steps n is Gaussian by virtue of the Central Limit Theorem (see section 1.2). In this case, we can choose $\theta = 2$ and find:

$$\left\langle |x|^2 \right\rangle^{1/2} \propto \sqrt{n} \qquad \forall \mu > 3 . \tag{2.36}$$

Hence, the Levy flights, i.e. $1 < \mu < 3$, are the interesting cases for which we can expect deviations from the standard diffusion process.

2.3.2   Generalised Central Limit Theorem and stability

Rather than discussing the general setting, we will focus on the Levy flight with a Cauchy-Lorentz distribution for the step size:

$$x_{n+1} = x_n + \delta x , \qquad L(\delta x) = \frac{1}{\pi b} \frac{1}{1 + (\delta x / b)^2}, \quad (b > 0) . \tag{2.37}$$

The evolution equation (2.18), which iteratively determines the probability distribution $L_n(x)$ becomes in the present case

$$L_{n+1}(y) = \int_{-\infty}^{\infty} dx \, L_n(x) \, \frac{1}{\pi b} \frac{1}{1 + (x - y)^2 / b^2} . \tag{2.38}$$

As detailed in subsection 2.2.4, such an equation can be solved in Fourier space: The Fourier transform of (2.38) is given by

$$\bar{L}_{n+1}(k) = \exp\{-b\,|k|\}\,L_n(k)\,. \tag{2.39}$$

With the usual initial condition,

$$L_0(x) = \delta(x)\,, \qquad \bar{L}_0(k) = 1\,,$$

the solution of (2.39) is found to be

$$\bar{L}_n(k) = \exp\{-b\,n\,|k|\} \quad \Rightarrow \quad L_n(x) = \frac{1}{\pi\,b\,n}\frac{1}{1+(x/b\,n)^2}\,. \tag{2.40}$$

Let us study stability, i.e. the limit of continuous time

$$n \to \infty\,, \qquad t = \delta t\, n = \text{finite}\,.$$

We introduce a new "diffusion type" constant d by

$$\frac{b}{\delta t} = d = \text{finite} \qquad \text{for } \delta t \to 0\,,$$

and obtain from (2.40):

$$L(x,t) = \lim_{n\to\infty} L_n(x) = \frac{1}{\pi\,d\,t}\frac{1}{1+(x/d\,t)^2}\,. \tag{2.41}$$

If we trivially re-write the last equation into

$$L(x,t) = \frac{1}{\pi}\frac{d\,t}{(d\,t)^2+x^2}\,,$$

it is straightforward to verify that indeed

$$\lim_{t\to 0} L(x,t) = \delta(x)\,.$$

We leave this exercise to the reader. Let us now study the *diffusion length* l_θ in (2.35):

$$(l_\theta)^\theta = 2\int_0^\infty dx\,\frac{1}{\pi\,d\,t}\frac{x^\theta}{1+(x/d\,t)^2}\,. \tag{2.42}$$

With the substitution $y = x/d\,t$, we find

$$l_\theta = g(\theta)\,d\,t\,, \qquad g(\theta) = \left[2\int_0^\infty dy\,\frac{1}{\pi}\frac{y^\theta}{1+y^2}\right]^{1/\theta}\,. \tag{2.43}$$

We indeed find *anomalous diffusion*: the diffusion length l_θ grows linearly with time for the Cauchy-Lorentz Levy flight, which needs to be contrasted

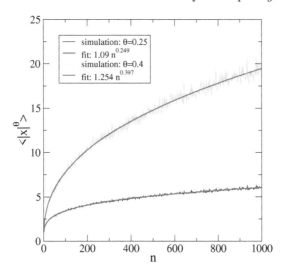

FIGURE 2.2: Fractional moment $\langle x^\theta \rangle (n)$ as a function of step time n for two values of θ.

with the result from Gaussian Random walks $l_\theta \propto \sqrt{t}$, which corresponds to standard diffusion.

Let us illustrate these findings by a numerical experiment. Using step-sizes drawn from the Cauchy-Levy distribution (2.37) (with $b = 1$), we generate a sequence of positions x_n, $i = 1 \ldots n_{\text{max}}$. We calculate N copies of this random walks with positions $x_n^{(i)}$, $i = 1 \ldots N$, and estimate the fractional moment by

$$\langle x^\theta \rangle (n) \approx \frac{1}{N} \sum_{i=1}^{N} \left[x_n^{(i)} \right]^\theta .$$

We expect that the result scales as

$$\langle x^\theta \rangle (n) \propto n^\theta . \tag{2.44}$$

Our numerical result for $N = 1000$ is shown in Figure 2.2, which shows $\langle x^\theta \rangle$ as a function for n for $\theta = 0.25$ and $\theta = 0.4$. Also shown is a power-law fit

$$\langle x^\theta \rangle (n) = a_0\, n^{a_1}$$

with a_0 and a_1 being fit parameters. A good agreement with the scaling law (2.44) is observed.

2.3.3 Computer experiments

Let us study a Levy flight where the step size is controlled by the distribution:

$$S(\delta x) = \frac{1}{2\sqrt{|\delta x|}\,(1 + \sqrt{|\delta x|})^2} \cdot \tag{2.45}$$

$$x_{n+1} = x_n + \delta x\,, \qquad x_0 = 0\,.$$

An analytic solution along the lines of the recursive equation such as (2.38) is not feasible in closed form. We will therefore resort to numerical simulations to find an estimate for the generalised diffusion length l_θ (2.35).

The distribution (2.45) has interesting characteristics: It will generate a Levy flight since the variance for the step size diverges (and, thus, evades the Gaussian case):

$$\int_0^c dx\, x^2\, S(x) = \int_0^c dx\, \frac{1}{2\sqrt{x}\,(1 + \sqrt{x})^2} \propto c^{3/2} \to \infty$$

for $c \to \infty$. Secondly, the probability distribution diverges at $\delta x = 0$. As it must be for a probability interpretation, the singularity is integrable, i.e. the probability to find a step size in a small interval $\delta x \in [0, \epsilon]$ is finite:

$$\int_0^\epsilon dx\, S(x) = \sqrt{\epsilon} + \mathcal{O}(\epsilon)\,.$$

In practice, both properties of $S(x)$ imply that the resulting Levy flight prominently features small and exceptionally large steps.

Let us briefly discuss how we generate the Levy flight in the computer experiment. To this aim, we will use the techniques explained in some detail in subsection 1.1.3. If $0 \le u < 1$ a random variable with uniform distribution, we introduce the maping:

$$\int_0^u = \int_0^{\delta x} S(y)\, dy, \qquad \delta x > 0$$

leading to

$$u = \frac{\sqrt{\delta x}}{1 + \sqrt{\delta x}} \qquad \Rightarrow \qquad \delta x = \frac{u^2}{(1 - u)^2} \cdot$$

We therefore propose to following procedure to construct one particular random variable δx:

1. Choose two uniformly distributed random numbers $u, w \in [0, 1]$.

2. Calculate $\delta x = \frac{u^2}{(1-u)^2}$.

3. If $w > 0.5$, $\delta x \to -\delta x$.

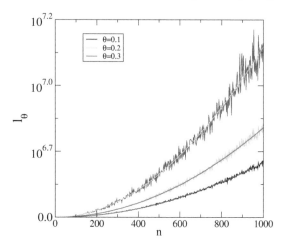

FIGURE 2.3: Diffusion length $l_\theta(n)$ as a function of step time n for three values of θ for a Levy flight with step size distribution (2.45).

We are now in the position to address the generalised diffusion length l_θ. For our flight here, it exists for $0 \le \theta < 0.5$. To this aim, we have generated $N = 16,000$ copies of a Levy-flight $x_n^{(i)}$ with "time" $n = 1 \ldots 999$ and $i = 1 \ldots N$. We have then estimated the diffusion length by

$$l_\theta(n) = \left[\langle x^\theta \rangle (n)\right]^{1/\theta} \approx \left[\frac{1}{N}\sum_{i=1}^{N}\left(x_n^{(i)}\right)^\theta\right]^{1/\theta} .$$

Typical results for $\theta = 0.1$, 0.2, 0.3 are shown in Figure 2.3. We find empirically that the numerical data, at least in the range $n = 1 \ldots 999$, are well described by the power-law

$$l_\theta = a_0(\theta)\, n^{a_1} . \tag{2.46}$$

Our findings for the exponent a_1 are listed in Table 2.1.
We do observe that the exponents are roughly θ independent as it was the case for the Cauchy-Lorentz Levy flight. Under the *assumption* that we can ignore the small trend in the data of Table 2.1 and that the exponent a_1 is indeed independent of θ (in the allowed θ range), we find by averaging

$$\bar{a}_1 \approx 1.9908 .$$

TABLE 2.1:   Exponents
diffusion length

θ	0.1	0.2	0.3
a_1	2.00012	1.99431	1.97806

This value is close to the value of 2. Note that the noise for higher θ values is higher. This implies that the less noisy data should get more emphasis on the average. How to do this in a systematic way will be discussed in great detail in the next section.

2.4 Random walks with potentials

So far, we have considered particles moving freely except for scattering between particles. We are now including an external force on the particles. Prominent examples are magnetic or electric forces or gravity.

2.4.1 The Langevin diffusion

Most of the previous studies of the Random Walk involved a discrete process, e.g. arising from a finite step size. Here, we will shed light on Random Walks starting from a continuous motion with a random force acting on the test particle. If $x(t)$ denotes the position of the particle, and hence, $dx/dt = u(t)$ its velocity, Newton's law of mechanics proposes the ordinary differential equation

$$m \frac{d^2 x}{dt^2} + \gamma \frac{dx}{dt} + \frac{dV(x)}{dx} = F(t) , \qquad (2.47)$$

where m is the mass of the particle, $\gamma > 0$ quantifies the damp that the particle experiences during its motion, $V(x)$ is an external potential and $F(t)$ is a random force on the particle that only depends on time.

In the first study, we return to the now familiar Brownian motion. Here, we describe this motion assuming that external forces are absent, i.e. we choose $V(x) = 0$, and the randomness is now induced by a random acceleration $a(t) = F(t)/m$ rather than random jumps in position due to collisions. Do we recover the familiar diffusion law (2.27)?

Note that $1/\tau_r = \gamma/m$ has the unit of an inverse time, which we will call *relaxation time*. With external forces absent, the above ODE becomes first order in the velocity:

$$\frac{du}{dt} + \frac{1}{\tau_r} u = a(t) . \qquad (2.48)$$

This ODE can be solved in closed form, e.g. by Variation of Constants method. Using the initial condition $u(t = 0) = u_0$:

$$u(t) = u_0 \, e^{-t/\tau_r} + e^{-t/\tau_r} \int_0^t e^{t'/\tau_r} a(t') \, dt' . \qquad (2.49)$$

So far, the motion has been deterministic with $a(t)$ a microscopic force. This force is not known in practice. To circumvent the lack of knowledge, we consider a distribution of random accelerations and are interested in the average motion of the particle. If we consider many particles at the same time (as there would be many air molecules in a macroscopic volume), we assume that for an individual particle, the interaction with the other particles is summarised in the random force $ma(t)$. In this case, the average over random accelerations would be equivalent to the sample average over the many particles.

The key achievement of this *statistical approach* is that we need not to specify $a_k(t)$, $k = 1 \ldots n$ with n the number of particles, but the information of the many particle interactions will be encoded in view parameters, which are experimentally accessible. This apparent loss of information, with which we get away and still can describe average motion, can be viewed as a high degree of universality in whatever the average motion will be.

What are the constraints for the random acceleration? Firstly, the average acceleration $\langle a \rangle$ should vanish. This means that there is no directed motion from the interaction with the other particles. Secondly, the interaction needs to respect *causality*: a random force at time t can only affect the particle speed $u(t')$ if the force acted in the past, i.e. if $t \leq t'$. Thirdly, we assume that the random force is dependent on the position of the particle. We can summarise these constraints by:

$$\langle a(t) \rangle = 0 , \tag{2.50}$$

$$\langle a(t)\, u(t') \rangle = C(t' - t)\, \theta(t' - t) \tag{2.51}$$

$$\langle a(t)\, a(t') \rangle = \langle a^2 \rangle \exp\left\{ -\frac{|t - t'|}{\xi} \right\} . \tag{2.52}$$

The latter property shows that the random force does not depend on the point in time when the noise occurs. We also assume that the noise is only correlated over time ξ. With property (2.50), we easily find that the average speed is independent of the random forces and obeys the microscopic damping law:

$$\langle u(t) \rangle = u_0\, e^{-t/\tau_r} .$$

This does not mean that the test particle comes to a hold: it is only that any information on the initial direction of a particle at time $t = 0$ is lost over time. The relevant time scale for this to happen is set by the relaxation time τ_r. This process is called *thermalisation*. This interpretation can be verified by studying the average of the velocity square:

$$\left\langle u^2(t) \right\rangle = \langle u_0^2 \rangle\, e^{-2t/\tau_r} + e^{-2t/\tau_r} \int_0^t dt_1 \int_0^t dt_2\, e^{(t_1 + t_2)/\tau_r}\, \langle a(t_1)\, a(t_2) \rangle . \tag{2.53}$$

Note that any linear term in $a(t)$ vanishes upon averaging because of $\langle u(0)\, a(t) \rangle = 0$ in view of (2.51). Let us study double integrals of type

$$I := \int_0^{t_3} dt_1 \int_0^{t_4} dt_2\, g(t_1 + t_2)\, f\left(|t_1 - t_2| \right) .$$

We can assume that $t_3 \leq t_4$. If this is not the case, we would swap the order of integration. We split the second integral and then swap the order of integration in the first integral

$$
\begin{aligned}
I &= \int_0^{t_3} dt_1 \int_0^{t_1} dt_2 \, g \, f + \int_0^{t_3} dt_1 \int_{t_1}^{t_4} dt_2 \, g \, f \\
&= \int_0^{t_3} dt_2 \int_{t_1}^{t_3} dt_1 \, g \, f + \int_0^{t_3} dt_1 \int_{t_1}^{t_4} dt_2 \, g \, f \\
&= \int_0^{t_3} dt_1 \left[\int_{t_1}^{t_3} dt_2 \, g \, f + \int_{t_1}^{t_4} dt_2 \, g \, f \right] ,
\end{aligned}
$$

where we have used in the last line that $f \, g$ is invariant under the swap $t_1 \leftrightarrow t_2$. The substitution $t_2 = t_1 + \tau$ finally yields:

$$
I = \int_0^{t_3} dt_1 \left[\int_0^{t_3 - t_1} d\tau \, g(2t_1 + \tau) \, f(|\tau|) + \int_0^{t_4 - t_1} d\tau \, g(2t_1 + \tau) \, f(|\tau|) \right].
\tag{2.54}
$$

Identifying

$$
t_3 = t_4 = t, \qquad g(x) = \exp\left\{ \frac{x}{\tau_r} \right\} \qquad f(x) = \langle a^2 \rangle \exp\left\{ -\frac{|x|}{\xi} \right\} ,
$$

we find for (2.53):

$$
\left\langle u^2(t) \right\rangle = \langle u_0^2 \rangle \, e^{-2t/\tau_r} + 2 \, e^{-2t/\tau_r} \int_0^t dt_1 \, e^{2t_1/\tau_r} \int_0^{t-t_1} d\tau \, e^{\tau/\tau_r} \, \langle a^2 \rangle \, e^{-\tau/\xi} .
$$

All integrals can be done in closed form. We will, however, use an approximation, which is very well justified in practice: we assume that the test particle is *slowly* moving and that the random accelerations occur very rapidly, i.e.

$$
\tau_r \gg \xi .
\tag{2.55}
$$

With this approximation, we then obtain the integral:

$$
\int_0^{t-t_1} d\tau \, e^{\tau/\tau_r} \, e^{-\tau/\xi} \approx \int_0^\infty d\tau \, e^{-\tau/\xi} = \xi .
$$

We therefore find

$$
\left\langle u^2(t) \right\rangle \approx \langle u_0^2 \rangle \, e^{-2t/\tau_r} + \left(1 - e^{-2t/\tau_r} \right) \langle a^2 \rangle \, \tau_r \, \xi .
\tag{2.56}
$$

In Mechanics, $E_{\text{kin}} = m \, u^2(t)/2$ is the *kinetic energy* of the test particle. Here, we find that for large times, $t \gg \tau_r$, the average kinetic energy becomes time independent:

$$
\left\langle E_{\text{kin}} \right\rangle = \frac{m}{2} \langle a^2 \rangle \, \tau_r \, \xi .
\tag{2.57}
$$

It is proportional to the relaxation time τ_r and the noise correlation time ξ.

Let us now study the average position of the test particle. Averaging the velocity $u(t)$ in (2.49), we find:

$$\left\langle \frac{dx}{dt} \right\rangle = \langle u_0 \rangle \, e^{-t/\tau} , \tag{2.58}$$

If we can exchange the order of time derivative and average, we could calculate the solution just by integrating. To analyse this, we replace momentarily the average by an estimator over n test particles:

$$\langle x(t) \rangle \approx \frac{1}{n} \sum_{k=1}^{n} x_n(t) .$$

The time evolution of each test particle is independent of the evolution of the other particles, and we obtain:

$$\left\langle \frac{dx}{dt} \right\rangle \approx \frac{1}{n} \sum_{k=1}^{n} \frac{d}{dt} x_n(t) = \frac{d}{dt} \left(\frac{1}{n} \sum_{k=1}^{n} x_n(t) \right) \rightarrow \frac{d}{dt} \langle x \rangle .$$

Hence, we obtain from (2.58) that the average position follows the "microscopic law":

$$\langle x(t) \rangle = \tau_r \langle u_0 \rangle \left(1 - e^{-t/\tau_r} \right) .$$

For large times, the centre of the "cloud" of particles is displaced by

$$\lim_{t \to \infty} \langle x(t) \rangle = \tau_r \langle u_0 \rangle .$$

Let us study the variance of the position since, after all, this informs the diffusion. For any test particle, we find its position as a function of time t by integrating (2.49):

$$x(t) = \tau_r u_0 \left[1 - e^{-t/\tau_r} \right] + \int_0^t dt_1 \, e^{-t_1/\tau_r} \int_0^{t_1} e^{t'/\tau_r} a(t') \, dt' . \tag{2.59}$$

From this, we find the average of its square

$$\langle x^2(t) \rangle = \tau_r^2 \langle u_0^2 \rangle \left[1 - e^{-t/\tau_r} \right]^2 \tag{2.60}$$

$$+ \int_0^t dt_1 \, dt_3 \, e^{-(t_1+t_3)/\tau_r} \int_0^{t_1} dt_2 \int_0^{t_3} dt_4 \, e^{(t_2+t_4)/\tau_r} \langle a(t_2)a(t_4) \rangle .$$

Note that if we square equation (2.59), we would encounter the mixed term

$$2 \tau_r u_0 \left[1 - e^{-t/\tau_r} \right] \int_0^t dt_1 \, e^{-t_1/\tau_r} \int_0^{t_1} e^{t'/\tau_r} a(t') \, dt' .$$

Upon averaging, this mixed term vanishes because of the causality constraint (2.51):

$$\langle u_0\, a(t') \rangle \;=\; \langle u(0)\, a(t') \rangle \;=\; 0 \qquad \text{since } t' > 0. \tag{2.61}$$

We now implement the approximation separating the slow from the fast time scales, i.e. $\xi \ll \tau_r$:

$$\langle a(t_2)a(t_4) \rangle \;=\; \langle a^2 \rangle\, \xi\, \frac{1}{\xi} \mathrm{e}^{-|t_2 - t_4|/\tau_r} \;\approx\; \langle a^2 \rangle\, \xi\, \delta(t_2 - t_4)\,, \tag{2.62}$$

where $\delta(t)$ is the Dirac δ-function, which we introduced already in (1.8). We hence need to calculate

$$\int_0^{t_1} dt_2 \int_0^{t_3} dt_4\, \mathrm{e}^{(t_2 + t_4)/\tau_r}\, \delta(t_2 - t_4) \;=\; \int_0^{t_1} dt_2\, \mathrm{e}^{2t_2/\tau_r} \int_0^{t_3} dt_4\, \delta(t_2 - t_4)\,.$$

Note that the δ function vanishes unless $t_2 = t_4$, so that we could replace t_4 by t_2 in the exponential function. Performing the t_4 integration, we note that this integral equals 1 if the arguments vanis once during integration and vanishes else (since the δ-function vanishes all the way through integration). We therefore find

$$\int_0^{t_1} dt_2 \int_0^{t_3} dt_4\, \mathrm{e}^{(t_2 + t_4)/\tau_r}\, \delta(t_2 - t_4) \;=\; \int_0^{\min(t_1, t_3)} dt_2\, \mathrm{e}^{2t_2/\tau_r}$$

$$=\; \frac{\tau_r}{2}\left[\mathrm{e}^{2\min(t_1, t_3)/\tau_r} - 1 \right].$$

Inserting this into (2.60), we complete the t_1 and t_3 integration:

$$\frac{\xi \langle a^2 \rangle \tau_r}{2} \int_0^t dt_1\, dt_3\, \mathrm{e}^{-(t_1 + t_3)/\tau_r}\left[\mathrm{e}^{2\min(t_1, t_3)/\tau_r} - 1 \right]$$

$$=\; \frac{\xi \langle a^2 \rangle \tau_r}{2} \int_0^t dt_1\, \mathrm{e}^{-t_1/\tau_r}\left[\int_0^{t_1} dt_3\, \mathrm{e}^{-t_3/\tau_r}\left(\mathrm{e}^{2t_3/\tau_r} - 1 \right) \right.$$

$$+\; \left. \int_{t_1}^t dt_3\, \mathrm{e}^{-t_3/\tau_r}\left(\mathrm{e}^{2t_1/\tau_r} - 1 \right) \right]$$

$$=\; \xi \langle a^2 \rangle \tau_r^2\left[t - \frac{3}{2}\tau_r + 2\tau_r\, \mathrm{e}^{-t/\tau_r} - \frac{1}{2}\tau_r\, \mathrm{e}^{-2t/\tau_r} \right].$$

We finally obtain the variance of the position (2.60):

$$\mathrm{var}(x) \;=\; \langle x^2(t) \rangle - \langle x(t) \rangle^2 \;=\; \tau^2\, \mathrm{var}(u_0)\left[1 - \mathrm{e}^{-t/\tau_r} \right]^2 \tag{2.63}$$

$$+\; \xi \langle a^2 \rangle \tau_r^2\left[t - \frac{3}{2}\tau_r + 2\tau_r\, \mathrm{e}^{-t/\tau_r} - \frac{1}{2}\tau_r\, \mathrm{e}^{-2t/\tau_r} \right].$$

There are two terms contributing to the variance of the position: the first term arises from the noise of the initial velocity and is, hence, proportional to $\mathrm{var}(u_0)$. The second term arises from the noise in the random acceleration of the particle.

The theory of Thermodynamics relates the average kinetic energy (2.57) of the test particle to the temperature T (Equipartition theorem) [15]:

$$\left\langle E_{\text{kin}} \right\rangle = \frac{1}{2} k_b T ,$$

where k_b is the Boltzmann constant. Comparing this with our result (2.57), we identify

$$\langle a^2 \rangle \tau_r \, \xi = \frac{k_B T}{m} . \tag{2.64}$$

This yields

$$\text{var}(x) = \tau^2 \, \text{var}(u_0) \left[1 - e^{-t/\tau_r} \right]^2 \tag{2.65}$$
$$+ \frac{k_B T}{m} \tau_r \left[t - \frac{3}{2}\tau_r + 2\tau_r \, e^{-t/\tau_r} - \frac{1}{2}\tau_r \, e^{-2t/\tau_r} \right] .$$

For small times t, i.e. $\xi \ll t \ll \tau_r$, we find:

$$\text{var}(x) = \text{var}(u_0) \left[t^2 - \frac{t^3}{\tau_r} \right] + \frac{k_B T}{m} \tau_r \frac{t^3}{3\tau_r^2} + \mathcal{O}(t^4) .$$

For large times, i.e. $t \gg \tau_r$, the leading behaviour is a linear rise of the variance with time:

$$\text{var}(x) \approx \frac{k_B T}{m} \tau_r \, t . \tag{2.66}$$

We recover normal *diffusive behaviour*. Comparing the latter equation with (2.14) and (2.15), we find for the diffusion constant D:

$$D = \frac{k_B T}{2m} \tau_r . \tag{2.67}$$

A linear relation between the temperature and the diffusion constant was first found by Einstein and independently William Sutherland. This relation is now known as Einstein relation in the kinetic theory in Physics.

To round off this section, we are going to study the correlation between noise and position, i.e.

$$\langle a(t_a) \, x(t) \rangle . \tag{2.68}$$

In the literature, sometimes you find put forward that this correlation vanishes. This is an additional assumption, which may prove in contradiction with present model. We would expect that the above correlation vanishes for $t_a > t$ due to causality, but for $0 \leq t_a \leq t$, the model in this section provides all ingredients to calculate this correlation. Indeed, using the trajectory (2.59), we find

$$C(t_a, t) := \langle a(t_a) \, x(t) \rangle = \int_0^t dt_1 \, e^{-t_1/\tau_r} \int_0^{t_1} e^{t'/\tau_r} \langle a(t_a) a(t') \rangle \, dt' .$$

where we have used (2.61). Using the approximation (2.62), we obtain

$$\langle a(t_a)\,x(t)\rangle \;=\; \langle a^2\rangle\,\xi \int_0^t dt_1\, e^{-t_1/\tau_r} \int_0^{t_1} e^{t'/\tau_r}\,\delta\!\left(t_a - t'\right) dt' \; .$$

Here, we can draw a first important conclusion: if $t < t_a$, then $t_1 < t_a$ since t is upper limit of the t_1 integration. If $t_1 < t_a$, then $t' < t_a$ for the same reason. In this case, the argument of the δ-function never vanishes and hence the δ-function is zero leading to

$$\langle a(t_a)\,x(t)\rangle \;=\; 0\;, \qquad \text{for } t_a > t$$

in agreement with causality. For $t_a \le t$, we simplify using the properties of the δ-function:

$$
\begin{aligned}
\langle a(t_a)\,x(t)\rangle &= \langle a^2\rangle\,\xi \int_0^t dt_1\, e^{-(t_1-t_a)/\tau_r} \int_0^{t_1} \delta\!\left(t_a - t'\right) dt' \\
&= \langle a^2\rangle\,\xi \int_{t_a}^t dt_1\, e^{-(t_1-t_a)/\tau_r} \;=\; \langle a^2\rangle\,\xi\,\tau_r\left[1 - e^{-(t-t_a)/\tau_r}\right].
\end{aligned}
$$

Using thermodynamic parameters (2.63), we finally find

$$\langle a(t_a)\,x(t)\rangle \;=\; \frac{k_B T}{m}\left[1 - e^{-(t-t_a)/\tau_r}\right] \qquad \text{for } t_a \le t\,. \tag{2.69}$$

There is a significant (and causal) correlation between noise and position, which approaches a constant for $t \gg t_a$.

2.4.2 Diffusion in external potentials

We now study the case of diffusion with a non-homogeneous potential $V(x)$ (2.47). We will assume that the motion is dominated by strong friction and set $m = 0$ in (2.47). This will also provide us with a new method to generate a set of random variables that are distributed according to a design probability distribution. The starting point is a discretised version of (2.47) (with $m = 0$):

$$dx \;=\; -\frac{dV(x)}{dx}\,dt \;+\; \eta(t)\,. \tag{2.70}$$

The change dx of position has two elements: an advection step, $\frac{dV(x)}{dx}\,dt$, according to classical mechanics and a random displacement η, which took place within time step dt due to collisions of the test particle with the surrounding medium. We will assume that the distribution of the displacements η are normally distributed:

$$P_\eta \;=\; \frac{1}{2\sqrt{\pi\,D\,dt}}\,\exp\{-\eta^2/4D\,dt\}\,.$$

Note that the width of the distribution is given by $\sqrt{D\,dt}$ leading to standard diffusive behaviour:

$$\text{var}(\eta) \propto dt\ .$$

To describe the system of the test particle exposed to random interaction with a medium, we introduce the probability distribution $P(x,t)$ for the position x for any given time $t > 0$.

Assume that the distribution at time t, i.e. $P(x,t)$, is available. Within time dt, the particle is moving to

$$y \;=\; x - \frac{dV(x)}{dx}\,dt \;+\; \eta(t)\ .$$

Hence, the probability that the particle is found at x at time t is given by $P(x,t)\,dx$. Hence, we find for the distribution at time $t + dt$:

$$P(y, t+dt) \;=\; \int_{-\infty}^{\infty} P(x,t)\,\delta\!\left(y - x + \frac{dV(x)}{dx}\,dt - \eta \right) \qquad (2.71)$$

$$\frac{1}{2\sqrt{\pi\,D\,dt}}\,\exp\{-\eta^2/4D\,dt\}\,dx\,d\eta\ .$$

It is the product of the probability to find the particle at x and the probability that the particle moves to y. Once, $\eta(t)$ is known, the motion to y is deterministic and given by equation (2.70). This is reflected by the δ-function in (2.71), which nullifies the integrand if its arguments do not vanish. Carrying out the η integration with the help of the δ-function, we find

$$P(y, t+dt) \;=\; \int_{-\infty}^{\infty} P(x,t)\,\frac{1}{2\sqrt{\pi\,D\,dt}}\,\exp\left\{ -\frac{\left(y - x + \frac{dV(x)}{dx}\,dt \right)^2}{4D\,dt} \right\}\,dx$$

We substitute for x:

$$2\sqrt{D\,dt}\,u \;=\; y - x$$

to obtain (using the shorthand notation: $V'(x) = dV/dx$):

$$P(y, t+dt) \;=\; \int_{-\infty}^{\infty} du\,\frac{1}{\sqrt{\pi}}\,P(y - 2\sqrt{D\,dt}\,u, t)$$

$$\exp\left\{ -\frac{\left(2\sqrt{D\,dt}\,u + V'(y - 2\sqrt{D\,dt}\,u)\,dt \right)^2}{4D\,dt} \right\}\,du$$

Expanding systematically in powers of $\sqrt{dt}$ up to and including order $dt^{3/2}$, we can easily perform the u integral. Upon this integration, the terms proportional to $dt^{1/2}$ and $dt^{3/2}$ vanish leaving us with:

$$P(y, t+dt) \;=\; P(y,t) + D\,\frac{\partial^2 P}{\partial y^2} + \frac{\partial}{\partial y}\left[P(y,t)\,V'(y) \right]\,dt \;+\; \mathcal{O}(dt^2)\ .$$

We now take the limit $dt \to 0$ to finally obtain a partial differential equation that describes the time evolution of the probability distribution (we rename $y \to x$):

$$\frac{\partial P(x,t)}{\partial t} = \frac{\partial}{\partial x}\left[D\frac{\partial P}{\partial x} + \frac{dV(x)}{dx}P\right]. \qquad (2.72)$$

This is the well-known Fokker-Planck equation [16] named after the physicists Adriaan Fokker and Max Planck. In the absence of any potential V, the equation collapses to the PDE (2.17), which we have studied in the context of the Wiener process in subsection 2.2.3.

Noticeably equation (2.72) has a time-independent, i.e. stationary solution with

$$D\frac{\partial P_{\mathrm{eq}}}{\partial x} + \frac{dV(x)}{dx}P_{\mathrm{eq}} = 0$$

implying

$$P_{\mathrm{eq}}(x) = c\,\exp\left\{-\frac{V(x)}{D}\right\},$$

where the integration constant c needs to be chosen to ensure proper normalisation of the probability distribution:

$$c^{-1} = \int_{-\infty}^{\infty} dx\,\exp\left\{-\frac{V(x)}{D}\right\} = 1.$$

We will show that any initial distribution $P(x,t=0)$ will "thermalise" to the stationary distribution $P_{\mathrm{eq}}(x)$. To this aim, we introduce the $Q(x,t)$ by:

$$P(x,t) = \exp\left\{-\frac{V(x)}{2D}\right\}Q(x,t). \qquad (2.73)$$

Inserting this equation into the Fokker-Planck equation (2.72) yields

$$-\frac{\partial}{\partial t}Q(x,t) = D\,H\,Q(x,t), \qquad (2.74)$$

$$H = -\frac{\partial^2}{\partial x^2} - \frac{1}{2D}\frac{d^2V}{dx^2} + \frac{1}{4D^2}\left(\frac{dV}{dx}\right)^2. \qquad (2.75)$$

The important observation is that H is a semi-positive operator:

$$H = R^\dagger R, \qquad R := i\frac{\partial}{\partial x} + \frac{i}{2D}\frac{dV}{dx}, \qquad R^\dagger := i\frac{\partial}{\partial x} - \frac{i}{2D}\frac{dV}{dx}.$$

Indeed for any sufficiently smooth and normalisable test function $\phi(x)$, we have

$$\int dx\,\phi^\dagger(x)\,H\,\phi(x) = \int dx\,\left[R\phi(x)\right]^\dagger R\phi(x) \geq 0.$$

In addition, H is a Hermitean operator and thus possesses a complete set of eigenfunctions $\psi_n(x)$ [1]. Because of the (semi-)positivity, its eigenvalues λ are semi-positive:

$$H\,\psi_n(x) \;=\; \lambda_n\,\psi_n(x)\,, \qquad \lambda_n \geq 0\,.$$

Because the Fokker-Planck equation has a stationary solution, there is at least one eigenfunction $\psi_0(x)$ with vanishing eigenvalue:

$$\psi_0(x) \;=\; \exp\left\{-\frac{V(x)}{2D}\right\}, \qquad \lambda_0 = 0\,. \tag{2.76}$$

At any time t, we can expand the function $Q(x,t)$ with respect to the orthogonal eigenfunctions:

$$Q(x,t) \;=\; \sum_n q_n(t)\,\psi_n(x)\,.$$

By inserting this into (2.75), the time dependence of the coefficients q_n is determined by

$$-\frac{dq_n(t)}{dt} \;=\; D\,\lambda_n\,q_n(t)\,.$$

The solution is given by

$$q_n(t) \;=\; q_n(0)\,\exp\{-D\,\lambda_n\,t\}\,.$$

Inserting this into (2.72) to recover the probability distribution $P(x,t)$ and using (2.76), we finally find

$$
\begin{aligned}
P(x,t) \;=\;& q_0(0)\,\exp\left\{-\frac{V(x)}{D}\right\} \\[2mm]
+\;& \exp\left\{-\frac{V(x)}{2D}\right\}\sum_{n=1}^{\infty} q_n(0)\,e^{-\lambda_n D\,t}\,\psi_n(x)\,.
\end{aligned} \tag{2.77}
$$

Since $\lambda_n > 0$ for $n \geq 1$, the contributions from the modes $\psi_{n\geq 1}$ are exponentially damped in time. Any initial distribution $P(x,0)$ exponentially approaches the stationary distribution $P_{\mathrm{eq}}(x)$, a process that we call "thermalisation".

2.4.3 Langevin simulations

In subsection 2.4.1, we have studied the motion of particle exposed to a (smooth) potential and a radom force. We then studied expectation values of e.g. kinetic energy or position for the case of vanishing potential. These expectation values were governed by a (time dependent) distribution function of the postion of the particle. This distribution function was normal, and, hence, describing a standard diffusion process.

---

[1] To simplify notation, we here assumed that there is countable, discrete set of eigenvectors

In this subsection, we will generalise the approach to more variables $x_1 \ldots x_n$. The goal will be to generate sets of variables $\{x_1 \ldots x_n\}$ that are distributed according to the probability

$$[dx] \, \exp\{-S(x_1 \ldots x_n)\} \,, \qquad [dx] := \prod_{k=1}^{n} dx_k \,. \qquad (2.78)$$

These sets can then serve to estimate observables in a Monte-Carlo simulation. To our knowledge, this method has been firstly proposed by Parisi and Wu [17]. A review can be found in [18].

To this aim, we promote the variables to functions $x_k(\tau)$ of the artificial time τ (Langevin time). The time evolution is dictated by the Langevin equations

$$\frac{dx_k(\tau)}{d\tau} = -\frac{\partial S}{\partial x_k} + \eta_k(\tau), \qquad (2.79)$$

where η_k is a random variable with a normal probability distribution. Details will follow below. For the numerical algorithm, we consider the evolution over a small time step $\Delta\tau$ and approximate:

$$x_k(\tau + \Delta\tau) \approx -\frac{\partial S}{\partial x_k} \Delta\tau + \eta_k \,, \qquad \eta_k = \int_{\tau}^{\tau+\Delta\tau} \eta_k(u) \, du \,. \qquad (2.80)$$

We model the noise by a Wiener process (see subsection 2.2.3) and assume that $\eta_k = (\vec{\eta})_k$ is a random variable with Gaussian distribution:

$$P(\vec{\eta}) = (4\pi\Delta\tau)^{-n/2} \exp\left\{-\frac{\vec{\eta}^2}{4\Delta\tau}\right\} \,.$$

With this setting, equation (2.80) generates a sequence of random vectors:

$$\vec{x}(0) \longrightarrow \vec{x}(\Delta\tau) \longrightarrow \vec{x}(2\Delta\tau) \longrightarrow \vec{x}(3\Delta\tau) \longrightarrow \ldots.$$

one can show along the lines of subsection 2.4.1 that the probability distribution of the random variable $\vec{x}$ asymptotically approaches the Boltzmann-Gibbs distribution

$$P(\vec{x}) \propto \exp\left\{-S(x_1 \ldots x_n)\right\} \,.$$

This is an important finding since it shows that the Langevin process is an option for generating ensembles following a Boltzmann-Gibbs distribution.

2.5   Exercises

Exercise 2.1 *Consider a symmetric ($p = q = 1/2$) Random Walk in d-dimensions with N steps in total, where $N > 0$ is an even number. Consider one such Random Walk as an event. For each event, we define the random variable R as follows: $R = 1$ when the walker returned to the origin at least once; choose $R = 0$ else.*

Carry out a large number of these Random Walks and estimate $\langle R \rangle$ for the dimensions $d = 1$ (walk on a line), $d = 2$ (walk in a plane) and $d = 3$ (walk in space).

If you want to find your way home, in what dimension is a Random Walk a viable strategy?

Exercise 2.2 *Study the properties of a symmetric ($p = q = 1/2$) Random Walk on a line with N steps in total, where $N > 0$ is an even number.*

Exercise 2.3 *Consider a random walk on three-dimensional lattice with step size 1. At each time step, the walker randomly chooses one of the six directions to step. The walker starts at the origin $\vec{x} = 0$, and the total number of steps is n_t. Let r denote the (Euclidean) distance from the origin after n_t steps.*

Assume that the probability distribution $P(\vec{x})$ for the position $\vec{x}$ after n_t steps is Gaussian. Show that the marginal distribution $p(r)$ is given by:

$$P(\vec{x}) = \frac{1}{(\pi a)^{3/2}} \exp\left\{-\frac{\vec{x}^2}{a}\right\}, \qquad p(r) = \frac{4\pi}{(\pi a)^{3/2}} r^2 \exp\left\{-\frac{r^2}{a}\right\}. \tag{2.81}$$

In a Computer experiment, choose $n_t = 500$, carry out $1,000,000$ random walks and record each time the end point distance r. Create a histogram of these values of r to estimate the marginal distribution $p(r)$.

Repeat the above experiment for $n_t = 1000$ and show both histograms in one graph.

Fit $b\,p(r)$ in (2.81) to each of your histograms (fit parameters are b and a). Convince yourself that $a(n_t = 1000) \approx 2\,a(n_t = 500)$.

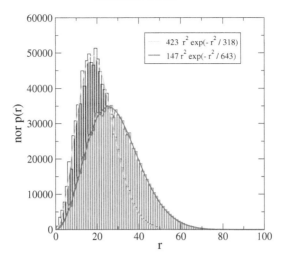

FIGURE 2.4: Estimate for the marginal distribution $p(r)$ for a three-dimensional random walk on a lattice for times $n_t = 500$ and $n_t = 1000$.

Comment: the parameter a should be proportional to Dt, D diffusion constant and t time, with $t = n_t$ in our case (see (2.16)). See Figure 2.4 for a solution.

Exercise 2.4 *Consider a random walk on a regular lattice in d dimensions. At each time step, the walker decides in which of the 2d direction they want to step. The walker starts at the origin and undertakes n_t steps in total. Consider a target point on the x^1-axis at distance d from the origin.*

Carry out $N = 100,000$ random walks and count the number of times ν that the walker accesses the target point. The ratio ν/N estimates the probability that the walker reaches the target point.

Choose $d = 2$ dimensions and $n_t = 2000, 4000, 6000$. For each choice, plot the estimated probability ν/N as a function of the distance from 1, 2, ..., 20.

Repeat the same suit of simulations for $d = 3$ dimensions and compare.

Comment: In $d = 2$, a strong time (i.e. n_t) dependence is observed. Note that for a given target distance, the probability always increases with increasing time since the walker has, at later times, more opportunity to hit the target. The numerical results are thus lower bounds for the $t \to \infty$ limit. In two dimensions, the expectation is that the probability to reach any point is one after an infinite amount of walking time.

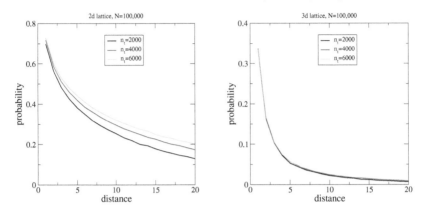

FIGURE 2.5: Probability of reaching a given target point with a "distance" from the origin within time n_t in two dimensions (left) and three dimensions (right).

In $d = 3$ dimensions, the result is different. Even at moderate times n_t, the probabilities for a given distance do not change much anymore, and we would expect a $\nu/N \propto 1/r$ behaviour (see (2.31)).

See Figure 2.5 for an illustration.

Chapter 3

Monte Carlo methods

In this chapter, we will introduce the basic concepts of Monte Carlo methods, which are widely used techniques, in particular to evaluate expectation values and other estimators over a given probability distribution. When the number of stochastic variables becomes large, an effective implementation of such methods relies on stochastic processes similar to those introduced in Chapter 2, leading to the so-called Markov chain Monte Carlo (MCMC), to which much of this chapter is dedicated. A crucial aspect of Monte Carlo methods is the correct assessment of statistical errors on the estimated quantities: in the case of non-trivial estimators, or in the presence of statistical correlations

DOI: 10.1201/9781315156156-3

among successive draws, a correct error estimate requires advanced statistical analysis techniques, which are introduced in the second part of this chapter

---

3.1 Objectives and concepts

In this section, we outline the main features of a Monte Carlo simulation, aimed at estimating the average of a function of stochastic variables over a given probability distribution, and the associated statistical uncertainty. The practical problem of actually sampling the distribution, as well as the case of more complex estimators, is discussed in the remainder of this chapter.

3.1.1 Simulation and error bar

One important application of Monte Carlo methods is to estimate the expectation values. Assume that random variables x are drawn from a probability distribution $P(x)$, then consider a mapping $f(x)$ of such variables. Our task is to estimate the mean value (or expectation value) of f:

$$\langle f \rangle = \int dx \, f(x) \, P(x) \,. \tag{3.1}$$

One sidetrack of this application is to compute integrals which can be put in the same form as the last equation. If the integration is over a high-dimensional space, i.e., $x \in \mathbb{R}^n$, $n \gg 1$, Monte Carlo methods are one of the most important numerical methods. This will be explored in detail in the next subsection.

The central idea is to generate a random sequence $f_n = f(x_n)$ and to estimate the expectation value (3.1) using

$$\langle f \rangle = \lim_{n \to \infty} \frac{1}{n} \sum_{k=1}^{n} f_k \,. \tag{3.2}$$

In practical applications, we sample the values f_n up to a large value of n. The exact expression (3.2) thereby turns into an approximation. If we are generating approximations, the estimate of the approximation error is as important as the estimate itself since otherwise the accuracy of our result cannot be judged. In the remainder of this subsection, we will develop such an error estimate, limiting ourselves to the case in which the x_n is drawn independently of each other; the more general case will be discussed later on.

Note that the random values $f_n = f(x_n)$ emerge from the random sequence $x_1 \ldots x_n$, which is distributed according to $P(x)$. We can interpret the random variables f as generated by a probability distribution $Q(f)$. We could

TABLE 3.1: Confidence intervals

error	σ	2σ	3σ
confidence	68.3%	95.4%	99.7 % .

numerically find this distribution by the histogram method discussed in sub-section 1.5. We will assume that the distribution $Q(f)$ does possess a second moment σ_f and that the central limit theorem 1.2.1 does apply. For sufficiently large n, the probability distribution of the average estimate (sample average)

$$\tilde{f} := \frac{1}{n} \sum_{k=1}^{n} f_k \qquad (3.3)$$

is then normal distributed with a standard deviation of $\sigma = \sigma_f/\sqrt{n}$. It is customary to quote one standard deviation as (absolute) error for the estimate $\tilde{f}$:

$$\langle f \rangle = \tilde{f} \pm \sigma .$$

In this case, the probability that $\tilde{f}$ approximates the true expectation value $\langle f \rangle$ within error bars is given by:

$$\int_{-\sigma}^{\sigma} df \, \frac{1}{\sigma\sqrt{2\pi}} \exp\left\{-\frac{f^2}{2\sigma^2}\right\} = 0.6826894920... .$$

If we want to quote a more conservative error bar, we can use a multiple of σ. The corresponding "confidence" probabilities are listed in Table 3.1.

How do we find σ? In practice, we have only the n random variables f_n at our fingertips. The task is to find σ_f since $\sigma = \sigma_f/\sqrt{n}$. We start with the definition

$$\sigma_f^2 = \left\langle (f - \mu)^2 \right\rangle = \left\langle f^2 \right\rangle - \mu^2 . \qquad (3.4)$$

The central idea is to replace the expectation values, and, since μ is also unknown, to use $\tilde{f}$ (3.3) as a substitute. Let us study the variable

$$s_f^2 = \frac{1}{n-1} \sum_{i=1}^{n} (f_i - \tilde{f})^2 , \qquad (3.5)$$

where the reason for the normalisation by $(n-1)$, instead of n, will be clear soon and is essentially related to the fact that the sample variables, f_i, fluctuate a bit less around their sample average than they would do around the true expectation value μ.

We consider now the expectation value $\langle s_f^2 \rangle_s$, where by $\langle \cdot \rangle_s$ we mean the average taken over the stochastic process which generates the sample, which in this case is particularly simple, since the variables f_i are all drawn from the same distribution $Q(f)$ (hence $\langle \cdot \rangle_s$ reduces to $\langle \cdot \rangle$ for averages at fixed i) and independently of each other. We find

$$
\begin{aligned}
\langle s_f^2 \rangle_s &= \frac{1}{n-1} \sum_{i=1}^{n} \left\langle \left[(f_i - \mu) - (\tilde{f} - \mu) \right]^2 \right\rangle_s \\
&= \frac{1}{n-1} \sum_{i=1}^{n} \left[\left\langle (f_i - \mu)^2 \right\rangle_s - 2 \left\langle (f_i - \mu)(\tilde{f} - \mu) \right\rangle_s + \left\langle (\tilde{f} - \mu)^2 \right\rangle_s \right].
\end{aligned}
$$

Using

$$
\left\langle (f_i - \mu)^2 \right\rangle_s = \sigma_f^2 \qquad \forall i,
$$

$$
\sum_{i=1}^{n} \left\langle (f_i - \mu)(\tilde{f} - \mu) \right\rangle_s = n \left\langle (\tilde{f} - \mu)^2 \right\rangle_s,
$$

we obtain

$$
\langle s_f^2 \rangle_s = \frac{1}{n-1} \left[n \sigma_f^2 - n \left\langle (\tilde{f} - \mu)^2 \right\rangle_s \right]. \tag{3.6}
$$

We finally need to work out the latter expectation value:

$$
\left\langle (\tilde{f} - \mu)^2 \right\rangle_s = \frac{1}{n^2} \sum_{i,k} \left\langle (f_i - \mu)(f_k - \mu) \right\rangle_s.
$$

Since the random variables f_i are independent, we find that the above expectation value on the right-hand side vanishes as long as $k \neq i$. Hence, the only contribution arises when $i = k$:

$$
\left\langle (\tilde{f} - \mu)^2 \right\rangle_s = \frac{1}{n^2} \sum_{i} \left\langle (f_i - \mu)^2 \right\rangle_s = \frac{1}{n} \sigma_f^2, \tag{3.7}
$$

which is consistent with the results of the central limit theorem. Inserting (3.7) into (3.6), we finally find the important result:

$$
\langle s_f^2 \rangle_s = \frac{1}{n-1} \left[n \sigma_f^2 - \sigma_f^2 \right] = \sigma_f^2. \tag{3.8}
$$

This paves the path to estimate the unknown variance σ_f. For large n, the variable s_f in (3.5) does not fluctuate much, and we might estimate

$$
s_f^2 \approx \langle s_f^2 \rangle_s = \sigma_f^2 \qquad (n \gg 1). \tag{3.9}
$$

3.1.2 Estimating expectation values

Let us put the formalism of the previous subsection to the test. We will generate random numbers from the exponential distribution (see subsection 1.1.4 for the method of how to do this)

$$
P(x) = \exp\{-x\}.
$$

We then map the random variables with function $f(x) = x^2$. We will seek an estimate (and error bars) for the expectation value:

$$\langle f \rangle = \int_0^\infty f(x)\, P(x)\, dx = \int_0^\infty x^2 \exp\{-x\}\, dx = 2 . \tag{3.10}$$

To this aim, we will generate n random variables $x_{1...n}$ and calculate the corresponding $f_k = f(x_k)$ "observables". For the estimate for $\langle f \rangle$, we simply calculate the average

$$\tilde{f} = \frac{1}{n} \sum_{k=1}^n f_k . \tag{3.11}$$

We also need to estimate the standard deviation σ_f of the probability distribution of the variables f. This will provide the error bar $\sigma_f/\sqrt{n}$ at one σ confidence level. Before we proceed with the numerical approach, let us calculate σ_f exactly. If $Q(f)$ is the probability distribution of the f random variable, its definition is

$$\sigma_f^2 = \int_0^\infty (f - \langle f \rangle)^2\, Q(f)\, df .$$

For $x \geq 0$, we substitute f for x:

$$x = \sqrt{f}, \qquad dx = \frac{1}{2\sqrt{f}}\, df$$

and find

$$P(x)\, dx = \exp\{-\sqrt{f}\} \frac{df}{2\sqrt{f}} = Q(f)\, df .$$

This leads to

$$\sigma_f^2 = \int_0^\infty \left(f - \langle f \rangle\right)^2 e^{-\sqrt{f}} \frac{df}{2\sqrt{f}} .$$

Rather than attempting this integral, it is convenient to reverse the substitution:

$$\sigma_f^2 = \int_0^\infty (x^2 - \langle f \rangle)^2 e^{-x}\, dx = 20 .$$

As detailed previously, an estimate for σ_f is provided by (see 3.5):

$$s_f^2 = \frac{1}{n-1} \sum_{i=1}^n (f_i - \tilde{f})^2 = \frac{1}{n-1} \sum_{i=1}^n f_i^2 - \frac{n}{n-1} \tilde{f}^2 . \tag{3.12}$$

Our numerical estimate $\tilde{f}$ is shown in Figure 3.1 as a function of n. Also shown are one standard deviation error bars. The exact result is indicated by the red line. Note that the exact result is for all n within the margins outlined by the error bar. To round off this section, the right panel of Figure 3.1 shows the estimate s_f for the standard deviation σ_f in comparison with the exact result $\sigma_f = 20$.

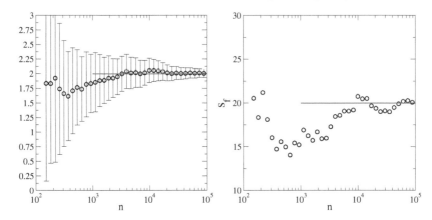

FIGURE 3.1: The estimate $\tilde{f}$ with error bars (one standard deviation confidence) as a function of n (left). The estimated s_f for the standard deviation σ_f (right).

3.2   Monte Carlo integration

We have outlined an efficient strategy to evaluate averages over a given distribution function, like that in Eq. (3.1). Such a strategy assumes that one is actually able to generate a sequence $x_1 \ldots x_n$, sampled according to the distribution $P(x)$. The construction of a numerical algorithm providing this sample is at the core of the Monte Carlo method, and often represents one of the major technical challenges.

In Chapter 1, we have described a simple method to draw a stochastic variable x according to the given distribution $P(x)$, starting from a (pseudo)random number generator providing stochastic variables u which are uniformly distributed in the interval $[0, 1]$. Such a method assumes that one is able to integrate and invert Eq. (1.11). This is not always feasible and, whereas valid alternatives exist for one or few stochastic variables, like, for instance, the von Neumann accept/reject method that we describe below, the task of creating an algorithm capable of sampling random variables according to a given distribution becomes unfeasible as the number of variables becomes very large; unfortunately, this is also the case for which Monte Carlo methods are expected to perform much better than other standard numerical integration techniques.

Trying to make the Monte Carlo approach successful and viable for a large class of complex problems, the first question is: are we really forced to sample according to a given distribution? After all, an integral like that in Eq. (1.11) can be interpreted in many ways, and one of them is the average of the function $f(x)P(x)$ over a flat distribution, which is much easier to reproduce; moreover, the scaling of the statistical uncertainty as the inverse square root of the sample size n, $\sigma = A/\sqrt{n}$, is a general result which is independent of the sampling distribution. However, as we discuss in detail below, the prefactor A can be strongly dependent on the sampling distribution, especially in the quite common case in which the probability distribution $P(x)$ is strongly peaked and significantly different from zero only in a very small region of the stochastic variables space. That forces, in practice, to sample as close as possible to the actual distribution $P(x)$.

A second important point is the following. What is really difficult, and usually not feasible, is to devise an algorithm whose only task is to make a single draw of the stochastic variables x according to $P(x)$. However, a viable alternative is to devise a stochastic process, in terms of a fictitious time, capable of generating a sort of random walk around the x space which, waiting enough time, will visit each volume dx with a frequency proportional to $dxP(x)$: the original average can then be traded off for a time average taken over the random walk. This simple idea reveals to be extremely powerful: the simplest and best framework to construct the stochastic process is provided by the theory of Markov processes and chains, which is concisely described in the following. A price one has to pay in this trade-off is that the estimates f_k taken along the stochastic walk will have some grade of correlation among themselves, so that the standard error analysis, based on the Central Limit Theorem, must be modified to take data autocorrelations into account. The price is well worth paying; however, the grade of autocorrelation between successive measurements emerges as a fundamental figure to judge the level of performance of a particular Monte Carlo algorithm.

3.2.1 The accept-reject method

The analytic approach outlined in Eq. (1.11) which, based on a change of variables, permits to obtain a given distribution starting from the uniform one, is practicable only in a few cases. A viable alternative is the so-called accept-reject method, which is usually associated with the name John von Neumann.

Suppose we do not know how to sample a given distribution $p(x)dx$, which for simplicity we assume to be restricted to the interval $[a, b]$ of the real axis, as in Figure 3.2. Suppose, however, that we are able to sample another distribution $g(x)$, defined over the same domain[1], such that a constant C exists for which $Cg(x) \geq p(x)$ for every x, as in Figure 3.2. Obviously, since both $g(x)$ and $p(x)$ are normalized distributions, we have $C \geq 1$, with $C = 1$ holding only when $g(x) = p(x)$.

---

[1]The domain could actually be larger, but then $p(x)$ could be redefined over the same domain and set to zero outside $[a, b]$, thus without changing the rest of the argument.

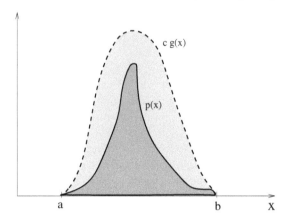

FIGURE 3.2: Accept-Reject method: $p(x)$ is the distribution we would like to sample, $g(x)$ is the distribution we know how to sample, C is a constant chosen so that $Cg(x) > p(x) \; \forall \; x$. The algorithm shoots uniformly over the region below $Cg(x)$, then only shoots hitting (with probability $1/C$) the region below $p(x)$ are accepted, i.e. their x coordinate is taken as an element of the sample.

The accept-reject algorithm is composed of the following two steps:

1. A pair of stochastic variables (x, y) is drawn, with x distributed according to $g(x)dx$ in the $[a, b]$ domain, and y uniformly distributed in the domain $[0, C\,g(x)]$;

2. If $y \le p(x)$, then x is accepted as an element of our sample; otherwise, it is rejected and we go back to point 1.

We are now going to prove that the accepted values of x are distributed according to $p(x)dx$. The probability of drawing x in the range $[x, x + dx]$ is $g(x)dx$. Once x is known, the probability of drawing y in $[y, y + dy]$ is $dy/(Cg(x))$; hence, the joint probability is given by

$$g(x)dx\frac{dy}{Cg(x)} = \frac{dxdy}{C}$$

i.e. in step 1. the pair is drawn with a uniform probability in the yellow area below the curve $Cg(x)$ in Figure 3.2.

In the accept-reject step 2, only pairs lying below the curve $p(x)$ are kept, so that the final sample of accepted points is uniformly distributed in the region below the curve $p(x)$, which has unit area. Hence, if we are interested in the distribution of x alone, this is given by

$$dx \int_0^{p(x)} dy = p(x)dx$$

proving that the accepted values of x are indeed distributed as required.

An important figure for this algorithm is the *acceptance* α, i.e. the probability of accepting x, whose inverse gives the average number of trials needed to perform one draw, which is a measure of the numerical effort required by the algorithm. The acceptance is given by the ratio of the area below $p(x)$ to that below $Cg(x)$, hence we have

$$\alpha = \frac{1}{C}$$

since both $g(x)$ and $p(x)$ are normalized. It shows that, even if $g(x)$ can be chosen quite arbitrarily, the closer it is to the original distribution $p(x)$, the better the algorithm will perform.

Example: we construct a simple algorithm (Creutz algorithm) to sample the distribution $p(x) = \mathcal{N}\sqrt{1 - x^2}\exp(\gamma x)$ over the interval $[-1, 1]$, where $\mathcal{N}$ is a proper normalisation constant, and estimate its acceptance when $\gamma \gg 1$.

Let us consider the distribution

$$g(x)dx = \frac{\gamma}{e^\gamma - e^{-\gamma}}e^{\gamma x}dx\,,$$

which is properly normalized over the same interval and can be sampled by drawing a random number $u \in [0, 1]$ and taking

$$x = \frac{1}{\gamma}\ln\left(e^{-\gamma} + u(e^\gamma - e^{-\gamma})\right)\,.$$

The constant C must be chosen so that

$$C\frac{\gamma}{e^\gamma - e^{-\gamma}}e^{\gamma x} \geq \mathcal{N}\sqrt{1 - x^2}e^{\gamma x}$$

over the whole interval. The minimum possible value is then

$$C = \mathcal{N}\frac{e^\gamma - e^{-\gamma}}{\gamma}\,.$$

For this choice, the acceptance step consists of comparing a random number in $[0, e^{\gamma x}]$ with $\sqrt{1 - x^2}e^{\gamma x}$ (notice that the actual normalisation factor $\mathcal{N}$ is not needed by the algorithm), so that the acceptance is given by

$$\alpha = \frac{1}{C} = \frac{\int_{-1}^{1} dx\,\sqrt{1 - x^2}\,e^{\gamma x}}{\int_{-1}^{1} dx\,e^{\gamma x}}$$

which we now try to estimate in the limit $\gamma \gg 1$. In that regime, it becomes very frequent to draw x very close to 1, in particular the probability of drawing x between $1 - \delta$ and 1 is approximately $1 - e^{-\delta\gamma}$, so most numbers are extracted in a small region of size $1/\gamma$ around 1; however, in that region the correction factor $\sqrt{1 - x^2}$ becomes very small and the acceptance becomes small as well. In this regime, the acceptance can be well approximated by

$$\alpha \simeq \frac{\int_{1-1/\gamma}^{1} dx\,\sqrt{1 - x^2}\,e^{\gamma x}}{\int_{1-1/\gamma}^{1} dx\,e^{\gamma x}} \simeq \frac{\int_0^{1/\gamma} dz\,\sqrt{2z}\,e^{-\gamma z}}{\int_0^{1/\gamma} dz\,e^{-\gamma z}} \simeq \frac{1}{\sqrt{\gamma}}\frac{\int_0^1 dt\,\sqrt{2t}\,e^{-t}}{\int_0^1 dt\,e^{-t}} \propto \frac{1}{\sqrt{\gamma}}$$

where $t = \gamma z = \gamma(1 - x)$. Hence, while the algorithm works reasonably well for moderate values of γ, the acceptance goes to zero as $1/\sqrt{\gamma}$ in the large γ regime, where it is necessary to devise a different algorithm.

3.2.2   The importance of a proper sampling

Suppose we need to make a Monte Carlo estimate of the average of the function $f(x)$ over the probability distribution $p(x)$, which is non-zero in the range $[a, b]$:

$$\langle f \rangle_p = \int_a^b f(x)p(x)dx \,; \qquad \int_a^b p(x)dx = 1 \,. \qquad (3.13)$$

As we have discussed above, we need to build a sample x_k, with $k = 1, \ldots n$, distributed according to $p(x)$ and then evaluate the average

$$\tilde{f} = \frac{1}{n}\sum_{k=1}^{n} f(x_k) \qquad (3.14)$$

over the sample. However, Eq. (3.13) can be rewritten in many different ways, for instance, we can rewrite

$$\langle f \rangle_p = \int_a^b f(x)p(x)dx = \int_a^b f(x)\frac{p(x)}{\tilde{p}(x)}\tilde{p}(x)\,dx = \left\langle \frac{fp}{\tilde{p}} \right\rangle_{\tilde{p}} \qquad (3.15)$$

i.e. we can sample x according to a different weight $\tilde{p}$, which is assumed to be a normalized distribution over $[a, b]$, and then measure the sample average of $fp/\tilde{p}$: the *wrong* distribution used in the sampling process is *reweighted* by inserting the ratio $p/\tilde{p}$ in the function to be averaged, obtaining exactly the same result as $n \to \infty$. Since $\tilde{p}$ can be any distribution, and since the Monte Carlo error is expected to scale as $1/\sqrt{n}$ independently of the chosen distribution, that seems, at least in principle, a very powerful technique to avoid problems when we do not know how to sample the original distribution $p(x)$. However, as we are now going to demonstrate, the factor appearing in front of $1/\sqrt{n}$ can be very different depending on the choice of $\tilde{p}$.

Let us consider the case, which is common to many practical situations, in which $p(x)$ ed $f(x)$ are as depicted in Figure 3.3: $f(x)$ is a regular and smooth function over the whole range, while the distribution $p(x)$ is non-zero only over a very small region of size δ, where $p(x)$ takes values of order $1/\delta$ in order to be properly normalized. Suppose that we decide to sample according to a uniform distribution, i.e.

$$\tilde{p}(x) = u(x) = \frac{1}{|b - a|} \,,$$

then the statistical error after n measures is expected to be $\sqrt{\sigma_u^2/n}$, where

$$\sigma_u^2 = \langle f^2 p^2 \rangle_u - \langle fp \rangle_u^2$$
$$= \frac{1}{|b - a|}\int_a^b f(x)^2 p(x)^2 dx - \frac{1}{(b - a)^2}\left(\int_a^b f(x)p(x)dx\right)^2 . \, (3.16)$$

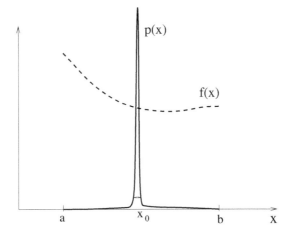

FIGURE 3.3: The importance of a proper sampling. If we need to compute the integral of $p(x)f(x)$ over the showed range, the best Monte Carlo strategy is to draw points around the region where the integral takes most of its contributions, which in this case means distributed as close as possible to $p(x)$.

Since the function $p(x)$ takes values of the order of $1/\delta$ over a range of order δ around x_0, the first integral will be of order $1/\delta$, if $f(x)$ is smooth around x_0, while the second integral is of order 1. Hence we have $\sigma_u^2 \sim 1/\delta$, i.e. it diverges as δ becomes smaller and smaller. From a practical point of view, the reason of this divergence is that most of the times we are sampling in a region where the function to be averaged is zero, apart from a few draws where it takes very large values, proportional to $1/\delta$: hence, most of the numerical effort goes for nothing, moreover fluctuations diverge.

On the other hand, if we sample according to $p(x)$, the statistical error will be $\sqrt{\sigma^2/n}$ where

$$\sigma^2 = \langle f^2 \rangle_p - \langle f \rangle_p^2 \tag{3.17}$$

and it is easy to verify that this quantity goes to zero proportionally to δ when $\delta \to 0$, if $f(x)$ is continous around x_0. Actually, were it possible, the best would be to sample according to $\tilde{p} \propto f(x)p(x)$, since in this case the variance would be exactly zero.

To summarise, despite the fact that the error goes as $1/\sqrt{n}$ for any sampling distribution, the square root of the variance in front of it can change abruptly, and the best strategy for variance reduction is to sample more where the integrand takes most contributions. This strategy is usually known as *importance sampling*.

3.3 Markov Chain Monte Carlo

Markov processes have been already introduced in Chapter 2. They are stochastic evolution processes, defined within a given set of stochastic

variables (which is called *the system* in the following) and in terms of a (usually fictitious) time variable, such that only the present state of the system
affects its future evolution. It is common to consider a discrete time variable,
so that the evolution proceeds step by step: in this case the process is called
Markov chain.

Markov processes resemble closely the evolution determined by a set of first
order differential equations, like Hamilton equations in phase space, or, in
the case of Markov chains, by a set of first-order recurrence relations[2]. In all
these cases, the common idea is: tell me where you are now, and I will tell
you where you are going in the future, no matter what you did in the past.
The only, obvious difference is that the evolution rule is not deterministic,
i.e. we are dealing with a stochastic process. Moreover, we assume that the
evolution rule stays unchanged during the stochastic process, i.e. it is *time
independent*. Actually, the simplest example of Markov process is the random
walk discussed in Chapter 2.

The idea of exploiting Markov chains to sample a given distribution is in some
sense dual to the idea at the basis of Statistical Mechanics (which is discussed
in Chapter 4). In the latter, one tries to rewrite time averages of some complex
system at thermal equilibrium in terms of averages taken over a given probability distribution, representing the so-called statistical ensemble. In Markov
chain Monte Carlo, the aim is instead to devise a stochastic dynamical process
such that averages taken over the fictitious Monte Carlo time coincide with
averages over a given probability distribution. There is no compelling reason
for adopting a Markov chain to do that, one could use a stochastic process
with memory of the past as well; however, the theory of Markov chains is much
simpler, thus allowing to predict their dynamics with relative ease, and leaves
at the same time enough freedom for choosing and optimizing algorithms for
a given problem.

In the following, we will assume, for simplicity, that the set of all possible
states of the system, i.e. of all the possible outcomes of the stochastic variables,
which is called *configuration space* in the following, is discrete, so that a given
state can be enumerated by an index a: we will denote such space by Ω.
Nothing dramatic happens when this is not true, just a change of some sums
into corresponding integrals. The Markov chain is then defined in terms of a
probability matrix W, which is usually called the *transition matrix*[3]:

Definition: $W_{ab} \equiv P(b \rightarrow a)$ gives the probability that the system, being in
state "b", moves to state "a" after one discrete time step.

The elements of the transition matrix will be encoded, more or less explicitly,
in the algorithm used to move from one state to the other in the computer

---

[2]Those do not include many famous recurrence sequences, like the Fibonacci one.

[3]For the sake of working with standard row-by-column multiplications in the composition
of Markov steps, we adopt a definition of the transition matrix which is the transposed of
what usually found in the literature.

implementation of the Markov chain. They have two obvious properties

$$0 \leq W_{ab} \leq 1 \ \forall \ a,b \in \Omega \ ; \qquad \sum_{a \in \Omega} W_{ab} = 1 \ \forall \ b \in \Omega. \qquad (3.18)$$

The first property simply states that W is a probability matrix, the second states that the stochastic process never stops, i.e. the system will surely reach some new state after each step.

It is quite easy to compute the probability of reaching a from b in k steps of the Markov chain: one has to consider the probability of a generic walk going from b to a, and then sum over all possible intermediate $k-1$ states, i.e.

$$P(b \rightarrow a, \ k \text{ steps}) = \sum_{c_1,c_2,\ldots c_{k-1}} W_{ac_{k-1}} \ldots W_{c_2,c_1} W_{c_1,b} = (W^k)_{ab}, \qquad (3.19)$$

therefore the k-steps transition matrix is simply given by the k-th power of the original matrix.

Let us now call p_a the target probability distribution that needs to be sampled. The idea is to generate, by means of the Markov chain, a stochastic walk $a^{(0)}, a^{(1)}, a^{(2)}, \ldots, a^{(k)}, \ldots$, where $a^{(0)}$ is a given starting state, such that each state a is visited with frequency p_a during the walk; in this way, states visited during the walk can be taken as a representative sample for Monte Carlo evaluation. That must be true independently of the starting state $a^{(0)}$, at least after waiting for an appropriate relaxation time. We can turn this requirement into a simple mathematical statement, we need that

$$\lim_{k \rightarrow \infty} (W^k)_{ab} = p_a \ \forall \ b \qquad (3.20)$$

so that, independently of the starting state b, after enough steps we end up in state a with probability p_a. Eq. (3.20) means that W^k, for large enough k, must tend to a matrix with all columns equal to p_a, so that the stochastic walk leads eventually to the desired sampling probability, completely losing memory of the starting position. How large must be k, in order to effectively achieve that, is a figure of merit of the chosen algorithm[4].

Before discussing how to build an appropriate W, which satisfies Eq. (3.20), let us introduce some additional properties that will be required in the following.

Definition: Ergodicity

A Markov chain is called *ergodic* if, for any pair of states $a,b \in \Omega$, there is an integer k such that $(W^k)_{ab} \neq 0$.

---

[4]A simple, obvious possibility would be to take $W_{ab} = p_a$ right from the beginning, i.e. to take a Markov chain that completely loses memory of the starting state b and selects the new state directly with probability p_a: such a transition matrix would work perfectly and indeed one has $W^k = W \ \forall \ k$ in this case. However, the point is that we are resorting to Markov chains just because we cannot sample p_a directly by simple means.

In other words, a Markov chain is ergodic if, starting from any state, we can end up in any other state of the system if we wait long enough. Typically, when the chain is non-ergodic the state space Ω gets divided into a set of disconnected regions, meaning that it is not possible to move from one region to the other via the given stochastic process, as if the Markov chain could be reduced to a set of simpler Markov chains, each describing the stochastic process within one of the disconnected regions. For this reason, an ergodic Markov chain is also called *irreducible*.

Definition: Period of a state

Consider a particular state "a" of the system, and the set of all integer numbers n such that $(W^n)_{aa} \neq 0$. The period of a, d_a, is defined as the greatest common divisor of all n's.

In other words, if we start from state a, future visits of a cannot take place but for integer multiples of d_a (not necessarily for all multiples of d_a).

Keynote: If a Markov chain is ergodic, every state has the same period, $d_a = d \ \forall \ a$, so that we can define the period d of the chain itself.

Reasoning: Let us consider a generic pair of states, a and b, and their associated periods, d_a and d_b. Since the chain is ergodic, it must be possible to go from a to b, and then back from b to a, in a finite number of steps. Let us consider one of these stochastic walks and let us call $\bar{k}$ the associated number of steps: by definition, $\bar{k}$ must be divisible by d_a.

Now, the given loop can be modified as we want in the following way: as we reach b, before moving back to a, we add a new loop of n steps going from b to b: by definition, d_b is the GCD of all the possible values of n. In this way, we have constructed a new loop of $\bar{k} + n$ steps going from a to a: $\bar{k} + n$ must be divisible by d_a, and since also $\bar{k}$ is divisible by d_a, n must be divisible by d_a as well. Therefore, we have proved that d_a is a divisor of all possible n such that $(W^n)_{bb} \neq 0$, so that $d_a \leq d_b$ by definition of GCD.

The above reasoning can be repeated, exchanging the role of a and b, to prove that $d_b \leq d_a$, so that finally we obtain $d_a = d_b$ for a generic pair of states.

Definition: Aperiodicity and regularity

An ergodic Markov chain is aperiodic if $d = 1$. An aperiodic, ergodic Markov chain is also called "regular".

Regular Markov chains are important because some of the theorems we are going to discuss in the following only apply to them. It is easy to ensure that the chain is aperiodic: one just needs a small probability of remaining in the same state at each step. Ergodicity is usually more difficult to prove or enforce.

3.3.1 Ensemble of Markov chains and some theorems on the spectrum of W

In order to investigate the properties of W, it is better to take a Statistical Mechanics attitude and discuss the behaviour of an ensemble of Markov processes, instead of a single one. To that purpose, we consider a very large (or infinite) set of identical copies of the same system (the ensemble), which are initially distributed over the configuration space Ω according to a starting distribution $\pi_a^{(0)}$, which is the fraction of the total number of copies placed in state a. We imagine to start a Markov chain, described by the matrix W, from each copy of the ensemble: at each step k the systems move from one state to the other, and the occupation number (population) of each state changes, so that we can define the distribution $\pi_a^{(k)}$ as a function of the step.

It is easy to understand how $\pi_a^{(k)}$ evolves step by step: the elementary process by which a copy ends up in state a at step $k+1$ starting from state b at step k happens with probability $W_{ab}\pi_b^{(k)}$; then we just need to sum over all possible starting states b:

$$\pi_a^{(k+1)} = \sum_b W_{ab}\,\pi_b^{(k)}\,. \tag{3.21}$$

Hence, the evolution of the distribution is simply given by a row by column multiplication by the matrix W, and we can give a simple expression for the probability distribution at step k:

$$\pi_a^{(k)} = \sum_b (W^k)_{ab}\,\pi_b^{(0)}\,. \tag{3.22}$$

Our requirement for the single process, Eq. (3.20), can now be rephrased as a requirement on the evolution of $\pi_a^{(k)}$. We want that, independently of the starting distribution $\pi_a^{(0)}$,

$$\lim_{k\to\infty} \pi_a^{(k)} = \lim_{k\to\infty} \sum_b (W^k)_{ab}\,\pi_b^{(0)} = p_a \tag{3.23}$$

where we have used Eq. (3.20) and the fact that $\pi_b^{(0)}$ is normalized, i.e. the distribution must converge to an *equilibrium distribution* equal to p_a. The word "equilibrium" stems from the fact that, if previous equations hold, then such distribution must be left unchanged by a single step of the chain. It is easy to show that the two conditions, Eq. (3.20) and Eq. (3.23) are equivalent: if Eq. (3.20) holds, then Eq. (3.23) trivially follows; on the other hand, if Eq. (3.23) holds, then Eq. (3.20) follows by concentrating the initial distribution in a single state, i.e. $\pi_{b'}^{(0)} = \delta_{b'b}$, where δ is the Kronecker delta symbol.

The rephrasing of the problem enables us to discuss it in terms of the spectrum of the matrix W, i.e. in terms of its eigenvalues λ and associated eigenvectors v_a:

$$\sum_b W_{ab}\,v_b = \lambda v_a^{(\lambda)}\,. \tag{3.24}$$

Indeed, it is clear that p_a must be an eigenvector of W with $\lambda = 1$, which is nothing else but the equilibrium condition. Then, in a few words, what we need to prove, in order to guarantee that Eq. (3.23) holds, is that there is a unique eigenvector with $\lambda = 1$, and that all other eigenvectors have $|\lambda| < 1$, so that any deviation of the starting distribution from the equilibrium one is washed away exponentially fast in the Markov chain time k. One can show that *regular* Markov chains satisfy such conditions, and in the following we discuss some details on how one derives such result, even if more rigorous presentations can be found elsewhere [19, 20]. However, such details may be skipped over by the uninterested reader, since the important point is to learn how to construct a Markov chain that satisfies the properties required to sample a given distribution.

Keynote: There is always at least one eigenvector with $\lambda = 1$.

Reasoning: The normalisation condition $\sum_a W_{ab} = 1$ implies that $\sum_a (W_{ab} - \delta_{ab}) = 0$, which is a condition of linear dependence among the rows of the matrix $W - \mathrm{Id}$, meaning that $\det(W - \mathrm{Id}) = 0$, so that the system $W_{ab} v_b = v_a$ has at least one solution.

Keynote: If λ is an eigenvalue of W, then $|\lambda| \leq 1$.

Reasoning: Let v_a be an eigenvector with eigenvalue λ. Then, considering the equation $\sum_b W_{ab} v_b = \lambda v_a$ and taking the absolute value of both members, we obtain

$$|\lambda||v_a| = |\sum_b W_{ab} v_b| \leq \sum_b |W_{ab} v_b| = \sum_b W_{ab} |v_b|$$

where we have exploited the fact that $W_{ab} \geq 0$. If we sum the above inequality over a, we obtain

$$|\lambda| \sum_a |v_a| \leq \sum_{a,b} W_{ab} |v_b| = \sum_b |v_b| \implies |\lambda| \leq 1$$

where the property $\sum_a W_{ab} = 1$ has been used.

Keynote: If v_a is an eigenvector with $\lambda \neq 1$, then $\sum_a v_a = 0$.

Reasoning: Considering the equation $\sum_b W_{ab} v_b = \lambda v_a$ and taking the sum over the first index we obtain

$$\lambda \sum_a v_a = \sum_{a,b} W_{ab} v_b = \sum_b v_b$$

from which the thesis follows, since $\lambda \neq 1$.

Keynote: If v_a is an eigenvector with eigenvalue $\lambda = 1$, and if W is an ergodic chain, then it is possible to take $v_a \geq 0 \ \forall \ a$, i.e. v_a can be normalized as a probability distribution over Ω.

Reasoning: Let us suppose that an eigenvector v with $\lambda = 1$ exists, which has both negative and positive elements. Then, the elements of Ω can be divided into two sets, those for which $v_a < 0$, and the other ones. Let us now consider the equation $v_a = \sum_b W_{ab} \, v_b$ and the associated inequality

$$|v_a| = |\sum_b W_{ab} \, v_b| \leq \sum_b |W_{ab} \, v_b| = \sum_b W_{ab} \, |v_b| \, ,$$

if the inequality is strict for at least one state then, by summing over a, we obtain an absurdity. However, in order to have only equalities, we need that for each case only addends having the same sign appear both on the left and on the right-hand side. Since we have supposed that v_a can take either signs, we need that W_{ab} only connects pairs for which v_a and v_b have the same sign. However, that implies that Ω can be divided into two sets such that it is not possible to go from one set to the other, in any number of steps, by the given Markov chain. This is not possible, since W was supposed to be ergodic.

The argument above does not take into consideration the possibility that the eigenvector v_a be complex, but little changes in this case. Repeating the above reasoning, one can write $v_a = \rho_a \exp(i\phi_a)$ and divide Ω into different subsets corresponding to states with the same ϕ_a. If $\lambda = 1$, then W cannot have matrix elements between states with different values of ϕ_a, i.e. it is not possible to go from one set to the other in any number of steps. That leads again to conclude that either W is non-ergodic, or ϕ_a is a constant independent of a, so that the eigenvector can be rescaled (by a complex number) in order that its elements become all real and positive.

Keynote: If W is ergodic, the eigenvector associated with $\lambda = 1$ is unique (apart from an overall normalisation constant).

Reasoning: Let us suppose that $v_a = \sum_b W_{ab} \, v_b$ and $v'_a = \sum_b W_{ab} \, v'_b$, where both v and v' are normalized, i.e. $\sum_a v_a = \sum_a v'_a = 1$, and that v and v' are different, i.e. $v_a \neq v'_a$ for at least some values of a. Then also $w_a = v_a - v'_a$ is an eigenvector with $\lambda = 1$, with at least some non-zero elements, however, at the same time we have $\sum w_a = 0$, meaning that w must have elements with differents signs. This is not possible because of the previous theorem, since W is ergodic.

Keynote: If W is regular, then $\lambda \neq 1 \implies |\lambda| < 1$.

Reasoning: Let us suppose instead that an eigenvalue $\lambda \neq 1$ with $|\lambda| = 1$ exists, and let v_a be an eigenvector associated with λ. As we have done above, we can write

$$|\lambda||v_a| = |\sum_b W_{ab}\, v_b| \leq \sum_b |W_{ab}\, v_b| = \sum_b W_{ab}\, |v_b|$$

but now the strict equality must hold for each a, otherwise, after summing over a, we would obtain $|\lambda| < 1$. That means that all elements v_b appearing in the sum $\sum_b W_{ab}\, v_b$ must have the same phase, i.e. they can be written as $v_b = \rho_b \exp(i\phi_b)$ with the same $\phi_b = \bar{\phi}(a)$ and, since we can write $\lambda = \exp(i\phi_\lambda)$ with $\phi_\lambda \neq 0$, the element v_a on the left-hand side must have a different phase $\phi_a = \bar{\phi}(a) + \phi_\lambda$.

Therefore, if we divide the states of Ω in classes corresponding to elements v_a having the same complex phase, we conclude that the probability that the Markov chain keeps a state belonging to a given class unchanged is exactly zero. That cannot be applied to states for which $v_a = 0$; however, we need that at least some states exist for which $v_a \neq 0$: that must be true; otherwise, we would have no eigenvector at all.

Now, since we have supposed $|\lambda| = 1$, the above reasoning can be applied to a generic power of W, i.e. for a generic number of steps of the Markov chain, unless $\lambda^{\bar{k}} = 1$ for some $\bar{k}$. That means that, if the system starts from a state with $v_a \neq 0$, the Markov chain will not bring it back to the same state but for integer multiples of $\bar{k}$, so that the period of the chain must be $\bar{k}$ at least (with $\bar{k} > 1$) if $\bar{k}$ exists, or infinite otherwise.

This is not possible, since by hypothesis the chain is regular (ergodic and aperiodic), hence its period must be one.

Let us summarise the results achieved above and derive some further conclusions. A regular Markov chain has a unique equilibrium distribution, let us call it $\pi_a^{(EQ)}$, which is left unchanged by the Markov chain evolution. All other eigenvectors correspond to eigenvalues λ such that $|\lambda| < 1$ and are characterized by the fact that the sum of their elements is equal to zero.

These facts are essential not only to be sure that equilibrium exists, but also to understand that we will actually converge to it, independently of the starting distribution. To see that, let us write the starting distribution as $\pi_a^{(0)} = \pi_a^{(EQ)} + \Delta\pi_a^{(0)}$, where $\Delta\pi^{(0)}$ represents its distance from the equilibrium one. Since both $\pi_a^{(0)}$ and $\pi_a^{(EQ)}$ are normalized probabilities, we have $\sum_a \Delta\pi_a^{(0)} = 0$, meaning that it can be expanded over the eigenvectors[5] with $\lambda \neq 1$; moreover, its elements are bounded by $|\Delta\pi_a^{(0)}| \leq 2$. We can write

$$\pi_a^{(k)} = (W^k)_{ab}\, \pi_b^{(0)} = (W^k)_{ab}\, (\pi_b^{(EQ)} + \Delta\pi_b^{(0)}) = \pi_a^{(EQ)} + \Delta\pi_a^{(k)} \qquad (3.25)$$

where $\Delta\pi_a^{(k)} = (W^k)_{ab}\, \Delta\pi_b^{(0)}$. Since $\Delta\pi^{(0)}$ lives in a space of eigenvectors with eigenvalues $|\lambda| < 1$, the magnitude of its elements will be contracted after each

---

[5] Our argument is not strictly rigorous, since we should show that a basis of eigenvectors exists for the transition matrix W, but this is not always the case.

iteration of W. Therefore, defining l_{exp} as the *sup* of $|\lambda|$ over all $\lambda \neq 1$, we have

$$|\Delta \pi_a^{(k)}| \leq 2\, l_{exp}^k = 2\exp(-k/\tau_{exp}) \tag{3.26}$$

where we have defined the so-called *exponential autocorrelation time*

$$\tau_{exp} \equiv -\frac{1}{\log l_{exp}}. \tag{3.27}$$

In conclusion, we will converge exponentially fast to the equilibrium distribution, with a relaxation time which depends on the starting distribution, but is at most equal to τ_{exp}. The initial part of the Markov chain, needed to achieve convergence towards equilibrium, is usually called the *thermalisation phase*. The exponential autocorrelation time is surely finite for systems with a finite number of possible states; however, it could be arbitrarily large for systems with infinite states, since l_{exp} could be arbitrarily close to 1. This possibility is not only academic, since we will face it as a real problem in numerical simulations of statistical systems close to a critical point.

3.3.2 Equilibrium condition and the detailed balance principle

The remaining question that needs to be answered is how to fix $\pi_a^{(EQ)} = p_a$, i.e. how to construct a regular Markov chain that converges to an assigned equilibrium distribution p_a. The task is much simpler than finding $\pi_a^{(EQ)}$ for a given W, which corresponds to solving an eigenvalue problem. We need instead to find a W having an assigned eigenvector of unit eigenvalue: as we will see, the problem has in general infinite solutions, a fact that can be properly exploited in order to optimize the Monte Carlo sampling.

The equilibrium condition, $\sum_b W_{ab}\, p_b = p_a$ can be rewritten, after some simple steps[6], as a set of equations, one for each state a

$$\sum_{b \neq a} W_{ab}\, p_b = \sum_{b \neq a} W_{ba}\, p_a \tag{3.28}$$

which have a very simple and intuitive meaning in terms of the invariance of the population of state a. The left-hand side represents the fraction of copies of the system which, being originally in a state other than a, move to a after one iteration of the Markov chain. The right-hand side represents the fraction of copies which, being originally in state a, move to a different state during the same step. The equation then represents an exact balance between the incoming and outgoing copies, which leaves the population of a, hence p_a, unchanged.

---

[6]Exploiting the normalisation condition for W, one can rewrite $p_a = p_a \sum_b W_{ba}$.

Eq. (3.28) reprents a very loose set of constraints for W_{ab}, which leaves a large freedom[7] and are not easy to be checked on a given algorithm. Therefore, since one is not interested in determining all possible solutions, but just some of them which can be suitable for Monte Carlo sampling, it is usual to adopt a more restrictive set of equations, which are a sufficient, but not necessary condition:

$$W_{ab}\, p_b = W_{ba}\, p_a \quad \forall \; a, b \, \in \, \Omega \, . \tag{3.29}$$

In practice, the balance between incoming and outgoing populations is not imposed as a whole on each state, but separately for each pair of states, checking that the total flux going from one state to the other is zero. This is known as the *detailed balance principle*, it guarantees the general balance and equilibrium condition (3.28), as can be checked by summing Eq. (3.29) over $b \neq a$, and is much more practical, since it is usually easier to check Eq. (3.29), having in mind two well-definite states of the systems, than to check a global condition like Eq. (3.28).

Of course, in this way, we are missing a large number of possible algorithms which would satisfy just the general balance condition; however, detailed balance is still loose enough to allow for a large variety of solutions and possible algorithm optimisation. In the following we are going to analyze two classical schemes of algorithms, known, respectively, as Metropolis and heat-bath, which satisfy the detailed balance principle.

3.3.3   The Metropolis algorithm

There is a whole class of algorithms respecting the detailed balance principle which takes its name from the first author (Nicholas Metropolis) of a famous paper, dated 1953, where one of the first applications of Monte Carlo simulations to the study of the statistical properties of matter was presented [21].

Let us describe the scheme of the typical Metropolis algorithm. In order to assign the elements of the matrix W, we need to describe the stochastic process by which we move from state a to state b in one elementary step of the chain:

1. Starting from a, a new, trial state $\tilde{b}$ is selected according to a *tentative* transition probability $A_{\tilde{b}a}$. The matrix A is properly normalized, so that it could define a Markov chain by itself; however, it can be completely unrelated to the equilibrium distribution we are looking for. We will just impose that A is symmetric, i.e. $A_{\tilde{b}a} = A_{a\tilde{b}}$, as in the original paper, however, even this constraint can be relaxed by properly generalizing the algorithm;

2. If $p_{\tilde{b}} > p_a$, then we take $b = \tilde{b}$;

---

[7]Let us assume, to fix ideas, that Ω has a finite number N of states. Then W, taking into account normalisation conditions, is written in terms $N^2 - N$ unknowns, which are poorly constrained by the N equations (3.28).

3. If instead $p_{\tilde{b}} < p_a$, then $b = \tilde{b}$ with probability $p_{\tilde{b}}/p_a$, and $b = a$ otherwise.

Last step is usually called the Metropolis or accept/reject step. Notice that if the move is rejected, i.e. if $b = a$, the Markov chain step is considered as done anyway, as if state a had been visited multiple consecutive times, which must all be counted when taking averages over the stochastic process.

Later we will better explicit the various steps described above by discussing a specific example. Now instead we verify that the algorithm satisfies detailed balance. In order to do that, we need to consider a generic pair of states, a and b, and compute the two transition probabilities, W_{ab} and W_{ba} associated with the stochastic process described above. In order to fix ideas, let us assume that $p_a \leq p_b$. Then the probability of moving from a to b is simply given by $W_{ba} = A_{ba}$, since the move will be accepted for sure; instead the probability of moving from b to a will be $W_{ab} = A_{ab} \, p_a/p_b$, since the move is accepted with probability p_a/p_b. Then, exploiting the symmetry of the matrix A, we have:

$$W_{ab} \, p_b = A_{ab} \frac{p_a}{p_b} \, p_b = A_{ba} \, p_a = W_{ba} \, p_a \tag{3.30}$$

which is indeed the detailed balance principle. Of course, a detailed balance is not enough and one should check, case by case, that the chain is also ergodic and aperiodic.

In some applications, it is not easy to devise a tentative transition probability A_{ab} such that it is symmetric. In those cases, one can resort to a generalisation of the algorithm, known as Metropolis-Hastings, after W.K. Hastings who proposed it in 1970 [22]. It reads as follows:

1. Starting from a, a new, trial state $\tilde{b}$ is selected according to $A_{\tilde{b}a}$;

2. One then considers the ratio

$$r = \frac{p_{\tilde{b}} \, A_{a\tilde{b}}}{p_a \, A_{\tilde{b}a}}$$

and one takes $b = \tilde{b}$ with probability r (or 1 if $r > 1$) and $b = a$ otherwise.

Detailed balance can be easily verified, as we have done for the standard case.

An important feature of the Metropolis-Hastings algorithm, which makes it suitable for a large variety of systems, is that we never need to know the exact value of the probability p_a (nor that of the tentative transition probability), but just the value of probability ratios, i.e. the algorithm is not affected by our ignorance about the overall normalisation factor entering p_a. This is important especially for complex systems, for which an exact computation of the normalisation constant could be difficult, think, for instance, of statistical systems discussed in Chapter 4, where in order to know the normalisation constant of the Boltzmann distribution one needs to compute the partition function.

Example:
Suppose we want to generate a sample of real variables according to the normal distribution, $p(x) \propto \exp(-x^2/2)$. We already know the Box-Müller algorithm, which allows to perform independent draws quite efficiently. However, this is a great example where to learn how to implement the Metropolis algorithm in practice, and which gives us the opportunity to discuss the possible subtleties one should consider when treating a system characterized by a continuous (instead of discrete) set of states, which in this case are the possible values of x.

We need to generate a sequence of draws $x_0, x_1, \ldots, x_k, \ldots$, starting from a given x_0 and assigning how we move from x_k to x_{k+1}. A possible algorithm is the following:

1. we choose $\tilde{x}$ with a uniform random distribution in the interval $[x_k - \delta, x_k + \delta]$;

2. if $\exp(-\tilde{x}^2/2) > \exp(-x_k^2/2)$ then $x_{k+1} = \tilde{x}$;

3. otherwise, we extract u randomly in $[0, 1]$ and if $u < \exp(-\tilde{x}^2/2 + x_k^2/2)$ then $x_{k+1} = \tilde{x}$, otherwise $x_{k+1} = x_k$.

It is apparent that, moving step by step, we can reach any point of the real line from any other point in a finite number of steps; moreover, at each step we have a non-negligible probability of remaining at the same point; therefore, the chain is ergordic and aperiodic. The tentative transition probability is clearly symmetric in this case: the particular choice can of course be modified at will, one could take it Gaussian, or anything else. The parameter δ can be chosen freely, it is usually taken fixed during the Markov chain and, as we will see, it characterizes the efficiency of the algorithm. It is interesting to notice that step 2. could be avoided by doing step 3. directly, since if condition 2. holds then the acceptance test 3. is surely passed: in this way one would avoid an *if* condition, with the risk of generating an unnecessary random variable u; what is better depends on efficiency considerations regarding the particular code implementation (and not the algorithm itself).

The algorithm seems a trivial implementation of the Metropolis scheme, however there are some details we should discuss more in deep. The stochastic variable is continuous and, in order to make the argument about detailed balance more sound, it is better to discuss in terms of transition probabilities between infinitesimal intervals. Let us consider two intervals, $[x, x + dx]$ and $[y, y + dy]$: the given distribution, p, assigns to them a probability $p(x)dx$ and $p(y)dy$, respectively. If we start from the first interval, the probability of selecting a point in the second interval is $A(y, x)dy = dy\,\theta(\delta - |x - y|)/(2\delta)$, where the θ-function is defined such that $\theta(x) = 1$ for $x \geq 0$ and $\theta = 0$ for $x < 0$. Notice that $A(y, x) = A(x, y)$, however this does not imply a symmetry for the tentative transition probability, since in general $dx \neq dy$. The parameter needed for acceptance of the transition $x \to y$ is therefore that of Hastings' generalisation:

$$r = \frac{p(y)dy\, A(x,y)dx}{p(x)dx\, A(y,x)dy} = \frac{p(y)}{p(x)} \tag{3.31}$$

which is indeed the one we have used. Now, assuming $p(y) < p(x)$, detailed balance is verified since

$$\begin{aligned} A(x,y)\, dx\, p(y)\, dy &= \frac{p(y)}{p(x)}\, A(y,x)\, dy\, p(x)\, dx \\ &= A(y,x)\, dy\, p(y)\, dx = A(x,y)\, dy\, p(y)\, dx\,. \end{aligned}$$

The above reasoning may seem a more cumbersome way to find the same simple result. However, the important point is that the simplification leading to $r = p(y)/p(x)$ takes place because *the measure used to define the probability distribution and the measure used to define the tentative transition probability are the same*. To better appreciate this point, you should consider the following variation of the exercise above. Suppose the real line has not a "flat" measure, i.e. that the measure of intervals be given by $\mu(x)dx$, where $\mu(x)$ is some positive function: that could be the case after some change of variables has been performed, or when one considers multivariate systems. Suppose we want to approach the same problem, i.e. to construct a Markov chain for an equilibrium probability distribution $p(x)$, meaning now that the probability of finding the stochastic variable in an interval $[x, x+dx]$ is $p(x)\mu(x)dx$. We can proceed as above, and choose the tentative new variable $\tilde{x}$ uniformly distributed in the interval $[x_k - \delta, x_k + \delta]$. What is the correct acceptance parameter now? Where does the non-uniform measure $\mu(x)$ enter? The point now is that the tentative transition probability has not changed, i.e. the probability of selecting the tentative change in $[y, y + dy]$ is still $A(y, x_k)dy$, as if the measure were flat, hence from Eq. (3.31) one obtains $r = (p(\tilde{x})\mu(\tilde{x}))/(p(x_k)\mu(x_k))$. However, we were able to select the tentative transition proportionally $A(y, x_k)\mu(y)dy$, the acceptance parameter would be $r = p(\tilde{x})/p(\tilde{x}_k)$.

In order to further discuss this point, we refer the reader to exercises 3.3 and 3.4.

3.3.4 Heat-bath Algorithms

We are going to describe a class of algorithms which can be useful for multivariate systems with a large number of stochastic variables. The general idea is to perform a Markov chain step in which most variables are kept fixed and just a few of them are changed; in particular, completely independent values are drawn for those variables, according to the conditional probability stemming from the original probability distribution for fixed values of the variables which are not touched. Therefore, the unchanged variables play the role, for the subset of modified variables, of a thermal heat bath which fixes their distribution function (hence the name of the algorithm). At each step of the Markov chain, the modified subset is typically chosen randomly, in order to ensure ergodicity.

An algorithm of this kind is viable only if the conditional distribution of the subset variables can be sampled easily and efficiently; from this point of view, it is not really important that the subsystem is small, but just that its distribution can be easily sampled. We are now going to discuss the algorithm more in detail and prove that it respects detailed balance.

Let us rewrite the index labelling the possible states in the form $a \equiv (\bar{a}, \alpha)$ where α parametrizes the possible states of the subsystem, and $\bar{a}$ those of the remaining set of variables. Therefore, the sum over all states can be rewritten as $\sum_a = \sum_{\bar{a}} \sum_{\alpha}$ and the probability distribution can be written as $p_a = p(\bar{a}, \alpha)$. We can define a probability distribution for the $\bar{a}$ variables, by integrating over the α variables

$$\bar{p}(\tilde{a}) = \sum_{\alpha} p(\tilde{a}, \alpha) \,. \tag{3.32}$$

The conditional probability, $P(\alpha|\bar{a})$, can be written in terms of p and $\bar{p}$:

$$p(\bar{a}, \alpha) = \bar{p}(\bar{a}) P(\alpha|\bar{a}) \tag{3.33}$$

from which we obtain

$$P(\alpha|\bar{a}) = \frac{p(\bar{a}, \alpha)}{\bar{p}(\bar{a})} = \frac{p(\bar{a}, \alpha)}{\sum_{\alpha'} p(\bar{a}, \alpha')} \,. \tag{3.34}$$

The elementary step of the Markov chain is built as follows:

1. $\bar{a}_{k+1} = \bar{a}_k$

2. α_{k+1} is independent of α_k and drawn according to $P(\alpha_{k+1}|\bar{a}_{k+1})$

so that the transition probability can be written as:

$$W_{(\bar{b}, \beta), (\bar{a}, \alpha)} = P(\beta|\bar{b}) \, \delta_{\bar{b}\bar{a}} \,. \tag{3.35}$$

Detailed balance is now checked by noting that

$$W_{(\bar{b}, \beta), (\bar{a}, \alpha)} \, p(\bar{a}, \alpha) = \delta_{\bar{b}\bar{a}} \, P(\beta|\bar{b}) \, P(\alpha|\bar{a}) \, \bar{p}(\bar{a}) \tag{3.36}$$

where Eq. (3.33) has been used: this is exactly equal to the result obtained by computing $W_{(\bar{a}, \alpha), (\bar{b}, \beta)} \, p(\bar{b}, \beta)$.

Example:

Suppose we need to devise a Markov chain for sampling N stochastic variables distributed according to

$$p(y_1, y_2, \ \ldots \ y_N) \, dy_1 \, dy_2 \, \ldots dy_N \propto \exp \left(-\prod_j y_j^2 \right) \prod_j dy_j \,.$$

A possible algorithm, based on the heat-bath scheme, is the following. Let $y_j^{(k)}$ be the values of the stochastic variables at step k of the Markov chain.

1. select i at random in $1, ..., N$

2. $y_j^{(k+1)} = y_j^{(k)}$ for $j \neq i$, while $y_i^{(k+1)}$ is extracted according to the Gaussian distribution proportional to $\exp(-Ay_i^2)$, where $A = \prod_{j \neq i} (y_j^{(k+1)})^2$.

3.3.5 Composition of Markov chains

It is usual to deal with problems where one can devise several algorithms (Markov chains), each having some good features, but failing on some particular aspect. In this respect, it is important to notice that Markov chains can be combined to create new Markov chains. Suppose W^A and W^B be two regular Markov chains having the same equilibrium distribution p_a, i.e. $\sum_b W_{ab}^A\, p_b = p_a$ and $\sum_b W_{ab}^B\, p_b = p_a$. We can define a new Markov chain W, in which at each step we select the step of Markov chain W^A with probability $w > 0$, and the one of Markov chain W^B with probability $(1 - w)$, that means

$$W_{ab} = wW_{ab}^A + (1 - w)W_{ab}^B. \tag{3.37}$$

It is trivial to check that p_a is an equilibrium distribution for W too. In order to verify that W is a good Markov chain, we just need to prove that it is regular. Now, taking a generic couple of states, we know that at least two paths exists, leading from one state to the other, both for W^A and W^B: each path is also a possible path for the newly defined chain W, since the probability that we always select W^A (or W^B) along each path is non-negligible. Also the number of paths leading from one state to itself includes at least all those one could find for W^A and W^B separately, therefore the G.C.D. of their lengths cannot increase, i.e. W will be aperiodic as well.

The argument above can be easily repeated for the composition of an arbitrary number of Markov chains: the choice of the different weights by which each chain is selected at each step is a matter of optimisation.

3.4 Advanced Error Analysis Techniques

In section 3.1.1, we have discussed how to estimate the statistical error on the Monte Carlo evaluation of some average over a finite sample. That was actually the basic approach, which can be safely used only when one is interested in computing the average of a given function of the stochastic variables, and assuming the sample is composed of statistically independent draws.

We are now going to explore two different ways in which the basic approach can fail, thus requiring a more careful thinking. The first is a consequence of the correlation existing between consecutive draws of a Markov chain: they are not statistically independent, each draw moving away from the previous one according to the assigned transition probability.

The second is related to the fact that sometimes one is interested in less trivial functionals of the probability distribution, which are not just simple averages of some function of the stochastic variables: combinations of different averages, like cumulants, are a typical example. The way to correctly estimate the error on the Monte Carlo evaluation obtained for such functionals is non-trivial: standard error propagation, when possible, should be done with much care. Moreover, in some cases, one should consider also a systematic error, in addition to the statistical one, affecting finite sample estimates, since it is not always guaranteed that the sample estimator evenly fluctuates around the true estimator, meaning that Monte Carlo measures might be affected by a *bias*.

Often these two kind of problems combine, and various kinds of advanced error analysis techniques have been developed to properly deal with them. We are going to expose some of them and, in order to make the discussion as simple as possible, our playground will be the normal distribution for a single stochastic variable.

3.5 Error estimate in the presence of autocorrelations

Let us consider a Metropolis algorithm for sampling a stochastic variable x distributed according to

$$p(x)\, dx \propto \exp\left(-\frac{(x-\mu)^2}{2\sigma^2}\right)\, dx \qquad (3.38)$$

with $\mu = 5$ and $\sigma^2 = 1$. The algorithm has been already outlined in section 3.3.3. We produce a sequence $x_0, x_1, \ldots$ of draws, where x_{k+1} is chosen starting from x_k as follows: $x^{(t)}$ is selected randomly in $[x_k - \delta, x_k + \delta]$, then $x_{k+1} = x^{(t)}$ if $p(x^{(t)}) > p(x_k)$, otherwise $x_{k+1} = x^{(t)}$ with probability $p(x^{(t)})/p(x_k)$ and $x_{k+1} = x_k$ otherwise.

On the left-hand side of Figure 3.4, we show a sequence produced by 15000 iterations of this algorithm, starting from $x_0 = 0$ and adopting $\delta = 0.1$. The starting point is quite far from equilibrium (it is five standard deviations from the average of the normal distribution), hence we need some iterations to achieve equilibrium: Figure 3.4 shows that, in our case, the thermalisation phase lasts for at least 1000 iterations. It is usually wise to be generous in estimating the duration of thermalisation: the statistical error will not be

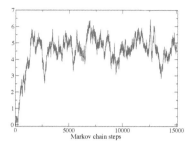 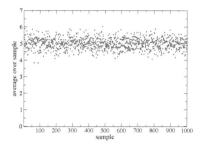

FIGURE 3.4: Left: sequence of 15000 Metropolis steps with $\delta = 0.1$ for sampling the normal distribution in Eq. (3.38). Right: sample averages $\tilde{x}$ for 1000 different samples built by the same algorithm, each made up of 10^4 thermalized steps.

much affected by discarding a larger number of iterations, while the damage of including draws which are still not properly distributed could be much worse. Therefore, we have decided to discard the first 5000 iterations, taking the last $N = 10000$ draws as our Monte Carlo sample. From now on, x_i, $i = 1, 10000$ will refer to such sample.

We will now estimate the average $\langle x \rangle$, as outlined in section 3.1.1, and compare it to the known result $\langle x \rangle = \mu = 5$. We have

$$\tilde{x} = \frac{1}{N} \sum_{i=1}^{N} x_i \quad ; \quad \sigma_{\tilde{x}}^2 = \frac{1}{N} \frac{1}{N-1} \sum_{i=1}^{N} (x_i - \tilde{x})^2 \tag{3.39}$$

and the result we get is

$$\langle x \rangle = 4.721 \pm 0.007 \tag{3.40}$$

which clearly indicates that something wrong is going on: we are about 40 standard deviations off the true average $\mu = 5$. Either the Markov chain algorithm is not reproducing the correct distribution, or our estimate of the standard deviation of the sample mean is largely underestimated.

In order to better understand this point, we have produced 1000 different samples as the one above (in practice, we have iterated the Markov chain for 10 million steps and divided it into 1000 different sequences), to see how the sample average $\tilde{x}$ is actually distributed. Results for the averages obtained from the different samples are reported in the right-hand side of Figure 3.4, from which we learn two things: sample averages are evenly distributed around $\mu = 5$, as expected; however, their standard deviation is very far from $\sigma_{\tilde{x}} = 0.007$ estimated above, and rather of the order of 0.1 or larger, i.e. at least one order of magnitude larger than our wrong estimate. Eq. (3.39) returns the wrong result because its derivation in section 3.1.1 assumes statistically independent draws, this is not true for a Markov chain Monte Carlo. Hence, we should work out again how the sample average is expected to fluctuate in this case.

The quantity we are interested in is

$$\sigma_{\tilde{f}}^2 = \langle (\tilde{f} - \langle f \rangle)^2 \rangle_s \qquad (3.41)$$

where in our case $f(x) = x$, while $\langle \cdot \rangle_s$ stands for the *average taken over the stochastic process which generates the sample*, a concept we have introduced at the beginning of this chapter. In the case of N independent draws, the probability distribution of the whole sample is simply the product of N independent probability distributions $\prod_{k=1}^{N} p(x_k)$; if instead the sample is built through a Markov chain, the sample generation probability is given by the Markov process itself.

Eq. (3.41) can be rewritten as follows:

$$\begin{aligned} \sigma_{\tilde{f}}^2 &= \left\langle \left(\left(\frac{1}{N} \sum_j f_j \right) - \langle f \rangle \right)^2 \right\rangle_s = \left\langle \left(\frac{1}{N} \sum_j (f_j - \langle f \rangle) \right)^2 \right\rangle_s \\ &= \left\langle \frac{1}{N^2} \sum_i \sum_j \delta f_i \, \delta f_j \right\rangle_s = \frac{1}{N^2} \sum_i \sum_j \langle \delta f_i \, \delta f_j \rangle_s \qquad (3.42) \end{aligned}$$

where $\delta f_i \equiv f_i - \langle f \rangle = f(x_i) - \langle f \rangle$ measures the deviation (fluctuation) of f at the i-th step from its average value. For independent draws, we have $\langle \delta f_i \delta f_j \rangle_s = \sigma_f^2 \delta_{ij}$ and one recovers the result of section 3.1.1. When this is not true, one defines the so-called *autocorrelation function*

$$C_f(i,j) \equiv \frac{\langle \delta f_i \, \delta f_j \rangle_s}{\sigma_f^2}, \qquad (3.43)$$

which serves as a measure of the correlation between two different draws of f and is normalized so that $C_f(i,j) = 1$ when $i = j$. We can state some useful properties of C_f and also predict, in the case of a regular Markov chain, its asymptotic behaviour in the limit of large separation between i and j.

First of all, $C_f(i,j)$ is by definition symmetric under the exchange of i and j and, for a time-independent Markov process, can only depend on the distance between i and j, i.e. $C_f(i,j) = C_f(|i-j|)$. Moreover, one has $|C_f(|i-j|)| \leq 1$, indeed we can write

$$2 \, \delta f_i \, \delta f_j = \pm (\delta f_i^2 + \delta f_j^2) \mp (\delta f_i \mp \delta f_j)^2$$

then taking the average $\langle \cdot \rangle_s$ of both sides to prove that

$$-2\sigma_f^2 \leq 2\langle \delta f_i \, \delta f_j \rangle_s \leq 2\sigma_f^2 .$$

Let us now compute $C_f(|i-j|)$ in terms of the transition matrix of the Markov process. In order to properly compute the autocorrelation function, we should find what is the probability of visiting state a at step i and then state b at step j (setting $j > i$ in order to fix ideas). Assuming the Markov chain has reached equilibrium, the probability of being in state a at step i is simply the sampling probability p_a. The probability of moving to b after $j - i$ steps is instead given by the transition matrix $(W^{|j-i|})_{ba}$. Therefore, the sought probability is $(W^{|j-i|})_{ba}\, p_a$ and we can write:

$$\langle \delta f_i\, \delta f_j \rangle_s = \sum_a \sum_b (W^{|j-i|})_{ba}\, p_a\, \delta f(a)\, \delta f(b)\,. \tag{3.44}$$

We can exploit the fact that, since the Markov chain is regular, for $|j-i| \gg 1$ $(W^{|j-i|})_{ba}$ is expected to converge exponentially fast to the equilibrium transition matrix, i.e.

$$(W^{|j-i|})_{ba} = p_b + R_{ba}(|j-i|) \quad \text{with} \quad |R_{ba}(|j-i|)| \sim O(e^{-|j-i|/\tau_{exp}}) \tag{3.45}$$

so that

$$\begin{aligned}
\langle \delta f_i\, \delta f_j \rangle_s &= \sum_a \sum_b p_b\, p_a\, \delta f(a)\, \delta f(b) + \sum_a \sum_b R_{ba}(|j-i|)\, p_a\, \delta f(a)\, \delta f(b) \\
&= \langle \delta f \rangle \langle \delta f \rangle + \sum_a \sum_b R_{ba}(|j-i|)\, p_a\, \delta f(a)\, \delta f(b) \tag{3.46} \\
&= \sum_a \sum_b R_{ba}(|j-i|)\, p_a\, \delta f(a)\, \delta f(b) \lesssim O(e^{-|j-i|/\tau_{exp}})
\end{aligned}$$

where we have used the fact that, by definition, $\langle \delta f \rangle = 0$. Therefore, the autocorrelation function is expected to vanish exponentially fast for large separation, with a decay time which is at most τ_{exp}, but can be shorter, and actually depends on the particular function f one is considering.

We can now go back to Eq. (3.42) and rewrite it as follows:

$$\sigma_{\bar{f}}^2 = \frac{1}{N^2} \sum_{i=1}^N \sum_{j=1}^N \langle \delta f_i\, \delta f_j \rangle_s = \sigma_f^2 \frac{1}{N^2} \sum_{i=1}^N \sum_{j-i}^N C_f(|j-i|) \tag{3.47}$$

where we have rewritten the double sum in a different way. For a finite sample, the range of the sum over the difference $j - i$ depends on i; however, if the sample is large enough, $N \gg \tau_{exp}$, we make a negligible error by extending the sum from $-\infty$ to $+\infty$, since $C_f(|j-i|)$ vanishes exponentially fast on both sides. Hence, the two sums become independent of each other and we can write

$$\begin{aligned}
\sigma_{\bar{f}}^2 &\simeq \sigma_f^2 \frac{1}{N^2} \sum_{i=1}^N \sum_{j-i=-\infty}^{+\infty} C_f(|j-i|) \\
&= \sigma_f^2 \frac{1}{N} \sum_{j-i=-\infty}^{+\infty} C_f(|j-i|) = \frac{\sigma_f^2}{N}(1 + 2\tau_{int}) \tag{3.48}
\end{aligned}$$

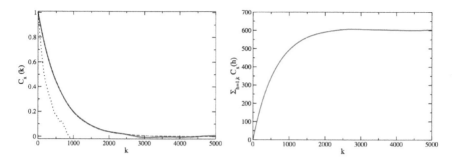

FIGURE 3.5: Left: autocorrelation function $C_x(k)$ for the Metropolis algorithm with $\delta = 0.1$, computed from a sample of 10 million steps; the dashed line is the function $\exp(-k/\tau)$ with $\tau = 600$, the dotted line is the autocorrelation function computed on a subsample of just 10 thousand steps. Right: integral of the autocorrelation function shown on the left. In particular, we show $\sum_{h=1}^{k} C(h)$ as a function of k: the asymptotic value yields the integrated autocorrelation time τ_{int}.

where we have introduced the so-called *integrated autocorrelation time*

$$\tau_{int} \equiv \sum_{k=1}^{\infty} C_f(k) \tag{3.49}$$

which for an exponentially vanishing function $C_f(k)$ coincides with its decay time, i.e. if $C_f(k) = e^{-k/\tau}$ (with $\tau \gg 1$) then $\tau_{int} \simeq \tau$. Eq. (3.48) has a simple interpretation: the variance of the sample mean is not σ_f^2/N but instead σ_f^2/N_{eff}, with $N_{eff} = N/(1 + 2\tau_{int})$, as if our sample were composed of N_{eff} effective independent draws. This is a reasonable result, since draws are correlated among themselves over a range of the order of τ_{int}.

Let us now go back to our original exercise. The left-hand side of Figure 3.5 shows the autocorrelation function of the variable x, for $|i - j|$ up to 5000 and for the Metropolis algorithm with $\delta = 0.1$ discussed above. In practice, the function has been computed from a sample of $N' = 10^7$ iterations of the Markov chain as follows:

$$C_x(k) = \frac{1}{\sigma_x^2} \frac{1}{N' - k} \sum_{i=1}^{N'-k} (x_i - \tilde{x})(x_{i+k} - \tilde{x})$$

where $\tilde{x}$ is the sample average evaluated on the same data set, and σ_x^2 has been estimated from the sample as usual. In principle one could also give an estimate of the statistical error on the determination of $C_x(k)$, but we skip the discussion about this point. The exponential decay of the autocorrelation function is clearly visible. On the right-hand side of Figure 3.5, we also show the "integral" of the autocorrelation function up to a maximum k, as a function of k: the integral clearly reaches an asymptotic value corresponding to $\tau_{int} \simeq$ 600; the function $\exp(-k/\tau_{int})$ also reproduces $C_x(k)$ quite well, as can be appreciated from the figure on the left-hand side.

Now, taking into account the result obtained for τ_{int}, we must revisit our previous estimate of the average on the sample with $N = 10000$ data, in particular of its error. That now becomes

$$\langle x \rangle = 4.721 \pm 0.007\sqrt{2\tau_{int} + 1} = 4.72 \pm 0.24 \qquad (3.50)$$

which is perfectly compatible with the exact average $\mu = 5$, being at slightly more than one estimated standard deviation from it. The error reported above is still not completely accurate, since the correct standard deviation of the sample average is $\sqrt{\sigma_x^2(2\tau_{int} + 1)/N} \simeq 0.35$, which is very close to the standard deviation observed for the distribution of sample averages on the right-hand side of Figure 3.4; the reason is that σ_x^2 itself has been estimated from the finite sample and is affected by a statistical fluctuations itself.

Actually, we have been cheating a little bit. The autocorrelation function and its integral have been estimated from a very large sample of 10 million draws, i.e. of size much bigger than τ_{int} itself, rather than from the original sample of 10 thousands draws: this is the reason we obtain so clean results for $C_x(k)$ and its integral in Figure 3.5. This is usually not possible in real life, where the sample size is constrained by computational budget limits.

Therefore, in Figure 3.5 we have also reported the autocorrelation function obtained from the original small sample of $N = 10000$ data. Results are quite different and less precise, and the integrated function would look ugly as a well, with a barely visible plateau[8]. In practice, we would have missed the correct autocorrelation time by roughly a factor 2, leading to an estimated standard deviation for the sample average of about 0.16 (instead of 0.24). This is not dramatic, the true average is still within 2 standard deviations. After all, in this game we are rolling dices, and we can just be a bit unlucky: two standard deviations could just be a manifestation of bad luck, 40 standard deviations (what we obtained before considering autocorrelations) is the clear manifestation of an evil bug somewhere.

It is interesting to notice that, in this particular simple case, the autocorrelation time of the Metropolis algorithm can also be estimated analytically. In order to simplify the calculation, we will do it in the limit of small δ, and consider a normal distribution with zero average and unit variance, the shift in the average μ being completely irrelevant to our purposes. The autocorrelation function after 1 Metropolis step can

---

[8]When the autocorrelation function starts fluctuating around zero, for $k > \tau_{int}$, its integral starts diverging like the distance covered by a random walk. Therefore, if the determination of the autocorrelation functions is not precise enough, the plateau could be barely visible.

be written as

$$
\begin{aligned}
\langle x_i\, x_{i+1}\rangle_s &= \frac{1}{\sqrt{2\pi}}\frac{1}{2\delta}\int_{-\infty}^{\infty} dx \int_{-\delta}^{\delta} dy\; e^{-x^2/2}\, x\left[(x+y)\,p_{acc} + x\,(1-p_{acc})\right] \\
&= \frac{1}{\sqrt{2\pi}}\frac{1}{2\delta}\int_{-\infty}^{\infty} dx \int_{-\delta}^{\delta} dy\; e^{-x^2/2}\, x\left[x + y\,p_{acc}\right] \\
&= 1 + \frac{1}{\sqrt{2\pi}}\frac{1}{2\delta}\int_{-\infty}^{\infty} dx \int_{-\delta}^{\delta} dy\; e^{-x^2/2}\, x\, y\, p_{acc}(x,y)
\end{aligned}
$$

where x is the value at the starting step, $x + y$ is the tentative new value at the next step and p_{acc} is the probability of accepting it, which depends both on x and y, as explained in the last line; moreover, we have already used the fact that $\langle x^2\rangle_s = \sigma_x^2 = 1$. Assuming $\delta \ll \sigma_x = 1$, we can now limit ourselves to the case of x and $x + y$ both strictly positive or negative: different combinations will give negligible corrections to our computation. Taking the positive case first, we have $p_{acc}(x,y) = 1$ for $y < 0$, while for $y > 0$ we can write

$$
p_{acc}(x,y) = \exp(-(x+y)^2/2 + x^2/2) \simeq \exp(-xy) \simeq 1 - xy
$$

where have we exploited the fact that $y \ll 1$ and that x is of $O(1)$. Therefore, we obtain

$$
\int_{-\delta}^{\delta} dy\; x\, y\, p_{acc}(x,y) \simeq \int_{-\delta}^{0} dy\; x\, y + \int_{0}^{\delta} dy\; x\, y\,(1 - x\, y) = -x^2 \frac{\delta^3}{3}\,,
$$

and exactly the same result is obtained for $x < 0$. Finally, we have

$$
\langle x_i\, x_{i+1}\rangle_s \simeq 1 - \frac{1}{\sqrt{2\pi}}\frac{1}{2\delta}\frac{\delta^3}{3}\int_{-\infty}^{\infty} dx\, e^{-x^2/2}\, x^2 = 1 - \frac{\delta^2}{6}\,.
$$

From that, assuming $\langle x_i x_{i+k}\rangle_s = \exp(-k/\tau)$, we derive

$$
\exp(-1/\tau) = 1 - \frac{\delta^2}{6} \quad \Longrightarrow \quad \tau \simeq \frac{6}{\delta^2} \tag{3.51}
$$

which for $\delta = 0.1$ gives $\tau \simeq 600$, which is perfectly consistent with what we have found numerically.

3.5.1   Data blocking techniques

A faster and more practical way to correctly estimate the statistical error on standard sample averages is based on the idea of repeating the naive computation, Eq. (3.39), on a modified sample obtained from the original one by averaging adjacent draws in a given block.

Let us consider again the sample of N draws x_i and, assuming N is even, let us define a new sample of blocked data $x_j^{(1)}$, $j = 1, N/2$, with

$$
x_j^{(1)} \equiv \frac{x_{2j-1} + x_{2j}}{2}\,; \tag{3.52}
$$

obviously we have $\overline{x^{(1)}} = \tilde{x}$, i.e. the average sample estimate is left unchanged by the blocking operation. We can now apply again the naive definition for the variance of the sample average to the blocked sample

$$
\begin{aligned}
\sigma^2_{(1)} &= \frac{1}{N/2} \frac{1}{N/2 - 1} \sum_{j=1}^{N/2} (x_j^{(1)} - \tilde{x})^2 \simeq \frac{1}{N/2} \frac{1}{N/2} \sum_{j=1}^{N/2} \frac{1}{4} (\delta x_{2j-1} + \delta x_{2j})^2 \\
&\simeq \frac{1}{N} \frac{1}{N} \sum_{j=1}^{N/2} (\delta x_{2j-1}^2 + \delta x_{2j}^2 + 2\delta x_{2j-1}\delta x_{2j}) \simeq \sigma^2_{(0)} + \frac{1}{N} \sigma_x^2 \langle \delta x_{2j-1}\delta x_{2j}\rangle \\
&\simeq \sigma^2_{(0)} (1 + C_x(1))
\end{aligned}
\tag{3.53}
$$

where we have defined $\sigma^2_{(0)}$ as the naive estimate of $\sigma^2_{\tilde{x}}$ on the original sample, $\delta x \equiv x - \tilde{x}$, and we have made use of the standard definition of autocorrelation function and of approximations valid for $N \gg 1$. We thus see that, in case of statistically independent draws, the naive estimate of the variance of the average remains unchanged for the blocked sample, while it grows in the case of positive autocorrelation, which is the usual case.

We can now iterate such a procedure, i.e. define a new sample $x_l^{(2)}$, $l = 1, N/4$, by blocking pairs of the sample $x_j^{(1)}$, and so on. After k steps of iteration, a sample $x_i^{(k)}$ of $N/2^k$ data[9], corresponding to averages over blocks of 2^k entries of the original sample, will be obtained, with an associated naive estimate $\sigma^2_{(k)}$. We expect that the sequence $\sigma^2_{(k)}$ be an increasing function of k, until a value of k is reached such that 2^k is of the order of the autocorrelation time of the original sample, so that adjacent blocks are effectively decorrelated and further iterations do not change the naive σ^2 estimate further. The saturation value of $\sigma^2_{(k)}$ can be taken as a correct estimate of the true variance of the sample average. Indeed, we have

$$
\sigma^2_{(k)} \simeq \frac{2^k}{N} \left(\frac{2^k}{N} \sum_{j=1}^{N/2^k} \left(\frac{x_{2^k(j-1)+1} + \cdots x_{2^k j}}{2^k} - \tilde{x} \right)^2 \right)
\tag{3.54}
$$

and, when 2^k is large enough so that different blocks can be considered as statistically independent, the quantity in big brackets can be taken as the correct variance of the average for samples of size 2^k, which is expected to scale as $1/2^k$ independently of the autocorrelation time (which just sets the prefactor), so that multiplying this quantity by $2^k/N$ will give the correct expected variance for averages over samples of size N.

We show two examples of the application of the blocking procedure in Figure 3.6, both cases refer to the Gaussian distribution, sampled via Metropolis

---

[9]We are assuming that N is divisible by 2^k. When this is not true, one can discard some entries of the sample (for instance, the last ones) so as to reduce N to the closest multiple of 2^k. Typically, that will not affect the estimate of $\sigma^2_{(k)}$ in a relevant way.

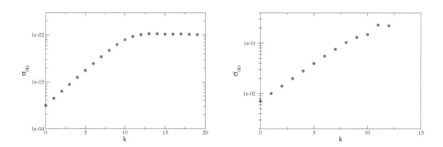

FIGURE 3.6: Naive error estimate for the sample average as a function of the blocking iteration and for two different sample sizes. Both samples have been drawn according to the Gaussian distribution with unit variance, via the Metropolis algorithm with $\delta = 0.1$. The sample size is $N = 10^7$ for the left figure, and $N = 10^4$ for that on the right.

with $\delta = 0.1$, as discussed above. On the left-hand side, the starting sample size is $N = 10^7$. The logarithmic scale on the vertical axis clearly shows that, in the first steps, the naive error grows exponentially with k: this is clearly understandable from Eq. (3.53), since in the presence of strong autocorrelation between adjacent blocks, $C(1) \simeq 1$ and the naive variance estimate approximately doubles at each blocking step, i.e. the naive error grows as the square root of the block size. Such exponential growth stops approximately around $k = 10$, after which a stable plateau is reached. Both the block size after which the growth stops, $2^{10} \sim O(10^3)$, and the ratio between the plateau value and naive error at $k = 0$ (slightly larger than 30) is compatible with the autocorrelation time that we have already found, $\tau \sim 600$; however, the blocking procedure can be a much more practical method. It should be clear that the particular choice of blocks sequence, scaling by a factor 2 at each step, is completely immaterial, and that any other sequence would work equally well. Let us stress that results might not be always so clear: the second example shown on the right-hand side of Figure 3.6, which is based on a sample of just 10^4 data, shows that in this case the plateau region is barely reached and the correct statistical error is not clearly identifiable; the reason is that the sample size is not so large, and in fact only $O(10)$ times larger than τ.

3.5.2   Markov chain optimization: looking for the highest possible efficiency

Previous discussion has shown that the efficiency of a given Monte Carlo algorithm, i.e. the numerical effort required to attain a given statistical accuracy,

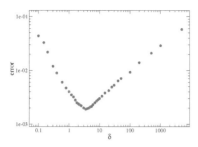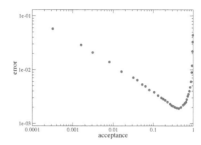

FIGURE 3.7: Left: Statistical error as a function of δ for samples of size $N = 10^6$ of the Gaussian distribution with unit variance. Right: the statistical error is reported as a function of the Metropolis acceptance.

can be strongly dependent on its parameters. For the case we have analysed, small values of δ, with $\delta \ll \sigma$, are not advisable: the Metropolis test acceptance is very high; however, we move very slowly at each Markov chain iteration and, as a consequence, autocorrelation times are large and of the order of $1/\delta^2$ (i.e. according to Eq. (3.48), statistical errors grow as $1/\delta$ at a fixed sample size N).

On the other hand, it is clear that very large values of δ are not advisable as well. Indeed, if $\delta \gg \sigma$ only a fraction of tentative moves proportional to σ/δ will have a chance to be accepted during the Metropolis step, so that, even if accepted steps will be very effective in updating the system, the autocorrelation time will be very large and proportional to δ, because of a very low acceptance proportional to $1/\delta$. At fixed N, statistical errors will scale as $\sqrt{\delta}$. We conclude that some optimal choice of δ must exist, which is expected to be of the order of the variance of the sampled distribution, σ. In order to further clarify the issue, we have produced different samples of equal size $N = 10^6$, hence requiring the same numerical effort, corresponding to different values of δ chosen in a wide range. On the left-hand side of Figure 3.7, we report the error on the sample average $\tilde{x}$ (estimated via the blocking procedure described above) as a function of δ. The logarithmic scale permits to better appreciate the behaviour of the error proportional to $1/\delta$ or to $\sqrt{\delta}$ for small or large values of δ, respectively; on the other hand, the minimum error, corresponding to the maximum efficiecency, is reached for $\delta \sim 3 - 4\ \sigma$.

The same data are shown on right-hand side of Figure 3.7 as a function of the Metropolis test acceptance, showing that the optimal acceptance in this case is around 40%. It is interesting to notice that the error does not change much around the optimal value of δ, so that no fine tuning of the parameter is really needed and values of δ which are reasonably close to the optimal one will work well.

3.6 Error estimate for non-trivial estimators: The Jackknife, and the Bootstrap

We are now going to analyse an issue which is not related to the problem of autocorrelation, although usually both problems have to be taken care of at the same time. We shall make use of the Gaussian distribution (with unit variance and zero average) as a laboratory also in this case; however, in order to clearly decouple the different kinds of problems, we will first start with sets of statistically independent data, produced via the Box-Müller algorithm.

Let us consider a sample of $N = 10^4$ data and the problem of estimating the quantity $\langle x^4 \rangle / (3 \langle x^2 \rangle^2)$ on such sample. The main difference with respect to previous examples is that now we are not interested in a simple average of some function of the stochastic variable over the sample, but in a combination of averages, a ratio in this case. Let us assume for the moment that the corresponding combination of sample averages, i.e.

$$E_N[x_1, x_2, \ldots, x_N] = \frac{\overline{x^4}}{3\,\overline{x^2}^2}, \tag{3.55}$$

is the correct estimator, in the sense that, if we consider many or an infinite number of draws of the same sample, then $\langle E_N \rangle_s = \langle x^4 \rangle / (3 \langle x^2 \rangle^2)$; even if later we will reconsider this hypothesis, by now we just want to understand how to correctly compute the statistical uncertainty σ_{E_N}.

The first naive way that is usually attempted in this situation is to first estimate the statistical errors on the various sample averages entering the ratio, then computing the overall statistical uncertainty by standard error propagation. We have tried to do that on our sample (which is not so useful to report graphically) obtaining

$$\langle x^4 \rangle = 2.96(10) \quad ; \quad \langle x^2 \rangle = 1.000(14)$$

i.e. a relative error 3.4 % on $\langle x^4 \rangle$ and 2.8 % on $\langle x^2 \rangle^2$. When taking the ratio, we are not allowed to sum relative errors in quadrature, as one could do for independent quantities, since both averages have been estimated on the same sample, so we have to take the straight sum of relative errors to finally obtain

$$\frac{\langle x^4 \rangle}{3 \langle x^2 \rangle^2} = 0.99(6). \tag{3.56}$$

Is that a correct estimate? It is largely compatible with the expected value, which for a Gaussian distribution with zero average is 1, hence for sure we have not underestimated the statistical error σ_{E_N}; however, we could have overestimated it.

In order to clarify this point, we have repeated the same estimate for 10^3 independent samples of the same size, to see how E_N is really distributed;

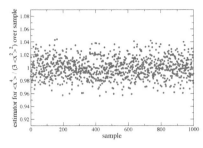 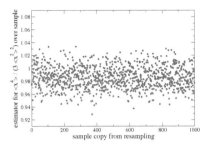

FIGURE 3.8: Left: Estimates of $\overline{x^4}/(3\,\overline{x^2}^2)$ over 10^3 independent samples with $N = 10^4$, in order to estimate the true variance of the estimator. Right: bootstrap estimates of the same quantity over 10^3 samples with $N = 10^4$ which have been resampled starting from the first sample of the left figure.

results are reported on the left-hand side of Figure 3.8. It is already visible by eye that 0.06 is an overestimate for σ_{E_N}, and indeed a direct computation yields $\sigma_{E_N} = 0.016$, i.e. almost a factor 4 lower. Such a bad estimate of the statistical error is equivalent to losing a factor 14 in numerical effort, so we should understand where we did go wrong and how to get a correct estimate. The wrong point is of course the use of standard error propagation: $\overline{x^4}$ and $\overline{x^2}^2$ are computed on the same sample and indeed we adopted a conservative standard sum of the relative uncertainties. However, actually, the existing correlation between the two sample averages acts in the opposite direction: if the sample is such that $\overline{x^4}$ has a fluctuation towards higher values, the same will typically happen for $\overline{x^2}$ too, so that fluctuations will partially cancel in the ratio E_N.

How to get the correct estimate for the standard deviation σ_{E_N} is non-trivial. In this particular case, error propagation could have been done more carefully, taking properly into account the covariance between $\langle x^2 \rangle$ and $\langle x^4 \rangle$; however, such solution is not viable for all possible estimators one has to deal with. Of course, also in this case we cannot think of repeating the sample extraction several times and measuring the fluctuations of E_N from sample to sample directly that would be a waste of time and information.

Instead, we have to devise some method which relies just on the single sample we have at our disposal. We will illustrate two different methods, known as the Jackknife and the Bootstrap. The common idea is that of extracting fictitious new samples starting from the original one, and to study how the estimator fluctuates over the new samples: this idea, which is known under the general name of *resampling techniques*, does not require much new numerical effort, since just the originally sampled system states are needed; that might seem a bit suspicious, since it sounds like creating new information from nothing, nevertheless it works very well.

3.6.1 The bootstrap

Let us consider a generic estimator over the finite sample made up of N independent draws. In our chosen simplified playground, i.e. the Gaussian distribution for one real stochastic variable, this is a generic function of N real variables, $F_N(x_1, x_2, \ldots, , x_N)$, which is usually symmetric for generic permutations of the sample variables and converges, for $N \to \infty$, to some functional $F[p(x)]$ of the probability distribution function $p(x)$ according to which the sample is drawn. The quantity E_N defined in Eq. (3.55) is one example, but one could think of any other combination of sample averages (the sample variance, for instance) or also of other less standard functions, like the median of the distribution.

Our problem is that of estimating how F_N fluctuates over different draws of the sample, in order to assign a statistical error to our estimate of F over the single sample. In particular, the statistical error is defined as

$$\sigma_{F_N} = \sqrt{\langle F_N^2 \rangle_s - \langle F_N \rangle_s^2} \tag{3.57}$$

where by $\langle \cdot \rangle_s$, as usual, we mean the average over the sample probability distribution, $P(x_1, x_2, \ldots x_N)$, which under the assumption of independent draws is defined as

$$P(x_1, x_2, \ldots x_N) \, dx_1 \ldots dx_N = p(x_1) \, p(x_2) \ldots p(x_N) \, dx_1 \ldots dx_N . \tag{3.58}$$

Our difficulties are that, on the one hand, for a generic F we do not have a simple estimator of σ_{F_N} itself, as it happens for simple average values, and that, on the other hand, we have to rely on one single sample, i.e. we cannot afford the sampling of the distribution over samples, Eq. (3.58): each new sample would require a new run of the Markov chain and a numerical effort equal to that needed for the original sample.

The idea of the bootstrap, originally proposed by Efron [23], is the following. Since the N elements of the single sample at our disposal are distributed according to $p(x)$, why do not we use them as our *dice box* to produce new *fake* samples? The extraction of the new j-th sample, $x_k^{(j)}$ with $k = 1, \ldots N$, will proceed as follows: we extract N integer random numbers $j_1, j_2, \ldots, j_N$, each uniformly distributed between 1 and N, then $x_k^{(j)} = x_{j_k}$, where x_{j_k} is the j_k-th element of the original sample. The new samples are fake in the sense that they are not genuinely drawn from the original distribution function $p(x)$, but just from a finite sample of it; however, they are produced at a very low cost, since no new Monte Carlo generation is needed.

For each new sample, we can define a new estimate of F_N

$$F_N^{(j)} \equiv F_N(x_1^{(j)}, x_2^{(j)}, \ldots x_N^{(j)}) \tag{3.59}$$

we can then consider M new fake samples, measure the variance of $F_N^{(j)}$ on them and take it as an estimate of σ_{F_N}, i.e. assume that

$$\sigma_{F_N} \simeq \sqrt{\frac{1}{M}\sum_{j=1}^{M}(F_N^{(j)})^2 - \left(\frac{1}{M}\sum_{j=1}^{M}F_N^{(j)}\right)^2}. \qquad (3.60)$$

It is important to stress that, for each new sample, the random integers j_k are extracted independently, i.e. it is allowed to have multiple repetitions of some elements of the original sample. Therefore, the resampled samples are not just a reshuffling of the original sample, which would lead to exactly the same estimates for F_N.

In order to illustrate how the bootstrap method works, we have considered the original sample with $N = 10^4$, on which the estimate in Eq. (3.56) was based, and exploited bootstrap resampling to produce 10^3 new samples of the same size[10]. On each of these samples, we have evaluated the estimator E_N in Eq. (3.55), obtaining the results reported on the right-hand side of Figure 3.8, where they can be directly compared to results obtained by a genuine draw of independent samples. Two things are worth noticing: *(i)* the bootstrap estimates fluctuate around a value which is a bit lower than 1, this is a reflection of the fact that the bootstrap samples are actually drawn from an approximate representation of the original distribution, consisting in a sample for which indeed we have $E_N \simeq 0.99$; *(ii)* nevertheless, fluctuations around the mean value are very similar in size to those obtained for genuine independent samples, so that we expect to obtain a reasonable estimate of σ_{E_N}, at least to the extent of accuracy needed for a statistical error.

In Figure 3.9 we report bootstrap estimates of σ_{E_N}, obtained according to Eq. (3.60), as a function of the number of resamplings M. One can see that just a few tens of resamplings are needed to converge to an estimate $\sigma_{E_N} \sim 0.017$, which is very close to the value $\sigma_{E_N} \simeq 0.016$ we have found before.

What we have found, in particular the number of resamplings which are needed to get a reasonable estimate of the statistical error, is quite general and independent of the particular problem and sample size under investigation. The interested reader can find more details about the bootstrap method in plenty of existing literature (see, e.g. Refs. [23, 24]).

Maybe it is worth concluding by discussing the origin of the name of the method: this is related to the famous sentence *pull oneself up by one's bootstraps*, which clearly summarises the kind of impossible task the method seems to achieve. Of course, there is no magic at all and, as we have already stressed above, the central point is the assumption that the real sample one has at disposal be a good and faithful empirical realisation of the full statistical pop-

---

[10]In principle, the size of the resampled samples could be different from that of the original one. Indeed, to the extent the original sample is considered as a good representation of the theoretical reference distribution, one could make use of it to produce samples of arbitrary size N' and study in this way the predicted fluctuations for the estimator $F_{N'}$.

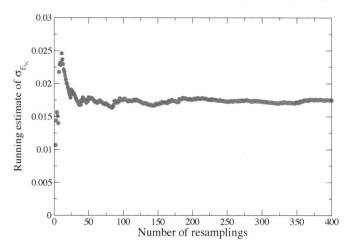

FIGURE 3.9: Bootstrap estimate of the statistical uncertainty on the estimator E_N in Eq. (3.55), as a function of the number M of resamplings appearing in Eq. (3.60).

ulation, i.e. of the theoretical distribution p. The assumption is better and better as the real sample size increases.

3.6.2   The jackknife

In this case, let us start from the origin of the name, which is linked to the well known tool of the same name, and is due to the many practical purposes to which this method can easily adapt. It was originally introduced by Tukey [25] as a tool to correct the presence of a bias in some estimators (a problem that we will discuss later in this chapter), then it was realized it could also be a practical way to obtain other useful information, e.g. on the variance of a given estimator.

Analogously to the bootstrap, also the jackknife is based on the production of several new samples starting from the original one, however, that happens now in a much more systematic way. In particular, one creates N new samples $y_k^{(j)}$ of size $N-1$, i.e. $j = 1, \ldots, N$ and $k = 1, \ldots, N-1$, by removing in each case one of the elements of the original sample:

$$\begin{aligned} y_k^{(j)} &= x_k \quad \text{for } k < j \\ y_k^{(j)} &= x_{k+1} \quad \text{for } k \geq j \,. \end{aligned} \qquad (3.61)$$

On each new sample, one makes an evaluation of the F_{N-1} estimator:

$$F_J^{(j)} \equiv F_{N-1}(y_1^{(j)}, y_2^{(j)}, \ldots y_{N-1}^{(j)}) \qquad (3.62)$$

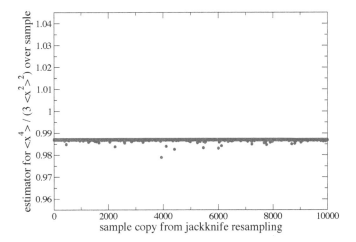

FIGURE 3.10: Estimates of $\overline{x^4}/(3\,\overline{x^2}\,^2)$ overall jackknife subsamples obtained from a sample with $N = 10^4$.

and then computes the variance of such estimates

$$\sigma^J_{F_{N-1}} \equiv \sqrt{\frac{1}{N}\sum_{j=1}^{N}(F^{(j)}_J)^2 - \left(\frac{1}{N}\sum_{j=1}^{N}F^{(j)}_J\right)^2}. \qquad (3.63)$$

How can that be related to the statistical uncertainty on F_N we are interested in? The relation in this case is more involved that in the bootstrap case. In Figure 3.10, we report the jackknife estimates $E^{(j)}_J$ for the estimator in Eq. (3.55), starting from the same starting sample with $N = 10^4$ already used as a playground for the bootstrap. It is well visible that fluctuations of F_{N-1} over the jackknife subsamples are strongly reduced, indeed we obtain $\sigma^J_{E_{N-1}} \simeq 1.7 \times 10^{-4}$. The reason is quite clear: the subsamples are very similar to each other, each pair differing only for the substitution of one single element. We will now give an euristic argument to establish a relation between σ_{F_N} and $\sigma^J_{F_{N-1}}$, thus obtaining the jackknife formula to evaluate the statistical uncertainty.

The information contained in σ_{F_N} is how F_N fluctuates as a consequence of the sample distribution explicited in Eq. (3.58), i.e. for independent fluctuations of each sample element x_k according to $p(x)$. As we have already said, the jackknife subsamples differ from each other for the substitution of just one element, therefore, also due to the symmetry of F under permutations of its arguments, we can consider $\sigma^J_{F_{N-1}}$ as the estimate of the fluctuations induced on F_{N-1} due to the fluctuation of just one of the sample elements, keeping the others fixed. Actually, we should consider the usual bias correction for the variance estimate, i.e. take $\sqrt{N/(N-1)}\,\sigma^J_{F_{N-1}}$ in its place.

It is quite reasonable to assume that the fluctuations of F_{N-1} when all variables fluctuate independently be the *incoherent* sum of those due to single variable fluctuations, i.e. that fluctuations sum in quadrature (this is actually a point where the argument and the method itself can go wrong). The reasoning is in some sense similar to that leading to the formula for error propagation in the presence of statistically independent variables, i.e.

$$\delta F_{N-1} = \sqrt{\sum_{k=1}^{N-1} \left(\frac{\partial F_{N-1}}{\partial y_k} \delta y_k \right)^2}.$$

Based on this assumption, and given the symmetry of F_{N-1} under permutation of its arguments, we can state that

$$\sigma_{F_{N-1}} = \sqrt{N-1} \sqrt{\frac{N}{N-1}} \, \sigma^J_{F_{N-1}}.$$

The final relation to σ_{F_N} is obtained taking for granted that the statistical error scales as $1/\sqrt{N}$, so that

$$\sigma_{F_N} = \sqrt{\frac{N-1}{N}} \, \sigma_{F_{N-1}} = \sqrt{N-1} \, \sigma^J_{F_{N-1}}. \tag{3.64}$$

Applying the formula above to our particular example, we obtain $\sigma_{E_N} \simeq 10^2 \, \sigma^J_{E_{N-1}} \simeq 0.017$, in very good agreement with the estimates obtained before.

Also for the jackknife method, we refer to the specialized literature for more details [26, 27]. In most cases, the bootstrap and the jackknife will work equally well. In general, the bootstrap is preferable when the starting sample size, N, is large, since usually a few tens of resamplings will suffice for the bootstrap, while the jackknife will require the estimate on N subsamples anyway. In addition, there are a few estimators for which the euristic argument given above is wrong and the jackknife does not work, while the bootstrap is still valid: a relevant counterexample is given by the median estimator.

3.6.3 Extension to the case of correlated data sets

We have illustrated the bootstrap and the jackknife resampling techniques for the case of statistically independent data sets. However, in most cases, one has to apply these techniques to data produced by Markov chain Monte Carlo: should we modify resampling to take into account autocorrelations, and how? In order to answer both questions, we have considered again the Metropolis algorithm with $\delta = 0.1$, and produced a sample of $N = 10^7$ draws according to the Gaussian distribution with zero average and unit variance. When applied to this sample in the same way described above, both the jackknife and the bootstrap return an estimate

$$\frac{\langle x^4 \rangle}{3\langle x^2 \rangle^2} = 1.0073 \pm 0.0005$$

which is around 14 standard deviations off the expected result. Therefore, we conclude that in this case, both techniques go wrong in estimating the fluctuations of E_N: it is clear that the reason must be the presence of auto-correlations among data; however, it is interesting to discuss where exactly the two procedures fail, and how to modify them.

In the case of bootstrap, the hypothesis that the original sample might be a good representation of the full distribution function is completely unaffected by the presence of autocorrelations among data. However, the procedure followed for resampling from this distribution should be reconsidered and matched to its purpose, which is that of reproducing the fluctuations that would be obtained for the estimator F_N if the original sample were drawn several independent times. Indeed, the standard bootstrap resampling, in which each new element is drawn randomly from the original set, reproduces the sampling of statistically independent data, so that the error estimate we get is wrong and actually the same we would obtain if F_N were measured on uncorrelated data.

We must modify the resampling procedures so that the new artificial samples contain the same degree of correlation as the original data set. The recipe to do that is simple and intuitive: instead of selecting each element randomly, we have to select random blocks of consecutive draws of the original sample, so that correlations (up to the chosen block size) present in the original data sample will be inherited by the artificial fake samples. Then, a study of the error estimate as a function of the block size will clarify what is the true error, since also in this case we expect the result to stabilize after the block size becomes larger than the autocorrelation time.

One can devise various different recipes to implement this idea. The original data sample could be divided in N/M blocks, of size M each, and then one could perform N/M random draws (with repetitions) of such blocks; or one could instead perform N/M random draws of single elements of the original data sample, and take each time the first M consecutive elements starting from the selected one. The data that we are going to illustrate have been obtained following the second recipe; however, it should be clear that the choice is immaterial, since the purpose is just that the new samples have built-in autocorrelations.

In the case of the jackknife, the error estimate goes wrong in a similar way. The quantity $\sigma^J_{F_{N-1}}$ defined in Eq. (3.63), i.e. how the estimator fluctuates when changing a single data entry, will give similar results (at fixed N) independently of the degree of autocorrelation existing among data. However, the assumption used to obtain the final estimate, Eq. (3.64), was that all entries in the data set fluctuate independently, so that single fluctuations giving rise to

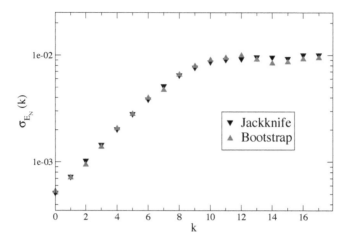

FIGURE 3.11: Bootstrap and jackknife estimates for the error on the estimator E_N in Eq. (3.55), as a function of the blocksize 2^k. The original data set ($N = 10^7$) has been drawn according to the Gaussian distribution with unit variance and zero average, using the Metropolis algorithm with $\delta = 0.1$.

the global fluctuation of F_{N-1} will sum up in quadrature (i.e. incoherently), giving rise to a factor $\sqrt{N-1}$. This assumption is not true when data are correlated: fluctuations will sum coherently over lengths of the order of the autocorrelation time, and incoherently over larger distances; as a consequence, the global fluctuation of F_{N-1} will be larger.

Also in this case, the solution is to divide the data set in N/M blocks of size M each, then applying a modified jackknife recipe in which N/M new samples are created, and in each of them an entire block is removed. Apart from that, everything proceeds in the same way, substituting N with N/M where appropriate. If the original data are not correlated, the final error estimate will remain unaltered: indeed, the fluctuations of the jackknife estimates, σ^J, will increase by roughly a factor $\sqrt{M}$ with respect to the standard jackknife, since M statistically independent data are removed at the same time; however, this factor is recovered when using N/M in place of N in Eq. (3.64). Instead, when data are correlated over a length much larger than M, σ^J will increase by a factor M with respect to the standard case (because of the almost complete correlation existing among the removed data), so that the final error estimate will increase by roughly a factor $\sqrt{M}$. Hence, also in this case, one has to repeat the modified jackknife for several, increasing block sizes, and look for a plateau of the corresponding error estimate as a function of M.

In Figure 3.11, we show results obtained for the error on $\langle x^4 \rangle/(3\langle x^2 \rangle^2)$, adopting the modified versions of the bootstrap and of the jackknife described above, as a function of the blocksize 2^k (in logarithmic scale). One can appreciate a very good agreement between the two procedures for each value of k, hence also on the final estimate, so that we can finally quote $\langle x^4 \rangle / (3\langle x^2 \rangle^2) = 1.007 \pm 0.010$, which is in good agreement with the theoretical expectation[11].

3.7 Biased Estimators

We are now going to discuss a different source of uncertainty affecting Monte Carlo computations, which is still related, as the statistical error, to the finite size of the sample, however, consists in a systematic deviation (bias) of the finite sample estimator from the true (infinite sample) one, hence must be classified as a *systematic error*. As we discuss in the following, such error, when present, is expected to vanish at least as fast as $1/N$, so that, for large samples, it is usually negligible with respect to statistical uncertainties, which scale like $1/\sqrt{N}$. Nevertheless, in many cases, it is easy to take care of it (the unbiased estimator for the variance is an example), and the Jackknife offers a general method to do that, at least partially.

Let us state the problem more precisely. Suppose we need to evaluate a generic estimator, consisting in a functional of the probability distribution of the stochastic variables, $F[p(x)]$, and let $F_N(x_1, x_2, \ldots, x_N)$ be its implementation over a finite sample (here each x represents the state of a system described by an arbitrary number of stochastic variables). The point is whether

$$\langle F_N \rangle_s = F \qquad (3.65)$$

or not. If Eq. (3.65) holds true, we say that the estimator F_N is unbiased, i.e. it returns values which fluctuate, sample by sample, around the true value F; however, such fluctuations are distributed around F so that they average to zero when repeating the experiment many times.

However, this is not always the case. As an explicit example, let us consider the case of a single real stochastic variable, and let $F = f(\langle x \rangle)$, where f is a generic function of the average value of x. A simple implementation of the finite sample estimator is $F_N = f(\bar{x}_N)$, where $\bar{x}_N$ is the sample average of x. Let $\delta \equiv \bar{x}_N - \langle x \rangle$, then, considering a Taylor expansion of f around $\langle x \rangle$, we can write (f' and f'' are the first and second derivatives computed in $\langle x \rangle$):

$$\langle f(\bar{x}_N) \rangle_s = \left\langle f(\langle x \rangle) + f'\delta + \frac{1}{2}f''\delta^2 + O(\delta^3) \right\rangle_s \qquad (3.66)$$

$$= f(\langle x \rangle) + f'\langle \delta \rangle_s + \frac{1}{2}f''\langle \delta^2 \rangle_s + O(\delta^3) = f(\langle x \rangle) + \frac{1}{2}f''\frac{\sigma_x^2}{N} + o(1/N)$$

---

[11]The reader might notice that the ratio between the naive error, 0.0005, and the real one, 0.010, points to an autocorrelation time of the order of 200, which is a bit lower than the one found before, $\tau \simeq 600$. However, in that case, the relevant autocorrelation time was that of the stochastic variable x itself, while for the quantity $\langle x^4 \rangle / (3\langle x^2 \rangle^2)$ the relevant autocorrelation times are those of x^2 and x^4, which are shorter.

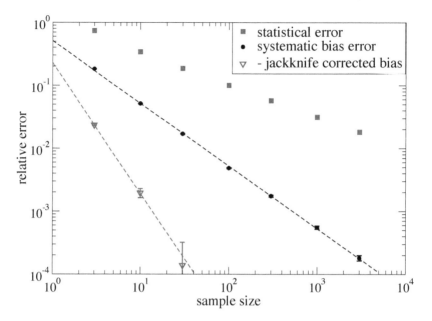

FIGURE 3.12: Bias errors, as a function of the sample size, before and after the Jackknife improvement, for the estimator $\exp(\bar{x}_N)$, where x is a normally distributed stochastic variable. Results are compared to statistical errors affecting the same estimator on the corresponding sample sizes. The dashed lines represent best fits (respectively to $1/N$ and $1/N^2$) for bias uncertainties with and without Jackknife improvement.

where we have used the fact that $\bar{x}_N$ is unbiased ($\langle\delta\rangle = 0$) and that by the central limit theorem $\langle\delta^2\rangle = \sigma_x^2/N$ apart from corrections of higher order in $1/N$. What we have found is actually a general result: bias errors always start as $1/N$ corrections.

We have already met an explicit example of the above result: the estimator of the variance over a finite sample underestimates the true variance by a factor $(N-1)/N = 1 - 1/N$. To see another explicit example, let us consider again the normal distribution with unit variance and average $\langle x\rangle = \mu$, that we have already used as a playground before, and the function $\exp(\langle x\rangle)$. We have fixed $\mu = 0$; however, its value is immaterial in what follows, especially if one considers relative deviations.

We have computed $\exp(\bar{x}_N)$ for samples of different sizes, produced via the Box-Müller algorithm, computing its average and statistical error by directly producing 10^6 different samples of the same size each time. In Figure 3.12, we report the quantity

$$\frac{\langle\exp(\bar{x}_N)\rangle_s - \exp(\mu)}{\exp(\mu)}$$

as a function of N (filled dots). It is quite clear that the systematic error vanishes indeed as $1/N$; a best fit to the function a/N^b returns $b = 1.004(9)$, $a = 0.518(13)$ and $\chi^2/\text{d.o.f.} = 9/4$, in agreement with Eq. (3.66) since $\sigma_x^2 = 1$ and $f'' = 1$; the relatively large value of the χ^2 test is justified by the fact that we are discarding $o(1/N)$ corrections in the fit ansatz.

Figure 3.12 also compares the bias errors with the statistical errors (filled squares) associated with samples of the corresponding size, which have been estimated directly from the variance of the sample estimators[12]. It is apparent that statistical errors largely dominate already for moderately large sample sizes.

Jackknife correction of the leading $1/N$ bias term

The example of the variance estimator shows that the bias can be corrected: in that case it is completely removed, i.e. multiplication by $N/(N-1)$ returns an unbiased estimator, essentially because we deal with a quadratic function of $\langle x \rangle^2$, so that corrections beyond $1/N$ are not present. In other cases, eliminating at least the $1/N$ term is already an appreciable result: the Jackknife method offers a general way to do that.

The idea is quite simple. For a given estimator F_N, the bias correction appears as a power series in $1/N$, whose coefficients are independent of N. Therefore, combining estimators defined on different sample sizes, part of the bias can be cancelled, in particular it is easy to prove that

$$\langle N F_N - (N-1) F_{N-1} \rangle_s = F + o(1/N)$$

i.e. the $1/N$ terms cancel out. Starting from a sample of N draws, it is always possible to derive the $N-1$ estimator by a Jackknife procedure. In particular, following the same notation adopted in paragraph 3.6.2, we can define

$$F_{N,jack} \equiv N F_N - \frac{N-1}{N} \sum_{j=1}^{N} F_J^{(j)} \qquad (3.67)$$

where $F_J^{(j)}$ is the $N-1$ estimator computed on the original sample after subtraction of the jth element.

In order to appreciate how that works, we repeated our experiment on the estimate of $\exp(\langle x \rangle)$ for a normally distributed variable, applying the Jackknife improved sample estimator and considering again 10^6 different samples for each sample size. Results for the relative bias are reported, with a minus sign in order to fit in the logarithmic scale, in Figure 3.12 as well[13]. It is apparent that bias corrections are now strongly reduced and become soon completely

---

[12]The statistical errors reported for the bias are also obtained from the variance of the sample estimators, but dividing in this case by $\sqrt{10^6}$, by the central limit theorem, since we are estimating the error on $\langle f(\bar{x}_N) \rangle_s$.

[13]The minus sign is expected: the definition in Eq. (3.67) cancels the $1/N$ term and changes the sign of the $1/N^2$ correction, which is positive in this case since the derivatives of the exponential function are all positive.

negligible with respect to statistical uncertainties; they are so tiny that it is difficult to correctly estimate them for $N \gtrsim 100$, even with a statistics of 10^6 samples. A best fit to the function a/N^b returns $b = 2.07(5)$, in agreement with expectations.

3.8 Exercises

Exercise 3.1 *Consider a three-state system and the following Markov chain acting on it*

$$W = \begin{pmatrix} 1 & \epsilon & \epsilon \\ 0 & 1 - 2\epsilon & \epsilon \\ 0 & \epsilon & 1 - 2\epsilon \end{pmatrix} \tag{3.68}$$

where W_{ij} is the probability of moving from j to i in one step. Is the chain regular? What is its equilibrium distribution? Give an estimate of its exponential autocorrelation time.

Exercise 3.2 *Consider a Markov chain corresponding to a random walk on a finite one-dimensional periodic lattice, i.e. a stochastic process acting on a lattice with k sites and periodic boundary conditions (a discrete ring), with an equal probability $1/2$ of jumping to the nearest neighbour on the left or on the right.*
Is the process ergodic? Is it regular? Can you add a small perturbation so as to make it regular? Compute the exponential autocorrelation time in this case.

Exercise 3.3 *Devise a Metropolis algorithm to uniformly cover the circle of unitary radius. Do it in cartesian coordinates x, y, first, and then in polar coordinates r, ϕ.*

Exercise 3.4 *Devise a Metropolis algorithm to uniformly cover the sphere of unitary radius. Do it in cartesian coordinates x, y, z, first, and then in spherical coordinates r, θ, ϕ.*

Exercise 3.5 *Define a Metropolis algorithm to sample the probability distribution*

$$p(x)dx = e^{-\gamma|x|}\,dx$$

and find the parameters which optimize its efficiency.

Exercise 3.6 *Define a Metropolis algorithm which uniformly explores the surface of the sphere in three and four dimensions.*

Exercise 3.7 *Consider again the cumulant ratio $\langle x^4 \rangle / (3 \langle x^2 \rangle^2)$ where x is a normally distributed Gaussian variable. Determine numerically the leading term of the bias error, comparing it with an analytical estimate. Then do the same with the Jackknife improved estimator.*

Exercise 3.8 *Discuss if and how the Jackknife correction of the bias would be affected by the presence of autocorrelations among data.*

Chapter 4

Statistical models

The statistical tools and techniques developed in previous chapters have wide applications in various branches of science (including social sciences), engineering and finance. Statistical mechanics, which is the subject of this chapter, is a relevant example of an application, and indeed one of the central topics of this book. Statistical mechanics is the branch of physics that derives thermodynamics properties of systems starting from the kinetics and the interactions of their elementary constituents. More specifically, the objective of statistical mechanics is to deduce macroscopic (i.e. measurable at our scale) thermodynamic observables of a system using a statistical approach to the dynamics of its microscopic constituents. In principle, it would be possible to apply the laws of (quantum) mechanics that govern the evolution of any physical system to the elementary constituents of matter to work out the resulting physical quantities. However, since a typical macroscopic system is made of $\mathcal{O}(10^{23})$ particles, solving exactly the associated equations is an impossible task even for the most advanced supercomputer. Statistics provides a better approach, enabling us to derive average properties of the system under investigation. Using statistics, we can in principle calculate the probability that a particle has a given speed or occupies a given state (characterised by its energy). In this context, statistical mechanics can be taken as a set of prescriptions to associate average properties of a given system to its thermodynamic observables once control parameters such as the temperature or the pressure are specified. Building on the set of rules provided by a statistical approach, not only can one prove that the expected properties of known systems (e.g. non-interacting pointlike particles giving rise to the ideal gas law) can be reproduced but also properties of more complex systems (e.g. spin models or superconductors) can be predicted.

The chapter is structured as follows. section 4.1 presents the basic elements of thermodynamics. In section 4.2, we will provide an introduction to statistical mechanics. Phase transitions are treated in a statistical mechanics approach in section 4.3. One of the most popular topics in statistical mechanics, the Ising model, is then discussed in section 4.4. Finally, in section 4.5, we shall provide an overview of other statistical models of wide use.

4.1 An introduction to thermodynamics

Statistical mechanics is the framework that derives thermodynamic properties starting from the laws of (quantum or classical) mechanics, which, due to the high number of particles involved, are treated not exactly, but statistically. In order to understand the foundations and the developments of statistical mechanics, we will provide here a concise description of thermodynamics. Historically, thermodynamics developed from the need to understand phenomena related to heat transmission and heat diffusion. In the beginning, the discipline developed as a set of empirical investigations. Later, mostly in the second half of the 1800, it was formalised into principles and laws. Most of the formalisation involved leading figures such as Carnot, Clausius, Kelvin and Maxwell.

4.1.1 Thermodynamic systems and associated states

Thermodynamics is also a branch of physics. Its objects of study are heat and related phenomena (e.g. heat transfer in between systems). The concept of system is central in thermodynamics. A system is a macroscopic (i.e. having a typical size that is much larger than the size of each of its elementary constituents) collection of particles we are interested in. An example is a gas in a container. A system can be open or closed. An open system can exchange particles and energy with an external environment. The external environment has to be intended as another system that is separated from the original one and to which we are interested in only because of the possibility of having these exchanges and the necessity to evaluate their effect on the original system. In a closed system, no exchange with the environment can take place. Going back to our example of the gas in a container, if the container is thermally insulating and hermetic, the system will be closed. If there is the possibility for exchanges of energy (e.g. through the compression made with a piston) or particles can escape because the container is not hermetic, the system is open. Another central concept is that of *observables*. An observable is any quantity that we are able to measure (e.g. the temperature, the pressure, the volume). Observables are classified either as *extensive* (i.e. having a value proportional to the number of constituents) or *intensive* (i.e. having a value that is independent of the number of constituents. Examples of extensive observables include the volume (indicated by V), the entropy (S), the enthalpy (H), the Gibbs free energy (G), the Helmholtz free energy (F) and the number of particles (N). Examples of intensive variables are the temperature (T), the density (ρ), the chemical potential (μ), the magnetic field (h). Some of those variables (e.g. T and P) are familiar to us through our everyday experience; the others will be defined below. Note that we can build intensive quantities by dividing extensive quantities by N or V. For instance, from the Helmholtz free energy F, we can definite the free energy per unit of volume f as $f = \frac{F}{V}$.

A system is said to be in **equilibrium** when there is no heat exchange among its parts. For instance, a pan of water on a hob is not at equilibrium, since heat will propagate from the hotter bottom of the pan to the colder top surface. Any system at equilibrium is in a *state*. A state can be represented as a point in a $\mathbb{R}^K$ space that describes all possible conditions in which the system can be. This state is defined by all physical observables one can measure on a system. In fact, it turns out that there are relations between those observables that reduce the number of variables that one can assign arbitrarily. A noticeable example of such a relation is the equation of state of an ideal gas

$$PV = NkT ,\qquad\qquad (4.1)$$

which states that, once we assign (for instance) the volume V, the pressure P and the number of particles N, then the temperature T is determined (k being a known constant, the Boltzmann constant).

Let's say that there are M relations between a set of k representative observables defining the $\mathbb{R}^K$ space of possible states. Once all those relations have been taken into account, $n = K - M$ independent observables can be identified. If no further relation can be identified, n takes the name of degrees of freedom. In other words, the number of degrees of freedom of a given thermodynamic system is an integer counting the observables whose value is

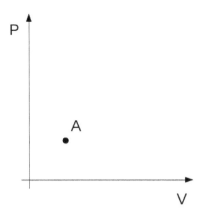

FIGURE 4.1: The space of all possible states for the ideal gas when the chosen independent variables are P and V. A represents a particular point in this space.

necessary and sufficient to specify the thermodynamic state of the system. While this set of observables can be chosen with a degree of arbitrariness (which would correspond to a change of variables), n is an intrinsic property of the system. Going back to our example of the ideal gas, it is possible to show that in this case $n = 2$. A possible graphical representation of the states of the system consists in a bidimensional plot in the variables P and V (see Fig. 4.1). Other choices (e.g. P and T) are also possible.

4.1.2 First look at differential forms

Relations between M observables in an $\mathbb{R}^K$ space (where $M < K$) describe a *differential manifold*. In any point of this differential manifold, we can define a *tangent space*, which is spanned by the linear independent vectors that are tangent at the given point to the hypersuperface fulfilling the constraints. The dimension of the tangent space in any point is equal to the number of degrees of freedom $n = K - M$. This is also the number of degrees of freedom (i.e. independent observables) that describe the system.

Let's fix a point A in the space of all states and let T be the corresponding tangent space. A linear function

$$\alpha : T \to \mathbb{R} \tag{4.2}$$

is called a *differential forms*. The ensemble of all α is vectorial space of the same dimension as T, which we indicate with T^*. T^* is called the *dual* of T. Differential forms provide the natural language to treat thermodynamics.

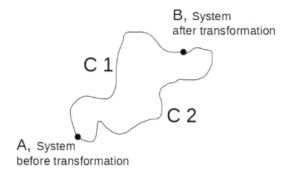

FIGURE 4.2: Two transformations connecting A and B, represented, respectively, by the curves C_1 and C_2.

A useful distinction is that between exact and non-exact differential forms. For this purpose, we introduce the integral of a differential form

$$\int_C \alpha \equiv \int \alpha(\vec{v}(t))\mathrm{d}t \tag{4.3}$$

where t $(0 \le t \le 1)$ parameterise the curve C and $\vec{v}(t)$ is the tangent vector to the curve C when the value of the parameter is t. This definition abstracts, for instance, the operation of computing the length of a curve as the integral of the velocity along that curve when we use the time as a parameter.

Now, let us consider two points A and B in the space of states. If two paths C_1 and C_2 exist that join the states A and B in the space of states of the system (see Figure 4.2) such that

$$\int_{C_1} \lambda \ne \int_{C_2} \lambda , \tag{4.4}$$

then λ is a non-exact or non-integrable form. Instead, if for any path joining A and B

$$\int_{C_1} \lambda = \int_{C_2} \lambda , \tag{4.5}$$

then λ is exact or integrable.

Parametrising a curve with a coordinate t naturally assigns a direction in which we move along the curve, the direction of increasing t. If for instance the two extremes A and B are identified, respectively, by $t = 0$ and $t = 1$, we will move from A to B. We can also make the opposite choice. It is easy to verify that

$$\int_{C_{A \to B}} \lambda = -\int_{C_{B \to A}} \lambda , \tag{4.6}$$

where $C_{A\to B}$ is the curve in which we start from A and end at B, while $C_{B\to A}$ is the very same physical curve, with a parametrisation that starts from B and ends at A. Using Eq. (4.6), it is easy to show that the integral of an exact form along a closed contour is zero:

$$\oint_C \lambda = 0 \, , \tag{4.7}$$

where the symbol $\oint$ indicates that the support C of the integral is a closed curve.

Similarly, it is possible to prove that a form is exact if its integral along all closed paths is zero. Putting both implications together, we have the following theorem:

Theorem 1 *A form is exact if and only if $\oint_C \Lambda = 0$ for any closed curve C.*

Finally, it is possible to prove that if λ is exact, a function ℓ exists such that its differential gives λ:

$$\lambda = d\ell \, . \tag{4.8}$$

4.1.3 First law of thermodynamics

In the following, infinitesimal quantities have to be meant in the framework of tangent space and associated differential forms: we make infinitesimal moves in the space tangent to the hypersurface describing the system and evaluate differential forms on vectors of this space (i.e. we measure the difference in values of thermodynamic observables for infinitesimal variations of the state). The infinitesimal mechanical work made by a system can be written as

$$\lambda = P \cdot dV \, . \tag{4.9}$$

We now call ω the infinitesimal heat provided by the system. This will provide us with all the elements to formulate a remarkable law of physics known as the first law of thermodynamics:

First law of thermodynamics in differential form: *An exact differential U exists whose infinitesimal variation is given by:*

$$dU = -\lambda - \omega \, . \tag{4.10}$$

We have specified that the law is formulated in differential form since an integrated version is also commonly found in elementary thermodynamics books. Either way, the first law of thermodynamics embodies the conservation of energy for a thermodynamic system and shows explicitly that heat is a form of energy.

The function U takes the name of internal energy. Note that integrals of exact differentials like U in thermodynamics are called *state functions*, since their difference between two states A and B depends only on the actual initial and final states A and B and not on any particular path that joins them. This is not true for observables like the work W (which is defined as the integral of λ and does depend on the path chosen between A and B) or the heat Q (integral of ω).

Note that above we have considered the internal energy of a system receiving only mechanical contributions (provided by λ) and thermal contributions (through ω). More in general, if more than one physical phenomenon contributes to the internal energy of the thermodynamic system, the differential form of the internal energy depends on the other variables. For instance, for a system that has a variable number of particles and a non-null magnetic moment, we write

$$dU = -\lambda - \omega + \mu dN - \vec{h} \cdot d\vec{M} \ , \tag{4.11}$$

where μ is the chemical potential, N the number of particles, $\vec{h}$ the external magnetic field and $\vec{M}$ the magnetisation.

4.1.4   Second law of thermodynamics

Differential forms provide a powerful formulation also for the second law of thermodynamics:

Second law of thermodynamics in differential form: *A function S exists such that*

$$dS = -\frac{\omega}{T} \ .$$

The function S is called the entropy and is again a well-defined observable for a given system. In the context of the second law of thermodynamics, the temperature T can be seen as *the inverse of the integrating factor of the entropy*[1].

We note that by combining the first and the second law of thermodynamics, we can write the differential of U as a linear combination of other differentials, i.e.

$$dU = -PdV + TdS + \mu dN - \vec{h} \cdot d\vec{M} + \dots \ . \tag{4.12}$$

This relation shows that U should naturally be regarded as a function of V, S, N, $\vec{M}$, ...

The existence of natural variables in which a given quantity is expressed is a crucial concept in thermodynamics. A procedure to define new observables through a specific change of variables known as the *Legendre transform* is defined in the next subsection.

---

[1]Given a (non-exact) differential form α, a function f is said to be its integrating factor if a function A exists such that $dA = f\alpha$.

4.1.5 Legendre transform and thermodynamic potentials

In general, consider $F = F(A, B)$. For the differential,

$$dF = \alpha dA + \beta dB \; , \tag{4.13}$$

with α and β functions of A and B. In particular,

$$\alpha(A, B) = \left(\frac{\partial F}{\partial A} \right)_B \tag{4.14}$$

Under the hypothesis that the Hessian (i.e. the symmetric matrix with all possible second derivatives) of F is different from zero, we can invert the dependence between α and A and write

$$A = A(\alpha, B) \; . \tag{4.15}$$

Let us consider now

$$G(\alpha, B) = F(A(\alpha, B), B) - A\alpha \; , \tag{4.16}$$

where now α and B need to be regarded as the independent variables. We have

$$dG = dF - \alpha dA - A d\alpha = -A d\alpha + \beta dB \tag{4.17}$$

We call G the Legendre tranform of F with respect to A. The pair (A, α) are called conjugate variables. Note that if we perform a Legendre transform of G with respect to α, we obtain F, i.e. the Legendre transform has identical square.

In addition, we have

$$\alpha(A, B) = \left(\frac{\partial F}{\partial A} \right)_B$$

$$A(\alpha, B) = \left(\frac{\partial G}{\partial \alpha} \right)_B$$

$$\beta(A, B) = \left(\frac{\partial F}{\partial B} \right)_A$$

$$\beta(\alpha, B) = \left(\frac{\partial G}{\partial B} \right)_\alpha$$

Given $U \equiv U(V, S, \dots)$, through an appropriate Legendre transform, we define the following thermodynamic potentials:

1. The Helmholtz free energy F

$$F(V, T, \dots) = U - TS \; ; \tag{4.18}$$

   at constant volume and temperature, the equilibrium state minimises F.

2. The Gibbs free energy G

$$G(P, T, \ldots) = F + TS \; ; \tag{4.19}$$

at constant pressure and temperature, the equilibrium state minimises G.

3. The enthalpy H:

$$H(P, S, \ldots) = U + PV \; . \tag{4.20}$$

at constant pressure and entropy, the equilibrium state minimises H.

Analogously, at constant volume and entropy, the internal energy U is minimised in the equilibrium state.

Note that all those thermodynamic potentials are state functions.

4.1.6 Maxwell's relations

Maxwell's relations are equalities between thermodynamic functions and their derivatives. We distinguish between two types of Maxwell relations: those derived directly from the expression of thermodynamic potentials (which we call type A) and those obtained by imposing that the thermodynamic potentials are state functions, i.e. that their differential is exact (type B). In the following, for simplicity we are going to omit the functions that are kept constant during the differentiation, while highlighting at the same time that in general is very important to specify them).

- **Type A Maxwell relations**
  Let us consider the internal energy in differential form:

  $$\mathrm{d}U = -P\mathrm{d}V + T\mathrm{d}S + \mu\mathrm{d}N - \vec{h} \cdot \mathrm{d}\vec{M} + \ldots \; .$$

  We can write

  $$P = -\frac{\partial U}{\partial V} \; , \qquad T = \frac{\partial U}{\partial S} \; , \qquad \ldots \; . \tag{4.21}$$

  Similar considerations can be made for other thermodynamic potentials, giving rise to further Maxwell's relations.

- **Type B Maxwell's relations**
  Given a twice-differentiable function $f(x, y)$, we have

  $$\frac{\partial}{\partial y}\left(\frac{\partial f}{\partial x}\right) = \frac{\partial}{\partial x}\left(\frac{\partial f}{\partial y}\right) \; .$$

  If we take $U \equiv U(V, S, N, \vec{M})$, we have, e.g.

  $$\frac{\partial}{\partial V}\left(\frac{\partial U}{\partial N}\right) = \frac{\partial}{\partial N}\left(\frac{\partial U}{\partial V}\right) \rightarrow \frac{\partial \mu}{\partial V} = -\frac{\partial P}{\partial N}$$

  Again, we can repeat this procedure for all thermodynamic potentials, generating a number of Maxwell's relations.

4.1.7 Heat capacities

We conclude our overview of thermodynamics by introducing two quantities that are often met in applications: the heat capacity at constant volume, C_V and the heat capacity at constant pressure, C_P. They are defined, respectively, as

$$C_V = \left(\frac{\partial U}{\partial T}\right)_V$$

and

$$C_P = \left(\frac{\partial H}{\partial T}\right)_P.$$

They are extensive quantities that describe the amount of heat required by a system to change its temperature by dT, and as such are related to the response of the system to temperature. Quantities of this type, which are expressed with second derivatives of free energies, are more in general referred to as susceptibilities, since they physically represent responses of the system to a change in a control variable. Specifically, the heat capacity describes the response of the system to a change of temperature. In particular, the higher the heat capacity, the more the variation of internal energy that is required to achieve a given increment in temperature.

4.2 From thermodynamics to statistical mechanics

Thermodynamics is the branch of physics that is interested in macroscopic properties related to heat generation, absorption and exchange. Thermodynamics is not concerned with how its laws emerge from the microscopic nature of a system. Statistical Mechanics instead focuses on the emergence of macroscopic properties from the microscopic ones.

Ideally, if we could write the equations of motions of elementary constituents, we should be able to determine the emergence of thermodynamic laws. However, in practice, in a typical thermodynamic system, we deal with order 10^{23} particles. The problem of solving so many coupled differential equations can not be tackled analytically and cannot be solved by any computer currently available and currently devisable. Assuming classical dynamics for the sake of the argument, only to store the positions and the momenta of a similar system in three dimensions, we would need to use a storage capacity of 10^{11} terabytes, with current computers having hard disks of the size of a terabyte and supercomputers of $10^4 - 10^5$ terabytes. Even in the most optimistic scenario, we would be a factor of 10^6 under resourced in terms of our storage ability for such a computation. Similar considerations hold for the raw computational power.

In order to bypass this issue, we can resort to statistical methods: we will treat the dynamical properties of microscopic constituents of the system as random variables extracted from a probability distribution. Therefore, instead of asking ourselves the question *what is the speed of the n-th particle of the system?*, we will be asking *what is the probability that a given particle has a velocity in the interval* $[v, v + dv]$*?*. Part of the problem is the identification of the probability distribution followed by the dynamical quantities of the system.

4.2.1   The micro-canonical ensemble

We start by considering a system in isolation. According to the laws of dynamics, the energy of the system is conserved. We use Hamiltonian mechanics as the framework for our approach.

In Hamiltonian mechanics, the degree of freedoms of the systen are the coordinates $\vec{q_i}$ and momenta $\vec{p_i}$ $i = 1, ..., 3N$ of the N particles that compose the system.[2] We represent the particles in the so-called *phase space*, which we indicate with the letter Γ. This is a $6N$ dimensional space whose coordinates are all the positions and momenta, ordered as $q_1, ..., q_{3N}, p_1, ..., p_{3N}$. Each $6N$-ple identifies a point in Γ space.

We now want to count physically equivalent states of a system, i.e. points that are close enough for their difference to be physically meaningless. Let us consider an infinitesimal volume element in Γ space, which can be written as

$$\mathrm{d}\Gamma = \mathrm{d}q_1 \cdots \mathrm{d}q_{3N} \mathrm{d}p_1 \cdots \mathrm{d}p_{3N} . \qquad (4.22)$$

We can identify a state of the system as a volume element in Γ which encloses all the points that are physically equivalent. if we call Δ the volume enclosed by equivalent states, the effective infinitesimal volume element becomes $d\Gamma/\Delta$. Given a state, incertitude and degeneracy define equivalent states. For instance, the indetermination principle would tell us that $\mathrm{d}p\mathrm{d}q \approx h$, where h is the Planck constant; moreover, identical (bosonic) particles are indistinguishable. We can use these observations to specify equivalent states, therefore effectively making the identification

$$\Delta = h^{3N} N! ,$$

where the factorial accounts for the indistinguishability of the particles in the system. Note that both h and $N!$ play a crucial role in our (semi)classical approach to Statistical Mechanics: h gives some natural dimension, therefore making the volume element dimensionless, as it should be for a pure number (remember that above we have set the task to count states of the system, and

---

[2]We are assuming that the system is three-dimensional.

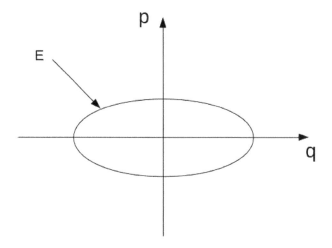

FIGURE 4.3: The hypersurface $H = E$ for an harmonic oscillator.

the number of states is a pure number). The factor of $N!$ solves a paradox related to the mixing entropy: without this factor, if we arbitrarily divide a gas into two subsystems, the entropy of the total system will not be given by the sum of the entropies of the two subsystems, which would be physically unacceptable. This will be further explored in exercise 4.6.[3]

Since our system is isolated, the Hamiltonian H is conserved by the dynamics. The number of states compatible with this condition is given by

$$\Sigma(E) = \int \frac{\mathrm{d}\Gamma}{h^{3N} N!} \delta(H(q_1, ..., q_{3N}, p_1, ..., p_{3N}) - E) \ ,$$

where the Dirac δ function selects a hypersurface of constant energy E.

Example 1 *The Hamiltonian of an harmonic oscillator is given by*

$$H = \frac{p^2}{2m} + \frac{1}{2} kq^2 \ ,$$

where k is the Hook constant and m is the mass. Therefore, for a system of a single harmonic oscillator, the hypersurface of constant $H = E$ is one-dimensional and in particular is an ellipsis (see Figure 4.3).

A central concept in Statistical Mechanics is the thermodynamic limit, which is the limit for the number of particles N of the system going to infinity taken at fixed intensive quantities (such as energy density, entropy per particle etc.).

---

[3]Note that exercise 4.6 is formulated in the canonical ensemble, which will be introduced in Sec. 4.2.2.

In this limit, an hypersurface Σ in $\mathbb{R}^{3N}$ has the same measure as the volume it encloses:

$$\Sigma(E) \simeq \Gamma(E) = \int_{0 \leq H(q_1, ..., p_{3N}) \leq E} \frac{d\Gamma}{h^{3N} N!} \tag{4.23}$$

This allows us to get rid of the δ function in an easy way, therefore enabling us to go from complicated surface integrals to a priori simpler volume integrals in order to calculate allowed states at fixed energy. All Statistical Mechanics considerations are always meant to be done (and hence valid) in the thermodynamic limit, which we will assume from now on.

A system with energy E is in one of the states corresponding to that energy. A relevant question is which states the system is in, or better what is the probability that the system is in a given allowed state. For this, we postulate the principle of equal a priori probabilities:

Definition 1 *The probability density $P(q_1, ..., q_{3N}, p_1, ..., p_{3N})$ to find the system in a volume $\frac{d\Gamma}{h^{3N} N!}$ centered at the state $q_1, ..., q_{3N}, p_1, ..., p_{3N}$ is the same for all choices of coordinates $(q_1, ..., q_{3N}, p_1, ..., p_{3N})$ such that $H(q_1, ..., q_{3N}, p_1, ..., p_{3N}) = E$.*

We then introduce the probability distribution $\rho(q_1, ..., q_{3N}, p_1, ..., p_{3N})$ as

$$\rho(q_1, ..., q_{3N}, p_1, ..., p_{3N}) = \begin{cases} c & \text{if } E \leq H \leq E + \delta E \\ 0 & \text{otherwise} \end{cases},$$

where c is a constant and δE is an infinitesimal variation of energy. Thermodynamic observables can now be introduced as average of microscopic observables in Γ space using the probability ρ. For instance, the internal energy U, which is the average energy of the system, is computed as

$$U = \langle H \rangle = \frac{\int d\Gamma \; H(q_1, ..., q_{3N}, p_1, ..., p_{3N}) \; \rho(q_1, ..., q_{3N}, p_1, ..., p_{3N})}{\int d\Gamma \; \rho(q_1, ..., q_{3N}, p_1, ..., p_{3N})}.$$

We introduce the entropy S as

$$S = k \log(\Gamma(E))$$

We now recall the first law of thermodynamics in differential form, which we can rewrite as

$$dS = \frac{P}{T} dV + \frac{1}{T} dU. \tag{4.24}$$

This enables us to define the *microcanonical temperature* as

$$\frac{1}{T} = \left(\frac{\partial S}{\partial U} \right)_V \tag{4.25}$$

We have now all the necessary ingredients for a procedure to derive the thermodynamic behaviour of a system from the microcanonical ensemble. This procedure is summarised by the following steps:

1. Compute $\Gamma(E) = \int_{0 \le H \le E} d\Gamma$;

2. Define $S(E) = k \log(\Gamma(E))$;

3. Compute $\frac{1}{T} = \left(\frac{\partial S}{\partial U}\right)_V = \left(\frac{\partial S}{\partial E}\right)_V$;

4. Invert to find $E(T) \equiv U(T)$;

5. Use Maxwell's relation to derive the other quantities.

4.2.2   The canonical ensemble

The micro-canonical ensemble is based on the principle of energy conservation for an isolated system. In thermodynamics, in general systems are not isolated, but are allowed to exchange heat with each other. This results in fluctuations in the energy.

This physical setting is captured by the canonical ensemble. The canonical ensemble can be derived by partitioning a microcanonical ensemble into two components. We refer to these components as subsystem 1 and subsystem 2. We will consider an asymmetric situation in which subsystem 2 is larger than subsystem 1. We will be interested in the thermodynamic behaviour of subsystem 1, and we will use subsystem 2 as a bridge towards the by-now familiar microcanonical case. In our notations, quantities with no subscript will refer to subsystem 1 (e.g. E is the energy of subsystem 1), quantities for subsystem 2 are over-lined (e.g. $\bar{E}$ for the energy of subsystem 2) and the subscript T will refer to quantities for the whole original system (e.g. E_T). Where needed for clarity, we will use the subscripts 1 and 2 to refer, respectively, to subsystem 1 and subsystem 2. For additive quantities, their values for the whole system is the sum of the values for the two partitions. For instance, for the number of particles

$$N_T = N + \bar{N} \ . \tag{4.26}$$

The number of particles in each subsystem is kept constant. We consider the case in which $N_T \simeq \bar{N} \gg N$, but at the same time both N and $\bar{N}$ are large enough for the two components to be considered in the thermodynamic limit. This setting is schematised in Figure 4.4. The whole system is isolated, but, taken singularly. the two subsystems 1 and 2 are not. The fact that the whole system is isolated means that 1 and 2 can only interact with each other. In terms of typical values of relevant observables, given the difference in number of particles and the fact that the overall system is isolated and at equilibrium, we expect that $\bar{E} \gg E$ and $\bar{V} \gg V$. Subsystem 2 is called *heat bath* or *(heat) reservoir*.

As a result of the exchanges between the two partitions, E and $\bar{E}$ both fluctuate. However, since $\bar{E} \gg E$, we can neglect the fluctuations of $\bar{E}$. We can

FIGURE 4.4: Setup for the derivation of the canonical ensemble from the microcanonical: we consider our target system as a subsystem of a much larger one, with which it can exchange energy with negligible influences on the fluctuation of the energy of the larger system itself. Both systems are large enough for the thermodynamic limit to provide an accurate description of their behaviour.

now apply the procedure for the derivation of the thermodynamic properties in the microcanonical ensemble for the overall system, break the energy into the sum of the contributions of the subsystems and focus on our subsystem of interest. We consider the entropy $S_2(\overline{E})$ of subsystem 2, for which we get

$$S_2(\overline{E}) = S_2(E_T - E) \simeq S_T(E_T) + \frac{\partial S_2}{\partial E_T}(E_T - E) \qquad (4.27)$$

$$\simeq S_T(E_T) + \frac{\partial S_T}{\partial E_T}(E_T - E) , \qquad (4.28)$$

where we have used the fact that the thermodynamic quantities in 2 are well approximated by those of the total system including 1 and 2. Defining the temperature of the total system from

$$\frac{1}{T} = \frac{\partial S_T}{\partial E_T} , \qquad (4.29)$$

which follows from the observation that this system is microcanonical, we can write

$$S_2(\overline{E}) = \text{constant} - \frac{E}{T} , \qquad (4.30)$$

where the constant term (which we will drop from now on) refers to microcanonical (and hence constant) quantities in the total system. However, given its large size, also the subsystem 2 is (approximately) microcanonical. Hence, we can write the volume in its Γ_2 space for given energy $\overline{E}$ as

$$\Gamma_2(\overline{E}) = e^{S_2/k} \propto e^{-\beta E} . \qquad (4.31)$$

For subsystem 1, we now consider an energy value E obtained by evaluating the Hamiltonian of the subsystem on its microscopic state corresponding to particle coordinates $q_1, ..., p_{3N}$. The total number of states that are compatible with this energy is given by $d\Gamma_1 \Gamma(\overline{E}) \propto d\Gamma e^{-\beta E}$, where the proportionality factor $\Gamma(\overline{E})$ is due to the fact that states in the two subsystems are independent, and hence, all states of subsystem 1 get an enhanced degeneracy by a factor $\Gamma(\overline{E})$. Since E is not fixed, in order to count states, this expression needs to be integrated over all configurations corresponding to allowed energies for the subsystem 1. This process exposes the quantity

$$Z = \int d\Gamma e^{-\beta H(q_1,...,p_{3N})} . \tag{4.32}$$

Note that in the previous expression, we have dropped the subscript 1 from $d\Gamma$, since for our argument we do not need to consider any longer subsystem 2; hence, there is no risk of confusion. Additionally, we have dropped the proportionality factor that relates $\Gamma_2(\overline{E})$ to $e^{-\beta E}$, as this will cancel out in the calculation of key thermodynamic observables (or, more generally, can be reabsorbed through a suitable normalisation condition).

The quantity Z defined in Eq. (4.32) takes the name of *partition function*. Its physical relevance is expressed by the relation

$$Z = e^{-\beta F} , \tag{4.33}$$

i.e. Z gives access to the Helmholtz free energy F. Since in turn F enables to derive all thermodynamic quantities, the thermodynamic behaviour of a system can be obtained in full if its partition function is known. For instance, given

$$F = -\frac{1}{\beta} \log Z , \tag{4.34}$$

we cam determine the entropy S as

$$S = -\left(\frac{\partial F}{\partial T}\right)_V \tag{4.35}$$

and then the internal energy U using

$$U \equiv F + TS . \tag{4.36}$$

As a matter of fact, access to U can also be obtained directly from Z, since

$$U = \langle H \rangle = \frac{1}{Z} \int d\Gamma \, H \, e^{-\beta H} = -\frac{1}{Z}\frac{\partial Z}{\partial \beta} = -\frac{\partial \log(Z)}{\partial \beta} . \tag{4.37}$$

Once U is known, the heat capacity C_V can be derived as

$$C_V = \left(\frac{\partial U}{\partial T}\right)_V = k\beta^2 \left(\langle E^2 \rangle - \langle E \rangle^2\right) \tag{4.38}$$

To show that the relation above holds, let us start by defining

$$\langle E \rangle = U = -\frac{\partial \log Z}{\partial Z} . \tag{4.39}$$

We now write $\langle E^2 \rangle$ as

$$\langle E^2 \rangle = \frac{1}{Z} \int dE\, E^2 e^{-\beta F} = \frac{1}{Z} \frac{\partial}{\partial \beta} \left(\frac{\partial}{\partial \beta} \int dE\; e^{-\beta E} \right) = \frac{1}{Z} \left(\frac{\partial^2}{\partial \beta^2} Z \right) . \quad (4.40)$$

Computing explicitly C_V gives

$$C_V = \left(\frac{\partial U}{\partial T} \right)_V = -\frac{1}{kT^2} \frac{\partial U}{\partial \beta} , \quad (4.41)$$

which, using the form of U, Eq. (4.39), gives

$$C_V = k\beta^2 \left(\frac{1}{Z} \left(\frac{\partial^2 Z}{\partial \beta^2} \right) - \left(\frac{1}{Z} \frac{\partial Z}{\partial \beta} \right)^2 \right) \quad (4.42)$$

Comparing the expression above with the expression for $\langle E^2 \rangle$ in Eq. (4.40) and of $\langle E \rangle$ in Eq. (4.39), we obtain Eq. (4.38). Note that the latter equation is a particular example of the more general *theorem of fluctuation and dissipation*, which related a dissipative quantity (i.e. a second derivative of the free energy (in this case C_V) to the fluctuations in an expression connected with a first derivative (U).

4.2.3   The grand canonical ensemble

The third ensemble we shall discuss is the grand canonical ensemble. In this setup, both the energy and the number of particles are allowed to fluctuate. We can define the chemical potential μ as

$$\mu = \frac{\partial F}{\partial N} , \quad (4.43)$$

where in the partial derivatives we keep fixed all other degrees of freedom (i.e. $T, V, \dots$). We now define the grand potential Ω as

$$\Omega = F - \mu N . \quad (4.44)$$

Let us consider a system for which $F \equiv F(T, V, N)$ (e.g. a gas). For the differential form of Ω, we then have

$$d\Omega = -PdV - SdT - Nd\mu , \quad (4.45)$$

from which we get

$$P = -\frac{\partial \Omega}{\partial V} \quad (4.46)$$

and

$$N = -\frac{\partial \Omega}{\partial \mu} \quad (4.47)$$

In the grand canonical ensemble, the expectation value of an observable O is defined as

$$\langle O \rangle = \frac{1}{\Xi} \sum_N e^{\beta \mu N} Z_N \langle O \rangle_N \; , \tag{4.48}$$

where the expectation value $\langle O \rangle_N$ is computed at fixed number of particles N (i.e. in the appropriate canonical system). We have introduced the definition

$$\Xi = \sum_N e^{\beta \mu N} Z_N \; , \tag{4.49}$$

where Ξ is the grand partition function and we have slightly changed the notation for the canonical partition function from Z to Z_N, to make explicit the fact that the latter is computed at a fixed number of particles N. As per assumption, N can fluctuate. To keep into account that in fact N is not fixed in our case, we need to sum over all the possible realisations of N.

Equation (4.48) can be derived by considering our system as a smaller portion of a larger canonical system and at the same time with a large average number of particles, so that the thermodynamic limit provides a good description of its properties.

The average number of particles, $\langle N \rangle$, is given by

$$\langle N \rangle = \frac{1}{\Xi} \sum_N N e^{\beta \mu N} Z_N = \frac{1}{\beta} \frac{\partial \log \Xi}{\partial \mu} \; . \tag{4.50}$$

Comparing the previous equation with Eq. (4.47) and highlighting the fact that in our setup the number of particles needs to be interpreted as an average property, we have

$$\Omega = -\frac{1}{\beta} \log \Xi + c \; , \tag{4.51}$$

where the constant c can be taken to be zero. We can then rewrite the grand partition function as

$$\Xi = e^{-\beta \Omega} \; . \tag{4.52}$$

The grand potential Ω is an extensive quantity. For a system in which Ω depends only on V, on T and on μ, since the only extensive independent variables among them is V, Ω must be proportional to V itself. This implies that we can write Ω as

$$\Omega = \omega(T, \mu) V \; , \tag{4.53}$$

or, equivalently, that

$$\frac{\partial \Omega}{\partial V} = \omega(T, \mu) \; . \tag{4.54}$$

From the differential form of Ω (4.45), we have that

$$\frac{\partial \Omega}{\partial V} = -P \; . \tag{4.55}$$

Putting together the two previous equations, we arrive at the expression

$$\Omega = -PV \ . \tag{4.56}$$

Therefore, for our system

$$\Xi = e^{\beta PV} \ . \tag{4.57}$$

Our presentation of the grand canonical ensemble concludes our basic overview of the derivation of thermodynamic expressions using Statistical Mechanics methods. At this point, it is worth noting that, despite the different assumptions, in the thermodynamic limit, the three ensembles we have discussed (microcanonical, canonical and grand canonical) reproduce the same thermodynamic relations for a given system. An explicit calculation for a specific system is proposed in exercise 4.3. Note, however, that the equivalence among ensembles, which relies on smoothness and invertibility, is lost at phase transitions points, where thermodynamic quantities are non-analytic. Phase transitions are a central topic in Statistical Mechanics. We will cover them in section 4.3.

4.2.4   From classical to quantum system

So far we have been dealing with classical systems, or, better, with systems in the classical approximation, with continuous energy levels, indistinguishable particles and volume of a state defined by the uncertainty principle. In this section, we discuss briefly quantum systems with a discrete energy spectrum in the canonical ensemble, as they will be playing a major role in the rest of the book.

The central quantity we need to compute is the partition function. Generalising our previous discussion, this is given by

$$Z = \mathrm{Tr}\, e^{-\beta H} = \sum_i \langle i | e^{-\beta H} | i \rangle = \sum_n \rho(n) e^{-\beta E_n} \ . \tag{4.58}$$

In this expression, $\mathrm{Tr}\, e^{-\beta H}$ is the trace of the operator $e^{-\beta H}$, which is computed summing the matrix elements $\langle i | e^{-\beta H} | i \rangle$ over all eigenstates $|i\rangle$ of the Hamiltonian H. If $H|i\rangle = E_i|i\rangle$, then $\langle i | e^{-\beta H} | i \rangle = e^{-\beta E_i}$. The rightmost expression in Eq. (4.58) sums over all energy levels n the *Gibbs factor* $e^{-\beta E_n}$ weighted by the density of states corresponding to level n (i.e. the number of states with energy E_n).

We note that the results

$$Z = \sum_n \rho(n) e^{-\beta E_n} \tag{4.59}$$

also holds for classical systems with a discrete energy spectrum, The latter are often models that extract the key features of quantum systems. Noticeable examples in this category are classical spin systems, which we will start introducing in section 4.4.

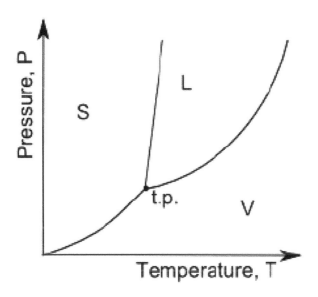

FIGURE 4.5: Schematic phase diagram for a substance displaying the solid, liquid and gas phases.

4.3 Phase transitions

In the previous sections of this chapter, we have introduced thermodynamic potentials and highlighted their use in deriving thermodynamic properties. In many cases, observables are obtained by taking derivatives of quantities such as partition functions. For this procedure to be allowed, there is an important assumption of analyticity. This assumption is valid except at phase transition points. Mathematically, phase transition points are defined just as singularity in the thermodynamic potentials. While this definition looks like a tautology in our context, phase transitions are in fact well-defined and relevant thermodynamic phenomena that play an important role in our world.

Physically, we say that a system undergoes a phase transition when we get a drastic change in their properties. To clarify this concept, let us take a familiar example. We know from our everyday experience that water can arise in three states: solid (ice), liquid and gas (vapour). While these three are by no means the only phases of matter, this behaviour is common to many substances we experience in our everyday life. A sketch of a substance that has the three phases discussed for water is presented in Figure 4.5. The diagram is drawn in the (T, P) plane and hence assumes that these two variables are the degrees of freedom of the system. The solid phase S is realised at low temperatures and high pressures. At high temperature and low pressures, the system is in the gas phase G. The liquid phase L opens up in some intermediate regimes. Phases are separated by phase boundaries, which for our prototype system are lines in the phase diagram. These lines mark phase transition points, and hence coincide with the singularities. When we cross a phase boundary, we go through a singularity of thermodynamic quantities. At phase boundaries, two phases coexist. There is a special point in the diagram in which three-phase boundaries meet. At this point, the three phases are stable and hence coexist. This point is called the triple point.

Note that phase boundaries can end. This happens in our hypothetical system (and for real systems such as water) at values of the temperature and pressures that we denote, respectively, as T_c and P_c. The subscript c is standard notation to identify points of phase transitions. The points at which phase transitions happen are called critical points. We will come back to the definition of critical points after we have discussed more in detail some other properties of phase transitions. The end of a phase boundary indicates that there is a way to connect the two corresponding phases without encountering a singularity. This suggests that the two apparently distinct phases are a manifestation of the same phase. Again, we will come back to this point later in this section.

The phase diagram we have presented is rather sketchy. Some systems which on the surface presents the three phases we have discussed are in fact more complex. For instance, water does not have only one solid phase, but there are 19 known crystalline forms of ice. Incidentally, we also remark another peculiarity of the phase diagram of water: while for many substances, the slope of the solid–liquid phase boundary is positive, as in Figure 4.5, for water this slope is negative. This fact corresponds to the well-known property that water is denser than ice (and in fact ice cubes float in a glass of water).

4.3.1 Classification of phase transitions

So far, we have discussed phase transitions as the appearance of singularities. This means that if we take a system near a critical point, its properties can change abruptly for infinitesimal changes in the control parameters. Let us clarify this with an example. Let us take a piece of ice at around 100 kPa (Kilo Pascal) of pressure (standard atmospheric pressure) at equilibrium. If we gradually raise the temperature keeping the system at equilibrium, when we reach a temperature of $T_c \simeq 273k$ the ice start to melt. Now, ice and water have a different density. This means that for arbitrarily small ϵ, the density of the system at $T + \epsilon$ and the density at $T - \epsilon$ are not the same as ϵ goes to zero, i.e.

$$\lim_{T \to T_c^-} \rho(T) \neq \lim_{T \to T_c^+} \rho(T) \, .$$

The singularity of ρ in this case is a discontinuity.

Since ρ is related to the volume, which in turn can be expressed as a first derivative of appropriate thermodynamic potentials (e.g. the Gibbs free energy), this discontinuity affects some first derivative of thermodynamic potentials. There are phase transitions for which the first derivatives of thermodynamical potentials are continuous, with the discontinuity appearing in higher-order derivatives. As a matter of fact, it proves to be a useful concept to classify phase transitions on the basis of the order of the derivative at which the discontinuity arises. Let's take the case of a canonical system, described by the Helmholtz free energy F (for other thermodynamic potentials, similar considerations apply). We say that the system undergoes a phase transition of order n at the values of the independent variables $(T_c, V_c, \dots) \equiv (x_1^c, \dots, x_k^c)$, with k the degrees of freedom of the system, if:

- All the partial derivaties $\frac{\partial^i F}{\partial x_{j_1} \dots \partial x_{j_i}}$ for $i < n$ exist at $x_1^c, \dots x_k^c$;

- There exist at least one partial derivative $\frac{\partial^n F}{\partial x_{j_1} \dots \partial x_{j_n}}$ that is ill-defined at $(x_1^c, \dots, x_k^c)$.

This definition is due originally to Ehrenfest, and for this reason is known as the *Ehrenfest classification* of phase transitions. The modern view on the Ehrenfest classification is that, while it does not capture all the complexity of phase transitions, provides a first point of entry to understand various possibilities that can happen in nature (and in models).

We have already seen the transition from ice to water, which is first order, as the first derivative of the (Gibbs) free energy is discontinuous. A discontinuous (or first order) phase transition characterises the whole phase boundary in the phase diagram of water, with the type and magnitude of the jump depending on the specific point on the phase boundary. Since the jump is a continuous function along the phase boundary, for a given point displaying a first-order phase transition (and therefore, a finite jump), there has to be an open neighbourhood of it along the phase boundary and containing this point such that for each point on this neighbourhood the jump is still finite. This means that the phase boundary cannot end on first-order phase transition points. Likewise, first-order phase transition points cannot appear in isolation, but must always arise as a part of a phase boundary. This conclusion does not apply only to our example, but is a general result.

We now want to understand what is the kinematic that creates a jump at a phase transition. To do this, let us work in the canonical system and in particular let us assume that we have computed the free energy and that we can express it or project it as a function of the internal energy. We know that the equilibrium state minimises the free energy. Therefore, at equilibrium and at a temperature T far from a phase transition, we expect to be in a situation such that sketched in Figure 4.6-1: there is one equilibrium state, and the internal energy is the one that minimises F. We call this value of internal energy U_1. Let us also assume that a phase transition arises for $T_c > T$. We

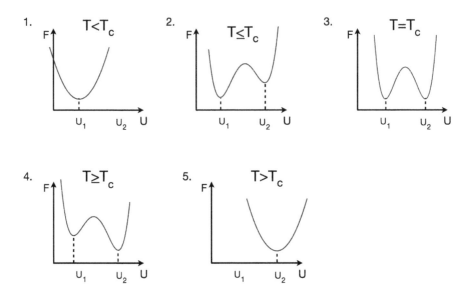

FIGURE 4.6: The sketches represent the free energy F as a function of the internal energy U for a given system as we cross a first order phase boundary. (1) For $T \ll T_c$, there is only one equilibrium state, which corresponds to the state with internal energy U_1. (2) For $T \leq T_c$, the equilibrium state has changed continuously, and a metastable state starts appearing at internal energy U_2. (3) At $T = T_c$ both states have the same minimal internal energy, and therefore are both equilibrium state; we are in a situation of coexistence of phases. (4) For $T \geq T_c$, the situation has reversed, with the state with internal energy U_2 being the equilibrium state and the one with internal energy U_1 being metastable. (5) Only the state corresponding to a smooth variation of U_2 is present at higher temperatures outside the critical region.

now increase the temperature to reach a value $T \lesssim T_c$. At this point, the function F will develop a local minimum U_2, with $F(U_1) < F(U_2)$. U_1 (which is not the original value, but has continuously changed with T) still provides the equilibrium state. However, the appearance of U_2 starts to signal that the physics is about to change: we are entering the so-called *critical region*, where gradually the system is affected by the nearby phase transition. This situation is sketched in Figure 4.6-2. We keep raising the temperature until we have $T = T_c$. Now, U_1 and U_2 (both continuously changed from the previous scenario) are two degenerate minima of F (Figure 4.6-3). As a consequence, the system has two equilibrium states and can be in either of them. Likewise, sub-portions of the systems can occupy either of the two states. This is the reason why boiling water forms bubbles. Changing a sub-portion of the system from either state does not cost free energy[4], therefore this operation provides an equally probable realisation of the system as the original one. However, moving from the state with internal energy U_1 to the one with internal energy U_2 does cost an amount of internal energy. Again, this is something that we see often in our everyday experience when, for instance, we melt ice: if we want to melt all ice, we need to keep providing heat to the system, and in this process the temperature will not change until all the ice is melted. What we are doing by melting more and more ice is to transfer parts of the system from one equilibrium state to the other, at the expense of providing heat. If we keep increasing the temperature, when we have $T \gtrsim T_c$, the state corresponding to U_2 becomes a global minimum (and hence is the proper equilibrium state), while the one corresponding to U_1 is a local minimum, and therefore metastable[5]. Finally, the state with internal energy U_1 disappears completely as we leave the critical region, for $T > T_c$ (Figure 4.6-5).

An important quantity that characterises first order phase transition is the latent heat per unit of mass, which we write as

$$l = (U_2 - U_1)/M \qquad \text{for } T = T_c, \qquad (4.60)$$

i.e. the amount of heat that the system has to be provided in order to convert a unit mass of our substance from the phase with $U = U_1$ to the phase with $U = U_2$. Since the internal energy is proportional to the mass, l is independent from M.

The quantity l also enters the expression for the slope of the phase boundary in the plane (T, P), which is given by the Clausius-Clapeyron equation

$$\frac{dP}{dT} = \frac{l}{T \, \Delta v}, \qquad (4.61)$$

where Δv is the difference of the volumes that the system occupies in the two phases divided by the mass itself (i.e. the specific volume difference).

We now leave for the moment first order phase transitions to discuss second order one. We start again from the observation that a first-order phase transition point must be an internal point on a phase boundary. This is obviously not the case for the end point of the liquid-gas phase boundary in Figure 4.5. Since there is no discontinuity after the line ends, exactly at that point, the discontinuity goes to zero, and the transition becomes of higher order, specifically of the second order. In fact, one finds experimentally that a key diverging

---

[4]This is true modulo surface effects, that we are neglecting.

[5]We refer to this state as *metastable* since we can prepare a system in that state, and the system will persist in it for a time that depends on the difference in free energy between the metastable and the equilibrium state. If we can only observe the system for a time that is significantly less that the expected decay time, we could mistake this state for the equilibrium state. The existence of metastable states gives rise to phenomena such as *superheating* and *supercooling*, whereby the system stays in the unstable phase for a small region of temperatures close to T_c.

quantity at this point is the isothermal compressibility of the fluid

$$\kappa_T = -\frac{1}{v}\left(\frac{\partial v}{\partial P}\right)_T. \tag{4.62}$$

The latter is a second-order derivative of the Gibbs free energy. We have defined those quantities as susceptibility. The divergence of a susceptibility creates a fundamental phenomenon: the emergence of scale invariance. In physics, scale is a fundamental concept, and we know that the phenomena we observe depend on the scale. The scale is generally defined via combinations of relevant quantities of the system. We may think that the critical point would naturally define a scale of some sort. However, this turns out not to be true: the system at that point does not know of any scale, whether we try to define a scale through Hamiltonian parameters (fundamental scales) or identify one using the critical parameters. If we take water and we bring it at the end point of the liquid-gas phase boundary, we can't see through it, i.e. exactly at that point and only at that point, the system becomes fully opaque. The explanation for this phenomenon, which is called *critical opalescence*, is that light of any frequency is scattered. This is compatible with scale invariance: the system has fluctuations at any scale, which interact with radiation at any scale. The property of dramatic fluctuations at all scales, which explain the divergence of relevant susceptibilities (in the case of a liquid, for instance, the all-scale fluctuations of densities justify the divergence of the isothermal compressibility) is a clear physical signature of second-order phase transitions. The emergence of those dramatic collective modes has earned them the name of *critical phenomena*[6]. For the sake of clarity, we note explicitly that a second order phase transition point does not have to appear necessarily as the end-point of a first order phase boundary, but can appear either as an isolated points or as a point of a phase boundary, i.e. phase boundaries can be also second order, or can change from first to second order at some values of the control parameters along them.

Transition of order third and above are in principle possible, but they are mostly found in a handful of models, and do not have known relevance in nature. A transition of order higher than two which we will discuss in some detail has infinite order.

Before moving to the next topic, it is worth to reflect on why singularity can appear. Again, we focus on the canonical ensemble. If we take the partition function (e.g. for a system with discrete energy levels but similar considerations hold for a system with a continuum spectrum), we formally write it as

$$Z(\beta) = \sum_n \rho(e)e^{-\beta E_n}, \tag{4.63}$$

---

[6] Following this naming, some authors refer to critical values of the parameters as, e.g. the critical temperature, the critical pressure and so on only in the case of second-order phase transition, naming the corresponding points of singularity for the first-order transitions *pseudocritical*. In this book, we shall use the adjective *critical* for the transition points of phase transitions regardless of their order

where we have made explicit the dependency on β. This sum has positive factors, so we expect a convergence radius which could be in principle finite or infinite. However, in case of a phase transition, we are looking for points of singularity rather than complements to disks. How can these isolated points (or lines, surfaces, etc., depending on the number of degrees of freedom) can arise? The answer is once again the thermodynamic limit: by summing over an infinite number of particles, critical domains can appear where the function was smooth (albeit possibly fast varying at some points) for a finite number of degrees of freedom. A consequence of this argument is that true phase transitions only make sense in the thermodynamic limit.

Elementary thermodynamics also provide a guide to the type of singularities that can arise. In fact, a celebrated result is that the free energy is always continuous. This means that the earliest discontinuity can only arise in the first derivative, and takes the form of a jump. This is the case of a first order phase transition, as discussed earlier.

4.3.2 The role of symmetries in phase transitions

By now, we know that phase transitions are identified by singularities, which can arise isolated point in the phase diagram of a system or at higher dimensional manifolds called phase boundaries[7]. The properties of the system vary dramatically as one crosses the phase transition, therefore we talk about two different phases for the system, above and below the critical point or critical separation manifold. In fact, the importance of those different phases is such that in some systems (e.g. water) they are given specific names (e.g. ice, rather than solid water). For such distinct behaviours, one would expect a well-defined separation boundary. Indeed, the properties of liquid water and steam are very different, and indeed we have a clear separation boundary. However, this boundary persists up to an end point, after which there is no clear distinction between the liquid and the gas phase. Therefore, should we consider the latter as separate phases or as the same phase? Both point of view are legitimate, but there is a powerful way to approach phase transitions (originally advocated by Landau) through symmetry considerations: the system has different phases when it has different symmetry properties for different regions of the phase diagram. Note that the shift from singularity to symmetry properties is crucial. For instance, liquid water is translationally invariant, and so is steam. Therefore, from the point of view of translational symmetry (and indeed all other spatial symmetries such as rotation, parity, etc.), there is nothing that distinguish water from steam. For this reason, with the Landau definition, water and steam are different manifestations of the same single phase. On the contrary, ice has a crystalline structure, which breaks translational invariance. Therefore, ice and liquid water can be distin-

---

[7]These isolated points are generally part of phase boundaries in an extended phase diagram, where new couplings and corresponding interaction terms are introduced.

guished by the different realisation of translational symmetry, and hence are two distinct phases.

The reader that is unfamiliar with this topic may be surprised by the use of the language that we have made above: why do we say that ice and water have a different realisation of translational symmetry, rather than ice is not translationally symmetric, while liquid water is? It is certainly matter of terminology, but in this case the terminology is crucially related to the physical behaviour of the system. Generally, when we talk about the symmetry of a system, we mean the symmetry of its Hamiltonian. Water and ice are the same substance, so they share the same Hamiltonian. Therefore, they would share the same symmetries. Why is then the structure of ice crystalline, while liquid water stays in an amorphous form? The answer is in the equilibrium state, which is different from the two of them. Basically, both realisations exist at all temperatures and pressures. However, they are the true equilibrium state in different regions of the phase diagram. This concept is not new, we have discussed is at length in the previous section, e.g. around Figure 4.6. What could be new is the realisation of the fact that these two states can have different symmetry transformation properties with respect to the Hamiltonian symmetries of the systems. In terms of translational symmetry, for instance, which leaves invariant the Hamiltonian, the equilibrium state of liquid water is invariant, while the equilibrium state of ice is not.

We are now in a position to introduce one of the most fruitful ideas in Statistical Mechanics (and Quantum Field Theory alike): the fact that a symmetry can be spontaneously broken. We say that a symmetry is spontaneously broken when the latter is an Hamiltonian symmetry of the system, but the equilibrium state (or, in the case of Quantum Field Theory, its vacuum state) is not invariant under that symmetry. On the contrary, when the equilibrium state is invariant under an Hamiltonian symmetry, we say that the symmetry is linearly realised. There are clearly remnants of the symmetry also in the case in which the latter is spontaneously broken. For instance, if we transform under the symmetry an equilibrium state that is not invariant, we must obtain another equilibrium state. As a consequence, in a phase in which a symmetry is spontaneously broken, the equilibrium state is not unique. Once again, the thermodynamic limit here plays an important role, as we know from Quantum Mechanics that groundstates are non-degenerate: we will need to take the limit for the number of particles going to infinity in order to prevent the typical tunneling phenomena that happen in Quantum Mechanics to produce a non-degenerate groundstate.

Equipped with these new ideas, we leave the specific examples we have discussed above and look at a more general case. Let's take a system with Hamiltonian H that is symmetric under transformations of the group G. G is either a discrete or a Lie group, which we take to be Abelian for simplicity. We recall that G is Abelian if and only for each $u_1 u_2 \in G$ we have

$$u_1 u_2 = u_2 u_1 \ , \tag{4.64}$$

where the product is the inner composition operation in G. As an example, we consider the group

$$\mathbb{Z}_N = \{e^{i \frac{2\pi}{N} m}\} \qquad m = 0, ..., (N-1) . \tag{4.65}$$

In an operatorial definition, H is symmetric under transformations if

$$\forall u_1 \in G \qquad u_1 H u_1^\dagger = H . \tag{4.66}$$

The equilibrium state $|0\rangle$ is symmetric if

$$u_1 |0\rangle = |0\rangle . \tag{4.67}$$

We now consider an operator ϕ that is not invariant under G, and in particular transforming in a non-trivial irreducible representation of G. For instance,

$$u_1 \phi u_1^\dagger \propto \phi = c(u_1)\phi , \tag{4.68}$$

with $c(u_1)$ the phase factor associated to u_1 (see Eq. (4.65)). We compute $\langle 0|\phi|0\rangle$ in the case in which $|0\rangle$ is invariant, and find

$$\langle 0|\phi|0\rangle = \langle 0|u_1^\dagger u_1 \phi u_1^\dagger u_1|0\rangle = c(u_1) \langle 0|\phi|0\rangle \Rightarrow \langle 0|\phi|0\rangle = 0 . \tag{4.69}$$

This result is more general: if $|0\rangle$ is invariant and ϕ is any operator that transform under a non-trivial irreducible representation of G, we must have that the expectation value of ϕ must be zero, i.e.

$$\langle \phi \rangle = 0 . \tag{4.70}$$

If instead an operator ϕ exists that is not invariant under a symmetry G of H and for which

$$\langle \phi \rangle \neq 0 , \tag{4.71}$$

then there is no other possibility than the symmetry being spontaneously broken. An operator ϕ that is not symmetric and enables to detect the broken phase is called an *order operator*, and its average $\langle \phi \rangle$ is an *order parameter*. At this point, it is worth to remark that the name *order parameter* derives from the fact that a non-zero value for $\langle \phi \rangle$ is associated with an ordered phase, which in turn is realised at low temperature. An example, which we will discuss in detail in the following section, is the alignment of magnetic moments in a ferromagnetic material, which arises below a critical temperature, but is destroyed by thermal fluctuation above the critical temperature. There are functions that have an opposite behaviour and are non-zero in the disordered phase. These functions, which are associated with topological (extended) objects rather than directly to degrees of freedom of the Hamiltonian, are called *dis order parameters*.

A (dis) order parameter inherits properties of the phase transition, in particular, concerning the order of the transition. We therefore expect that for a first

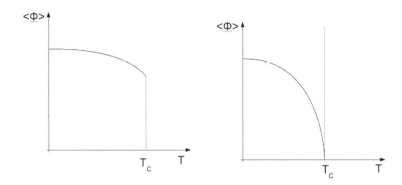

FIGURE 4.7: The behaviour of the order parameter as a function of a temperature, for a first-order phase transition (left) and a second-order phase transition (right).

order phase transition the order parameter will be jumping from a finite value to zero when approaching T_c from the broken phase, while for a second-order phase transition the order parameter will be continuous and equal to zero at T_c. These behaviour are sketched in Figure 4.7.

Order parameters are powerful tools for the study of phase transitions, beyond the fact that they act as a detector for the transition. In fact, it is argued that near the transition point an effective expression for the free energy can be written in terms of the order parameter and the relevant symmetry of the transition. This free energy has the symmetry that drives the phase transition, and the dynamics results in symmetry breaking below T_c and symmetry restoration above T_c. The behaviour of the free energy in terms of $\langle \phi \rangle$ is sketched in Figure 4.8 for a first-order phase transition and in Figure 4.9 for a second-order phase transition. The resulting free energy takes the name of *Landau free energy*. In the Landau free energy there are details that are system-dependent (such as the value of the temperature T_c), while other features (e.g. the form) are universal, i.e. depend only on the symmetry properties of the system. As a consequence, systems with transitions driven by the same symmetry have the same form of the Landau free energy. This is the first manifestation of a property called *universality*, which holds for second order phase transitions. Universality states that systems with the same phase transition driving symmetry and the same dimensionality share quantities such as the critical exponents, i.e. the exponents that describe the divergence of susceptibilities. Systems with the same universal quantities (which, it is worth repeating, depend only on the dimensionality of the systems themselves and the relevant symmetry of the phase transition) are said to be in the same *universality class*.

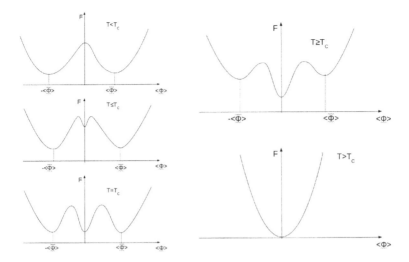

FIGURE 4.8: Cartoon of the behaviour of the free energy in terms of the order parameter as the temperature varies across a first-order phase transition. On the left, at the top we have the case of a system in a broken phase, where the free energy has multiple minima corresponding to the non-degenerate equilibrium states, for which the value of the order parameter is different. As we approach the critical temperature (middle figure), the system will develop a local minimum at zero, which has, however, higher free energy. At criticality (bottom), the minimum at zero is degenerate with the minima of the free energy corresponding to the broken symmetry equilibrium states. As we raise the temperature above T_c (top right), the minima of F realised in the broken phase become local minima, with the one at zero becoming global and therefore the genuine equilibrium state. Finally, at higher temperatures, the only minim of F is the one corresponding to $\langle\phi\rangle = 0$.

The concept of universality is central in the study of phase transitions, as it enables us to predict relevant properties of more complicated systems studying simpler and where possible exactly solvable models in the same universality class.

4.3.3 An application: the Van der Waals gas

The Van der Waals gas offers an instructive example that enables us to see in action most of the concepts relating to phase transitions that we have discussed so far in this section. We start with the familiar equation of state of an ideal gas,

$$Pv = kT , \qquad (4.72)$$

with $v = V/N$ the specific volume of the system, defined as the volume occupied by a single particle. The idealisation of this system consists in the double assumption that particles are point-like and that there is no interaction among these microscopic constituents of the system. Both

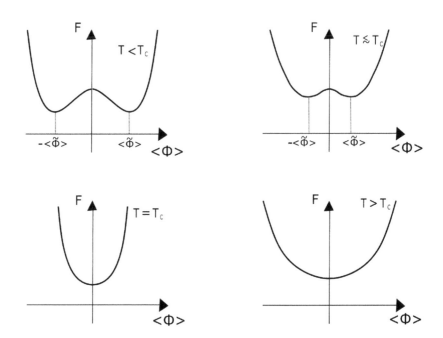

FIGURE 4.9: Cartoon of the behaviour of the free energy in terms of the order parameter as the temperature varies across a second-order phase transition. At low temperature (top left), we have degenerate minima of F corresponding to the non-degenerate broken symmetry equilibrium states. As we raise the temperature (top right), these minima get closer together, and the collapse at T_c (bottom left), in a local quartic behaviour near the minimum. At higher temperature, F has a single minimum at $\langle \phi \rangle = 0$

assumptions are not realised for real gases. To get closer to reality, we can model particles as spheres of volume v which interact among each other with an attractive potential that goes like $1/v^2$. The ideal specific volume should then be replaced by $v - b$, since this is the actual volume that each particle can occupy. Likewise, the ideal pressure should be replaced with $P + a/v^2$, where P is the measured pressure. The Van der Waals equation takes then the form

$$\left(P + \frac{a}{v^2} \right)(v - b) = kT \ . \tag{4.73}$$

The system has an inflection point, which can be obtained by imposing the equation of state and the conditions

$$\frac{\mathrm{d}P}{\mathrm{d}v} = 0 \ , \tag{4.74}$$

$$\frac{\mathrm{d}^2 P}{\mathrm{d}v^2} = 0 \ , \tag{4.75}$$

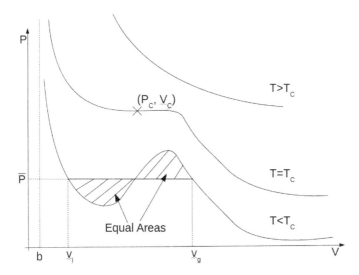

FIGURE 4.10: Behaviour of the Van der Waals pressure as a function of the volume at various temperatures. The three possible behaviours are shown: $T < T_c$, $T = T_c$ and $T > T_c$. For the first case, we also show Maxwell's construction of the equilibrium curve.

where T is kept as a parameter. The solution for the inflection point gives the values

$$P_c = \frac{a}{27b^2} \quad , \quad v_c = 3b \quad , \quad kT_c = \frac{8a}{27b} . \tag{4.76}$$

We have used the subscript c not by coincidence: we will shortly show that these values correspond to the critical end point of the liquid-vapour phase boundary. Indeed, the above critical temperature separates two distinct regimes of the system: a regime for which P as a function of v has a local maximum and a local minimum, realised for $T < T_c$, and a regime in which P is monotonically decreasing as a function of v ($T > T_c$). Figure 4.10 shows these possible behaviours. We note that, as expected, in all cases the pressure diverge if we try and squeeze v to its minimal value b.

Among the three typical behaviours, the one for which $T < T_c$ shows an unexpected feature: $P(v)$ cannot be inverted everywhere, since the function is not monotonic. Physically, we can rightfully suspect that this behaviour is not the one that the system follows. This intuition, however, requires an explanation, which was provided originally by Maxwell: the non-invertibility region corresponds to metastability, with the equilibrium state described by the horizontal line such that the two shadow areas in Figure 4.10 are equal. If we follow this horizontal path, we see that the specific volume as a function of the pressure has a jump-type discontinuity, i.e. we are in the presence of a first-order phase transition. If we now look at the direct figure, which shows P as a function of v, we see that we can change v without changing P. If we

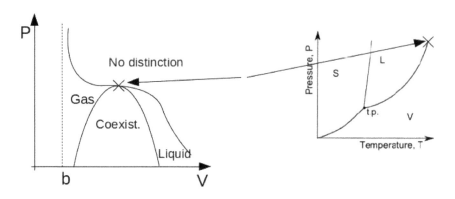

FIGURE 4.11: Phase structure of the Van der Waals gas (left) and the connection with the liquid-gas part of the phase diagram of Figure 4.5 (right).

move so that v is reduced, we are changing the state of the system from gas to liquid, and in the process the pressure of the gas does not change until we have completed the conversion, after which the system becomes incompressible, as shown by the drastic rise of P. We conclude then that the Van der Waals equation for $T < T_c$ supplemented with the Maxwell's construction provides a good description of the liquid-gas phase transition. As we move towards T_c, the local maximum and minimum in the metastable part of the pressure curve move towards each other, until they merge in the horizontal inflection points (v_c, P_c) at T_c. At this point, the first-order phase boundary disappears. Therefore, this point is the critical end point of the phase diagram. This can be checked explicitly by computing the isothermal compressibility and showing that it diverges for the critical values of P, v and T.

The phase structure of the Van der Waals gas is summarised in Figure 4.11, where the line separates the four regions of the (v, P) plane in which we have the gas phase, the liquid phase, coexistence of the two phases and no distinction among them. We can ask whether there is an observable that can distinguish between temperatures below T_c, where the two phases are different, and temperature above T_c, acting as an order parameter. While such an observable can be constructed in terms of the density along the critical line, the interpretation as an order parameter is not possible, since there is no symmetry that distinguish the two phases, as indicated by the fact that the phase boundary ends.

Indeed, from the point of view of critical phenomena, there is little more we can learn at an introductory level from the Van der Waals model. To further our understanding, we shall now move to a different physical phenomenon, namely, spontaneous magnetisation at low temperatures and its loss at higher temperatures, and we shall study a cornerstone model that captures the crucial features of magnetic materials: the Ising model.

4.4   The Ising model

The Ising model, which we will study in this section, displays a phase transition whose phenomenology is somewhat more remote from our experience than the liquid-gas phase transition, described by the Van der Waals model. However, as anticipated in the previous section, unlike in the Van der Waals model, in the Ising model there is a genuine symmetry whose breaking and restoration drive the phase transition. As a result, the Ising model provides us with a prototype system where the benefits of an approach based on the existence of a genuine order parameter can be exposed. In addition, the Ising model is based on a Hamiltonian that is simple enough to be tractable in notable cases, and yet it is in no way so simplistic to be irrelevant for practical applications. As a matter of fact, In two dimensions, and in the absence of an external field, the model can be solved explicitly, and the solution (which requires a non-trivial amount of work and for this reason won't be provided in its gory details in this book) enables us to determine relevant quantities such as the critical exponents associated with the phase transition. Due to its undoubted physical relevance coupled with the availability of an exact solution, the two dimensional Ising model is often used as a laboratory for testing new approaches to more complicated systems. Therefore, knowing its properties in sufficient detail turns out to be useful to understand a wealth of statistical mechanics and numerical methods that find applications to more complicated systems. Last but not least, thanks to the concept of universality, the physics of the Ising model proves to be more general and profound that the deceptively simple Hamiltonian on which it is based may suggest at first sight.

4.4.1   Ferromagnetism and the Ising model

The Ising model[8] was originally introduced to describe spontaneous magnetisation in *ferromagnetic materials*. Phenomenologically, these materials are observed to possess spontaneous magnetisation (and hence to act as macroscopic magnets) at temperatures below a critical temperature T_c. Abobe T_c, these materials become *paramagnetic* (i.e. the magnetisation is proportional to the external magnetic field). The ferromagnetic phase $(T < T_c)$ can be characterised by the existence of *Curie domains*. regions of macroscopic size in which the microscopic magnetisation points in the same direction (see Figure 4.12).

---

[8]The model is named after the German physicist Ernst Ising, who first studied it following a suggestion of his Ph.D. supervisor, Wilhelm Lenz, of the celebrated Lenz law of electromagnetism.

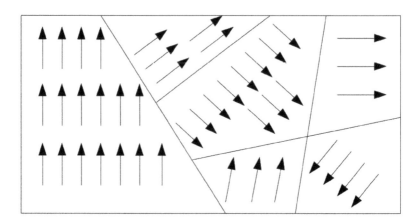

FIGURE 4.12: A schematic representation of Curie domains. The arrows indicate the direction of the elementary magnetisation and the continuous lines the boundaries of the Curie domains.

We define the magnetisation per site as

$$m = \frac{M}{V} \, , \tag{4.77}$$

with M the total magnetisation, and the magnetic susceptibility χ as

$$\chi = \left(\frac{\partial m}{\partial h} \right)_{h=0} \, ,$$

where h is the external magnetic field.

Writing a model that explains ferromagnetism entails proposing a fundamental Hamiltonian from which one can derive physical observables having the observed behaviour. In general terms, the relevant observables are derived from the partition function Z as follows:

$$F = -kT \log Z$$

$$\downarrow$$

$$U = -\frac{\partial}{\partial \beta} \log Z \qquad S = -\left(\frac{\partial F}{\partial T} \right)_V \frac{U-F}{T} \qquad M = -\left(\frac{\partial F}{\partial h} \right)_T$$

$$\downarrow \qquad\qquad\qquad \downarrow \qquad\qquad\qquad \downarrow$$

$$C_h = \left(\frac{\partial U}{\partial T} \right)_h \qquad C_X = T\left(\frac{\partial S}{\partial T} \right)_X \qquad (X = h, m) \qquad \chi = \frac{1}{V}\left(\frac{\partial m}{\partial h} \right)_T$$

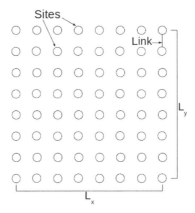

FIGURE 4.13: A two-dimensional Ising system of sizes L_x and L_y. Sites and links are indicated.

The Ising model is formulated on a grid in D dimensions. The geometrical objects of the grid we are interested in are its points and the links that connect two neighbour points. For thermodynamic considerations, we will be interested in infinite grids. However, operationally, it is convenient to perform calculations on finite grids and then take the limit for the grid size going to infinity. The dimension of the grid in the i-th direction is indicated by L_i. Generally, all L_i are taken to be equal to L, i.e. the grid is taken to be a square if $D = 2$, a cube for $D = 3$ and a hypercube for $D > 3$. A sketch of a two-dimensional lattice is provided in Figure 4.13.

On each point i of the grid, a spin variable $\sigma_i = \pm 1$ is defined. We define a configuration of the system as a unique assignment of all σ_i. Configurations are enumerable. We indicate a given configuration with the notation $C_k \equiv \{\sigma_i\}_k$, where k is an integer labelling the given configuration. On a finite lattice, the number of total configurations is 2^N, with N is the number of lattice sites.

In addition to the elementary degrees of freedom, a crucial ingredient of a model is its dynamics, which is specified by a Hamiltonian. For the Ising model, the Hamiltonian is

$$H = -J \sum_{\langle ij \rangle} \sigma_i \sigma_j - h \sum_i \sigma_i \,,$$

where $\langle ij \rangle$ is the link joining i and j, and the first term is summed over all the links. The Hamiltonian depends on two real parameters, the spin self-coupling J, whose intensity determines how strongly spins are coupled to their nearest neighbours, and the external field h. Since the dynamics favours the minimisation of H, $J > 0$ aligns nearest-neighbour spins, while $J < 0$ anti-align them. If we take $h = 0$, it is then clear that $J > 0$ corresponds to the ferromagnetic case. A system with $J < 0$ is referred to as anti-ferromagnetic.

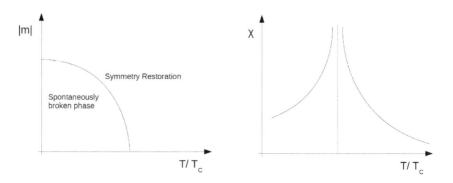

FIGURE 4.14: Typical behaviour of the magnetisation per site m (left) and of the magnetic susceptibility χ (right) for a ferromagnetic material.

Although anti-ferromagnetic systems are also physically relevant, here we will focus our attention on the ferromagnetic case ($J > 0$). In this case, and still for $h = 0$, H has two degenerate minima:

$$H = H_{min} = -J N_{links} = -J\frac{q}{2}N \quad \text{if} \quad \left.\begin{array}{l} \sigma_i = 1 \\ \sigma_i = -1 \end{array}\right\} \qquad \text{for all } i \,, \qquad (4.78)$$

where we have introduced q, the coordination number, defined as the number of nearest neighbours of a point.

The existence of two degenerate minima (or groundstates) for H is a consequence of the symmetry of the Hamiltonian under the action of the discrete group $\mathbb{Z}_2 = \{1, -1\}$ when $h = 0$. Indeed it is easy to verify that for $h = 0$ H is left invariant when all the spins are multiplied by $z = -1$ (the only non-trivial element of the $\mathbb{Z}_2$ group). This means that, given a configuration, its z-transformed configuration provides the same value for H.

A relevant role in the system is played by observables that under the transformation $\sigma_i \to z\sigma_i$ are either even (i.e. invariant, like H is) or odd (i.e. those who change sign). In the latter class is the magnetisation m, which, for a given configuration, is identified with the average of the spin over the sites:

$$m = \frac{\sum_i \sigma_i}{V} \,. \qquad (4.79)$$

m enters H through the coupling h, which has the role of providing a direction along which m is energetically encouraged to align. Indeed, if we take the limit $J \to 0$, H is minimised by $\sigma_i = h/|h|$, where $|h|$ is the absolute value of h. Since the term $h \sum_i \sigma_i = hM = hmV$ in the Hamiltonian is not invariant under $\sigma_i \to z\sigma_i$, it is said to *break* the $\mathbb{Z}_2$ symmetry.

We present in Figure 4.14 the behaviour of the magnetisation and the magnetic susceptibility for a ferromagnetic material, together with its interpretation in

FIGURE 4.15: The ising lattice in 1D: open (left) and closed chain (right).

terms of symmetry breaking and restoration coming from the Ising model. Of course, whether the predictions of the model match the observed behaviour (and therefore the physical interpretation in terms of symmetry is correct) needs to be verified with an explicit calculation. Therefore, in order to make progress with our understanding of the system, we shall compute physical observables and compare results with our expectations. We start our tractation from the easier one-dimensional case.

4.4.2 Exact solution of the 1D Ising model

By now the reader is accustomed to the fact that Statistical Mechanics calculations need to be performed in the thermodynamic limit. Usually the latter is reached by formulating the problem for a system of finite size and then carefully extrapolating for a diverging number of degrees of freedom. An aspect that is relevant for a finite system is the treatment of boundary effects. While boundary effects disappear in the thermodynamic limit, when computing observables for a finite size, one needs to take care of the boundary. As a matter of fact, a judicious prescription for the boundary (e.g. respecting the translational symmetry of an infinite system) greatly helps with the technicalities of the calculation. Restricting the argument to the one-dimensional case for the time being, in the left side of Figure 4.15 we show a typical linear lattice with L sites. We number the sites from 0 (further on the left) to $L - 1$ (further on the right). While all the sites $1, \ldots, L - 2$ have two neighbours, one on the left and the other on the right, the sites 0 and $L - 1$ have only one neighbour. The calculations at finite L simplifies without affecting the final result if we can treat all the sites on equal ground. One possibility is to use periodic boundary conditions. In this setup, one closes the chain identifying the left neighbour of $i = 0$ with $i = L - 1$ and the right neighbour of $L - 1$ with 0. Geometrically, this corresponds to formulating the system on a discretised ring or equivalently on a closed chain (Figure 4.15 right).

In the following, let us consider a system with periodic boundary conditions. The Hamiltonian H can be rewritten as

$$H = \sum_i \left(-J\sigma_i\sigma_{i+1} - \frac{h}{2}\sigma_i - \frac{h}{2}\sigma_{i+1} \right) = \sum_i H_i , \qquad (4.80)$$

which exposes a symmetry $i \leftrightarrow i+1$. Let us stress that, due to periodic boundary conditions, $i = L$ needs to be identified with $i = 0$. The partition function is given by

$$Z = \sum_{\{\sigma_i\}} e^{-\beta H} = \sum_{\{\sigma_i\}} \prod_i e^{\beta \sigma_i \sigma_{i+1} + \beta \frac{h}{2}(\sigma_i + \sigma_{i+1})} . \qquad (4.81)$$

We now introduce the 2×2 matrix T

$$T = \begin{pmatrix} T_{1,1} & T_{1,-1} \\ T_{-1,1} & T_{-1,-1} \end{pmatrix} , \qquad (4.82)$$

with the element $T_{\sigma,\tau}$ given by

$$T_{\sigma,\tau} = e^{\beta J \sigma \tau + \frac{\beta h}{2}(\sigma + \tau)} , \qquad (4.83)$$

with $\sigma = \pm 1$, $\tau = \pm 1$. In other words, $\log T_{\sigma\tau}$ gives the contribution to H of two neighbour sites when one of the spin is σ and the other is τ. The whole matrix T (called the *transfer matrix*) accounts for the four possible combinations of σ and τ values. Explicitly:

$$T = \begin{pmatrix} e^{\beta J + \beta h} & e^{-\beta J} \\ e^{-\beta J} & e^{\beta J - \beta h} \end{pmatrix} \qquad (4.84)$$

An inspection of the expression for Z in Eq. (4.81) shows that

1. Z can be written as the product of factors involving nearest-neighbour spins only;

2. each element of the product is an element of T, whose realisation depends on the current configuration $\{\sigma_i\}_k$;

3. as a consequence of the previous point, at fixed configuration, the product is of N elements of T, which is naturally ordered by writing consecutively factors referring to consecutive links;

4. at fixed configuration, the second index of an element of T in the above product must match the first index of the element of T associated with the following link, both indices referring to the spin associated with the common site;

5. since we have imposed periodic boundary conditions, the first index of the first factor of T (the one associated with the link connecting site 0 and site 1) is the same as the second index of the last factor of T (associated to the link connecting site $L-1$ with site 0);

6. finally, summing over all configurations corresponds to summing over all indices of T.

By virtue of the last point, we can write

$$Z = \sum_{i_0, \ldots, i_{L-1}} T_{i_0 i_1} T_{i_1 i_2} \ldots T_{i_{L-1} i_0} \tag{4.85}$$

where $i_0 = \pm 1, \ldots, i_{L-1} = \pm 1$. Since T is independent on the particular link, we get

$$Z = Tr\left(T^N\right). \tag{4.86}$$

In order to compute the trace, we find explicitly the eigenvalues of T, which are given by

$$\lambda_\pm = e^{\beta J}\left[\cosh(\beta h) \pm \sqrt{\sinh^2(\beta h) + e^{-4\beta J}}\right], \tag{4.87}$$

where we note that $\lambda_+ > \lambda_-$. In terms of $\lambda_\pm$, Z is given by

$$Z = Tr\left(T^L\right) = \lambda_+^L + \lambda_-^L = \lambda_+^L\left(1 + \left(\frac{\Lambda_-}{\Lambda_+}\right)^L\right) = \lambda_+^L + \ldots, \tag{4.88}$$

where the ellipses indicate subleading terms in the thermodynamic limit $L \to \infty$. Keeping only the leading contribution, we can write

$$Z \simeq \Lambda_+^N, \tag{4.89}$$

from which

$$F = -kT \log Z \simeq -LkT \log \Lambda_+, \tag{4.90}$$

where the latter approximation is exact in the thermodynamic limit. As for the magnetisation,

$$m = \lim_{L\to\infty} \frac{-1}{L}\left(\frac{\partial F}{\partial h}\right) = \frac{\sinh(\beta h)}{\sqrt{\sinh^2(\beta h) + e^{-4\beta J}}} \tag{4.91}$$

A sketch of m at various T values is provided in Figure 4.16. At $\beta = \infty$ ($T = 0$) m is the sign function of argument h. For any finite β

$$\lim_{h\to 0} m = 0, \tag{4.92}$$

i.e. the model never realises an equilibrium state with spontaneous magnetisation, but it is always in the disordered phase. Hence, the model in one dimension does not describe the physics of spontaneous magnetisation, which is the conclusion to which Ising arrived at in his Ph.D. work. The reason why we keep studying the Ising model, especially in the context of critical phenomena, is that in higher dimension the result is radically different, as we are going to show below.

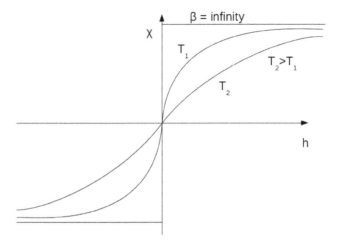

FIGURE 4.16: The magnetisation m for the 1D Ising model at various values of the temperature.

4.4.3   Mean-field solution of the 2D Ising model

The very same techniques that have been exposed to solve the Ising model in one dimension can be used to provide a solution of the model in two dimensions. Although the generalisation might look somewhat straightforward, indeed solving the Ising model in two dimensions with transfer matrix techniques is a highly non-trivial exercise, which was first successfully completed by Onsager in 1941 [28]. Even before the Onsager solution, it was known that the two-dimensional case is qualitatively different from the one-dimensional system in so far a phase transition from an ordered to a disordered phase is present. A proof of the existence of the phase transition was provided by Peierls [29].

In this section, we approach the problem with an approximate solution technique known as *mean field*. The advantage of this technique is that it provides insights on the physics of the model with a relatively simpler setup and more straightforward calculations. However, as we will see, the solution fails to describe correctly quantitative properties of the phase transition. Understanding the reason of the failure of mean field as a general framework in this case and the conditions under which it would provide an exact description of a phase transition are interesting topics *per se* that, however, go beyond the scope of this book. We refer the interested readers to specialised textbooks such as [30, 31].

The key idea behind the mean field approximation is to replace the spin–spin interaction with an "environmental" variable giving on average the same effect as the original interaction, but on single spin degree of freedom rather than on pairwise products. This can be achieved performing the replacement

$$h \to h_{eff} = h + h_{spins} \,, \tag{4.93}$$

where we have introduced the effective variable h_{spins}, which describes the magnetic effect of all the other spins on a single degree of freedom in the form of an internal magnetic field.[9] Together with the external magnetic field h, the latter generates an effective magnetic field that interacts with the single spins. The whole system can henceforth be described as the coupling of non-interacting spins with this effective external field, i.e.

$$H = (h_{ext} + h_{spins}) \sum_i \sigma_i \,. \tag{4.94}$$

In order to determine an expression for h_{spins}, we start from the original Hamiltonian,

$$H = -J \sum_{\langle ij \rangle} \sigma_i \sigma_j - h \sum_i \sigma_i \,. \tag{4.95}$$

We then decompose each single spin σ_i as its average (i.e. the magnetisation m) and the fluctuations around it (indicated with $\delta\sigma_i$),

$$\sigma_i = m + \delta\sigma_i \,. \tag{4.96}$$

From this expression, we trivially obtain

$$\delta\sigma_i = \sigma_i - m \,. \tag{4.97}$$

Replacing Eq. (4.96) into the quadratic term of the Hamiltonian gives

$$-J \sum_{\langle ij \rangle} \sigma_i \sigma_j = -J \sum_{\langle ij \rangle} (m + \delta\sigma_i)(m + \delta\sigma_j) \tag{4.98}$$

$$= -J \left(\sum_{\langle ij \rangle} m^2 + m \sum_{\langle ij \rangle} (\delta\sigma_i + \delta\sigma_j) + \sum_{\langle ij \rangle} \delta\sigma_i \delta\sigma_j \right) \,. \tag{4.99}$$

The mean field approximation assumes that the fluctuations $\delta\sigma_i$ are much smaller than m, so that the quadratic term $\delta\sigma_i \delta\sigma_j$ can be neglected. Under this assumption,

$$-J \sum_{\langle ij \rangle} \sigma_i \sigma_j = -J \left(\sum_{\langle ij \rangle} m^2 + m \sum_{\langle ij \rangle} (\sigma_i + \sigma_j) - m \sum_{\langle ij \rangle} (m + m) \right) \tag{4.100}$$

$$= -J \left(-\sum_{\langle ij \rangle} m^2 + m \sum_{\langle ij \rangle} (\sigma_i + \sigma_j) \right) \,, \tag{4.101}$$

---

[9]The homogeneity of h_{spins} is a consequence of the translational invariance of the system in the thermodynamic limit.

where we have used (4.97). Now, we can perform some of the sums, which gives:

$$\sum_{\langle ij \rangle} m^2 = N m^2 \frac{q}{2} ,$$ (4.102)

$$\sum_{\langle ij \rangle} \sigma_i = \frac{q}{2} \sum_i \sigma_i = \frac{q}{2} N m ,$$ (4.103)

where N is the number of lattice sizes. Inserting these results into the Hamiltonian of the system, we find that in the mean field approximation

$$H = N J m^2 \frac{q}{2} - (Jqm + h) \sum_i \sigma_i ,$$ (4.104)

which allows us to perform the identification $h_{eff} = Jqm + h$ (note however, in respect of our earlier discussion, that the term $NJm^2q/2$ is also relevant). This simplified form for H allows us to compute explicitly the partition function:

$$Z = \sum_{\{\sigma_i\}} e^{-\beta H(\{\sigma_i\})} = e^{-N\frac{q}{2}\beta Jm} \sum_{\{\sigma_i\}} e^{\beta(Jqm+h)\sum_i \sigma_i}$$ (4.105)

$$= e^{-N\frac{q}{2}\beta Jm} \prod_i \sum_{\sigma_i = \pm 1} e^{\beta(Jqm+h)\sigma_i}$$ (4.106)

$$= e^{-N\frac{q}{2}\beta Jm} \left(2\cosh[\beta(Jqm + h)] \right)^N ,$$ (4.107)

from which we derive the free energy

$$F = -\frac{1}{\beta} \log Z = JN\frac{q}{2}m^2 - \frac{N\log 2}{\beta} - \frac{N}{\beta} \log(\cosh[\beta(Jqm + h)]) .$$ (4.108)

The equilibrium state is characterised by the value of m for which F has a minimum, i.e.

$$\frac{\partial F}{\partial m} = 0,$$ (4.109)

which implies

$$m = \tanh(\beta(Jqm + h)) .$$ (4.110)

The latter, which is often referred to as Equation of State, is an equation of self-consistency for m. In order to assess whether the system undergoes spontaneous magnetisation, we are interested in the limit $h \to 0$. Hence, we restrict ourselves to the case $h = 0$.

The equation can be solved graphically. A graphical representation is reported in Figure 4.17, which shows that for $T = 0$ there are two solutions for which $m = \pm 1$ and one solution with $m = 0$ (pane (A)). An analysis of the extremal condition for F shows that the solution with $|m| = 0$ is a local maximum; hence, at equilibrium $m = \pm 1$. As a feature, the presence of three solutions

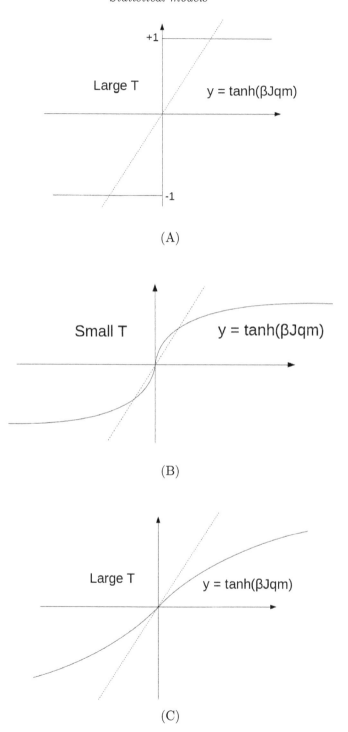

FIGURE 4.17: Graphical solution of the self-consistency condition (4.110) for $h = 0$ and (A) $T = 0$, (B) $T < T_c$ and (C) $T > T_c$.

with the minimum of F obtained at two values of m equal in modulus but opposite in sign and the maximum at $m = 0$ persists for low enough temperatures (pane (B)), while at high enough temperatures only the solution $m = 0$ is present, providing in this case a minimum for F with trivial magnetisation. This analysis shows that a critical value T_c exists separating a broken symmetry phase where $|m| \neq 0$ from a symmetric phase in which $m = 0$. This critical value is attained when the line $y = m$ is tangent to $y = \tanh(\beta Jqm)$, which provides the condition

$$\beta_c Jq = 1 . \tag{4.111}$$

From this, we read straightforwardly

$$T_c = 1/(k\beta_c) = Jq/k . \tag{4.112}$$

As expected, at the critical temperature physical observables are non-analytic. We begin to show explicitly this fact by considering the behaviour of m. Since near T_c m is small, we can use the expansion

$$\tanh^{-1} x \simeq x + \frac{x^3}{3} , \tag{4.113}$$

which is correct up to order x^3. After taking the inverse hyperbolic tangent on both sides of Eq. (4.17) and making use of this expansion, for $T < T_c$ we obtain

$$m + \frac{m^3}{3} = \beta Jqm , \tag{4.114}$$

from which we can see explicitly the three solutions for m. Considering that the equilibrium solution has $|m| \neq 0$, we get

$$m^2 = 3(\beta Jq - 1) = 3\frac{T_c - T}{T} \simeq 3\frac{T_c - T}{T_c} . \tag{4.115}$$

It is convenient to introduce the reduced temperature t as

$$t = \frac{T_c - T}{T_c} , \tag{4.116}$$

which measures the distance of T from T_c in units of T_c. In terms of t, Eq. (4.115) reads

$$m^2 = 3t , \tag{4.117}$$

which gives the two solutions

$$m = \pm (3t)^{\frac{1}{2}} . \tag{4.118}$$

Above T_c (where $t < 0$), Eq. (4.115) has only the solution $m = 0$. The left pane of Figure 4.14 represents schematically the behaviour of m as a function of T.

The magnetic susceptibility χ, defined as

$$\chi = \left(\frac{\partial m}{\partial h} \right)_{h=0} , \tag{4.119}$$

can also be obtained from the equation of state (4.110). Taking the derivative with respect to h of both sides and then setting $h = 0$ gives

$$\frac{\partial m}{\partial h} = (1 - \tanh^2(\beta m q dJ)) \left[\beta + Jq\beta \frac{\partial m}{\partial h} \right] . \tag{4.120}$$

Making use of the equation of state itself at $h = 0$ and of the definition of χ, we find

$$\frac{\chi}{\beta} = (1 - m^2) \left(1 + \frac{\chi}{\beta_c} \right) , \tag{4.121}$$

where we have also used the expression $\beta_c = \frac{1}{Jq}$. We now consider the case $t < 0$, whereby $m = 0$. In this case, after some simple algebra, Eq. (4.121) gives

$$\frac{\chi}{\beta_c} = (-t)^{-1} . \tag{4.122}$$

For the case $t > 0$, we use the result $m^2 = 3t$ (see Eq. (4.117)) and find (again, after some algebraic manipulations)

$$\frac{\chi}{\beta_c} = (2t)^{-1} . \tag{4.123}$$

We note that for both $t \lessgtr 0$, we have

$$\chi \propto |t|^{-1} , \tag{4.124}$$

however, the divergence is twice as fast for $t < 0$. Indeed, if we define the amplitude $A_\pm$ as

$$A_\pm = \lim_{t \to 0^\pm} (t\chi(t)) , \tag{4.125}$$

we find

$$A_+ = \beta_c/2 \quad \text{and} \quad A_- = \beta_c . \tag{4.126}$$

The behaviour of χ as a function of t is shown in Figure 4.14 (right). At $t = 0$ and for non-zero h, the equation of state becomes

$$m = \tanh [(m + h\beta_c)] . \tag{4.127}$$

Taking the inverse hyperbolic tangent on both sides and using the expansion (4.113), we obtain

$$m^3 = 3h\beta_c , \tag{4.128}$$

which shows that at $t = 0$ $m \propto h^{1/3}$.
Finally, we analyse the behaviour of the heat capacity C_V across the phase

transition, as predicted by the mean field theory. We derive the internal energy U as

$$U \equiv \langle H \rangle = \frac{JqN}{2}m^2 - JqNm^2 = -\frac{1}{2}JqNm^2 \,, \tag{4.129}$$

from which we then obtain C_V as

$$C_V = \frac{\partial U}{\partial T} \,. \tag{4.130}$$

Since for $T > T_c$ $m = 0$, in this case we obtain

$$U = 0 \,, \tag{4.131}$$

which gives

$$C_V = 0 \,. \tag{4.132}$$

Using $m^2 = 3t$, for $T < T_c$ we find

$$U = \frac{3}{2}JqNt = -\frac{3}{2}JqN\left(\frac{T_c - T}{T_c}\right) \,, \tag{4.133}$$

yielding to

$$C_V = \frac{3}{2}\frac{JqN}{T_c} = \frac{3}{2}Nk \,. \tag{4.134}$$

Hence, the intensive quantity $c_V = C_V/N$ has a discontinuity across the phase transition:

$$\lim_{t \to 0^+} c_V = \frac{3}{2}k \quad \text{and} \quad \lim_{t \to 0^-} c_V = 0 \,. \tag{4.135}$$

This behaviour is sketched in Figure 4.18.

4.4.4   Dualisation and critical temperature

A powerful insight on the two-dimensional Ising model can be obtained through the Kramers-Wannier duality. The latter is a transformation that maps the original Ising model into an Ising model at a different temperature. In particular, if $Z(T)$ is the partition function of the original Ising model, one can prove that a mapping exist such that

$$f(Z(T)) = f(Z(T^\star)) \,, \tag{4.136}$$

where F is a function of Z and $T^\star$, the temperature of the dual system, is a function of the original T. In particular, we shall see that small T corresponds to large $T^\star$ and viceversa.

As an aside, dualities are very powerful techniques that allow to solve a system via the mapping to another system whose solution is known. The general form of Eq. (4.136) is

$$f(Z(T)) = g(Z^\star(T^\star)) \,, \tag{4.137}$$

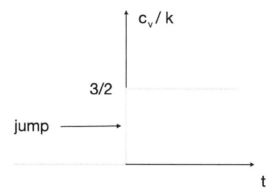

FIGURE 4.18: The behaviour of the specific heat c_V as a function of the reduced temperature t (in blue).

where the identity is between a function f of the partition function of the original system, $Z(T)$, and a function g of the partition function Z^* of the dual system. The latter is computed at a temperature T^* that is in a known relation with T. The interchange of the low-temperature regime with the high-temperature regime through duality is a general property. The Kramers and Wannier transformation provides a simple and yet very instructive example on how dualities can be used in practice to extract quantitative information on a system.

From the point of view of the general description of duality transformations, the Ising model is a special case, since it is *self-dual* (i.e. it maps into itself, albeit at a different temperature). At first sight, one may think that by mapping a system into itself there is nothing to gain: if we start from a problem we can not solve, we end up exactly with the same problem, and hence, no progress can be made. However, one has to consider that the temperatures of the two systems are different and in a known relation. This will allow us to find, under the assumption that a single-phase transition exists, the critical temperature.

We start from the partition function of the Ising model,

$$Z = \sum_{\{\sigma\}} \prod_{\langle ij \rangle} e^{\alpha \sigma_i \sigma_j} , \qquad (4.138)$$

where we have simplified the notation by introducing the variable

$$\alpha = J/kT . \qquad (4.139)$$

We shall now make use of the identity

$$e^{\alpha \sigma_i \sigma_j} = \cosh \alpha \, (1 + \sigma_i \sigma_j v) , \qquad v = \tanh \alpha , \qquad (4.140)$$

which can be verified by direct substitution, considering that $\sigma_i \sigma_j = \pm 1$. Hence,

$$\prod_{\langle ij \rangle} e^{\alpha \sigma_i \sigma_j} = (\cosh \alpha)^{2N} \prod_{\langle ij \rangle} (1 + \sigma_i \sigma_j v) . \tag{4.141}$$

Now we can expand the product and we get a power series in v, where the exponent indicates the length of a path in the lattice whose elementary steps are links. A generic path is not necessarily connected. Paths that are not closed do not contribute to the partition function, since positive and negative contributions cancel exactly when we sum over the σ. Hence, the only paths that contribute are the closed paths. As a consequence, if we indicate with $n(r)$ the number of paths of length r, Z can be rewritten as a sum over all closed paths connecting sites:

$$Z = 2^N (\cosh \alpha)^{2N} \sum_r n(r) v^r . \tag{4.142}$$

Incidentally, in the high temperature regime, this can be seen as a well-defined expansion controlled by r, since

$$\lim_{T \to \infty} v = 0 . \tag{4.143}$$

Hence, the expression (4.142) is referred to as the high temperature series.

In the Ising model, we conventionally think in terms of spins. An alternative picture takes as fundamental variables the links (i.e. the product of two spin variables), since links are the objects that ultimately enter in the expression of the energy. In particular, when measured with respect to the ground state, the energy of a given configuration gets contribution only by bonds with anti-aligned spins. In terms of links with misaligned spins at their end, we can rewrite the Hamiltonian of the model as

$$\mathcal{H} = -2NJ + 2JN_{+-} , \tag{4.144}$$

where N_{+-} is the number of anti-aligned neighbour spin pairs. In this picture, in which we take the link as a fundamental variable, we can imagine the energy as concentrated in the middle of each activated link, with an activated link being a link that joins two neighbour misaligned spins. We can mark for instance the middle point of an activated link with an X. We now say that two links are *neighbour* if either they touch or they are parallel and separated by one lattice spacing. For a given configuration, if we join all the neighbour links that are -1, we obtain closed paths that separate spins having value -1 from spins having value $+1$ (or viceversa). These paths take the name of *domain walls*, since they separate regions (or domain) with different spin values. In general, a given configuration can have several disconnected domain walls. N_{+-} in Eq. (4.144) is the sum of the perimeters of all domain walls in that configuration (see Figure 4.19).

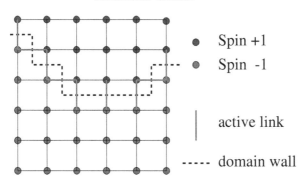

FIGURE 4.19: Ising configuration with "active links" and corresponding domain wall.

Using the expression (4.144) for the Hamiltonian, we can rewrite the partition function by summing over all possible domain wall configurations as

$$Z = e^{2\alpha N} \sum_r m(r) e^{-2\alpha r} , \qquad (4.145)$$

where $m(r)$ is the number of domain wall configurations of total boundary perimeter r.

As the next step, we define the dual lattice as the lattice formed by the central points of the squares in the original lattice. Closed loops of links belonging to the dual lattice can be interpreted as domain wall boundaries in the original lattice and closed loops of links in the original lattice can be superimposed to domain wall boundaries in the dual lattice. Therefore, for each domain wall line in the dual lattice, there is a loop of links in the original lattice and for each domain wall in the original lattice there is a closed loop of links in the dual lattice. This proves the identity

$$m(r) = n(r) \qquad (4.146)$$

at fixed r.

We now introduce α^* from

$$\tanh \alpha^* = e^{-2\alpha} , \qquad (4.147)$$

which can be proved to be equivalent to the relation

$$\sinh(2\alpha^*) \sinh(2\alpha) = 1 . \qquad (4.148)$$

With this substitution, and identifying $m(r)$ with $n(r)$ following the argument provided above, we can rewrite the partition function as

$$Z(\alpha) = e^{2\alpha N} \sum_r m(r) e^{-2\alpha r} = (\tanh \alpha^*)^{-N} \sum_r n(r) v^{*r} , \qquad (4.149)$$

with

$$v^* = \tanh \alpha^* \ . \tag{4.150}$$

On the other hand, we can also write

$$Z(\alpha^*) = 2^N \left(\cosh \alpha^* \right)^{2N} \sum_r n(r) v^{*r} \ . \tag{4.151}$$

By comparison

$$Z(\alpha^*) = 2^N \left(\sinh \alpha^* \cosh \alpha^* \right)^N Z(\alpha) \ , \tag{4.152}$$

or equivalently

$$\left(\sinh \left(2\alpha^* \right) \right)^{-N/2} Z(\alpha^*) = \left(\sinh \left(2\alpha \right) \right)^{-N/2} Z(\alpha) \ , \tag{4.153}$$

where the identity

$$2 \cosh x \sinh x = \sinh \left(2x \right) \tag{4.154}$$

has been used, together with Eq. (4.148) for symmetrising the result in (α, α^*). If we know that a phase transition exists for the original system and we assume that the transition temperature is unique, from (4.148) we get

$$\sinh^2(2\alpha_c) = 1 \ , \tag{4.155}$$

which, being $\alpha \geq 0$, implies

$$\sinh(2\alpha_c) = 1 \ . \tag{4.156}$$

It is now easy to show that

$$\alpha_c = \frac{1}{2} \log \left(\sqrt{2} + 1 \right) \ , \tag{4.157}$$

which coincides with the result obtained from the Onsager's solution.

4.4.5   Comparing mean-field with the exact solution

When compared to the mean-field value $\beta_c = (Jq)^{-1}$, the Kramers-Wannier result for β_c, Eq. (4.157), makes it manifest that the mean-field approximation fails at the quantitative level. Before analysing the reasons, it is worth going into a detailed comparison of the results in the two cases.

Among the crucial quantities that characterise a phase transition are critical exponents, i.e. the exponents that describe the singularity in relevant physical observables in the limit(s) $t \to 0^{\pm}$. In general terms, we write the leading behaviours in the $t \to 0$ limit of the physical quantities we have computed in subsection 4.4.3 as follows:

$$m \underset{t \to 0^+}{\propto} |t|^{\beta} \ ; \qquad \chi \underset{t \to 0}{\propto} |t|^{-\gamma} \ ; \qquad C_V \underset{t \to 0}{\propto} |t|^{\alpha} \ . \tag{4.158}$$

TABLE 4.1: Comparison of mean-field values for the inverse critical temperature and for the critical exponents with those obtained in the exact solution for the two-dimensional Ising model on a square lattice.

Quantity	Mean Field	Exact Value (from Onsager's solution)
$\frac{J}{kT_c}$	$\frac{1}{q} = \frac{1}{4}$	$\frac{1}{2}log(\sqrt{2}+1) \approx 0.44$
α	0 (no divergence)	0 (log divergence: $C_v \propto log\lvert 1 - \frac{T_c}{T}\rvert$)
β	1/2	1/8
γ	1	7/4
δ	3	15

These relations define the critical exponents α, β and γ. In addition, we define δ from the leading behaviour of m as a function of h when $t = 0$:

$$|m| \underset{h \to 0}{\propto} |h|^{1/\delta} \ . \tag{4.159}$$

In Tab. 4.1, we provide a comparison of β_c and of the critical exponents defined above as obtained in the mean-field approximation and from the exact solution (for the latter, we refer the reader to, e.g. [30]). The conclusion is that mean-field is a useful tool for predicting the qualitative features of the model, but the quantitative results it returns can not be trusted as being correct. An interesting question is whether there are cases in which mean field theory is quantitatively correct. In order to answer this question, we would need to revisit the nature of the approximation. Since once more a full understanding of the argument we are giving requires the introduction of tools and concepts that are outside the scope of our tractation (e.g. Landau Theory and scaling laws), here we only sketch the argument, referring to textbooks such as [30, 31] for deeper insights.

We start by revisiting the mean-field approximation, which can be recast into

$$\langle \delta \sigma_i \delta \sigma_j \rangle = 0 \ , \tag{4.160}$$

i.e. into the fact that spin fluctuations are uncorrelated. The quantitative failure of mean field hence means that spin fluctuations are correlated. On a general ground, one can write the correlation function

$$\Gamma(r) = \langle \delta \sigma_i \delta \sigma_j \rangle \propto e^{-\frac{r}{\xi}} r^{D-2+\eta} \ , \tag{4.161}$$

where $r = |i - j|$ (i.e. r is the distance between i and j). ξ is a quantity with dimensions of a length, which has the natural interpretation of the distance over which spins are correlated (indeed for $r \gg \xi$, $\Gamma(r) \simeq 0$) and η is a critical exponent known as *anomalous dimension*[10]. The correlation length ξ is characterised by its own critical behaviour:

$$\xi \underset{|t| \to 0}{\propto} |t|^{-\nu} , \tag{4.162}$$

where ν is the corresponding critical exponent. The astute reader would have already spotted the reason of the failure of mean field theory: with ξ diverging at criticality, the approximation $\xi \to 0$ is hard to justify. The opposite question hence arises: why with such a bad approximation we obtain nearly physical results? In fact, things are more subtle than they look. $\Gamma(r)$ has the same dimension of m^2. Hence, in order to evaluate the effect of the fluctuations, we would need to compare them with m^2. Using the approximation of quadratic fluctuations, one finds that that mean-field provides the exact result for $D > 4$, while for $D < 4$ one has to keep into account the fluctuations. Being the limiting case, $D = 4$ is called the *upper critical dimension*. We will touch again on the upper critical dimension in subsection 4.4.7.

We conclude with a note on the critical exponents. So far, we have introduced six critical exponents. Having six (or more) independent quantities is not theoretically appealing. Is it then possible that our critical exponents are not all independent? The answer is indeed that they aren't. The relations between them are provided by the scaling laws

$$\begin{array}{lll} \text{Fisher Law:} & \gamma = \nu(2 - \eta) \ ; \\ \text{Widom Law:} & \gamma = \beta(\delta - 2) \ ; \\ \text{Rushbrooke Law:} & \alpha + 2\beta + \gamma = 2 \ ; \\ \text{Josephson Law:} & \nu d = 2 - \alpha \ . \end{array} \tag{4.163}$$

Note that both the mean-field and the exact critical exponents of the Ising model fulfil these scaling laws. The scaling relations above can be derived by assuming that, under a rescaling of the unit of length r of the form

$$r \mapsto r/b , \tag{4.164}$$

with b the scale factor, the singular part of the free energy density behaves like an homogeneous function:

$$F(t, h) = b^{-D} F(t b^{D_t}, h b^{D_h}) . \tag{4.165}$$

---

[10]In fact, naive dimensional analysis would fix $\eta = 0$. A non-zero value for η is possible because of the existence of the scale T_c, which is dynamically generated (i.e. appearing from the dynamics despite not being present explicitly anywhere in the microscopic Hamiltonian) and can appear in dimensionless ratios such as $(T/T_c)^a$, where a is a generic exponent.

Using ξ as the scale factor and considering that $\xi \propto |t|^{-\nu}$, one finds the critical exponents in terms of the quantities D_t and D_h:

$$\alpha = 2 - \frac{D}{D_t} \; ; \qquad \beta = \frac{D - D_h}{D_t} \; ;$$

$$\gamma = \frac{2D_h - D}{D_t} \; ; \qquad \delta = \frac{D_h}{D - D_h} \; ; \qquad (4.166)$$

$$\nu = \frac{1}{D_t} \; ; \qquad \eta = 2 + D - 2D_h \; .$$

The previous relations show that there is a clear sense in which ν and η are fundamental quantities (they are the only two that respectively depend only on D_h and D_t, which in turn are related to the homogeneity hypothesis of f), while the other critical exponents can be seen as derived.
We note that, while we have used the Ising model as a template, the results presented in this section about the validity of mean-field theory and the scaling relations among critical exponents are generally applicable.

4.4.6 Monte Carlo simulation and phase transition

A general and powerful route for obtaining information about a phase transition is through Markov Chain Monte Carlo simulations. Using a Monte Carlo method, one can generate representative samples for the system and then study thermodynamic quantities as ensemble averages. With an apparent simplicity that nevertheless gives rise to profound physical features, the Ising model offers the perfect reference to understand the practicality of implementing Markov Chain Monte Carlo methods. In addition, it provides an ideal playground to familiarise with data analysis tools and enable us to discuss widely used methods to extract the value of the critical temperature and of the critical exponents (which, in our context correspond to solve the physics of the phase transition for a model).
In this section, we present two simple and yet effective Monte Carlo algorithm implementations for the specific case of the Ising model, the Heat Bath and the Metropolis algorithm. We will postpone to Chapter 5 the discussion of more advanced Monte Carlo techniques. For simplicity, we consider an Ising model over a rectangular grid of size $L_x \times L_y$. We also set $J = 1$ (a different value of J corresponding to a rescaling of β) and $h = 0$.
The steps of the Metropolis algorithm can be implemented as follows:

For a site i and the spin σ_i

1. Calculate
$$b = \sum_{j \in \langle xy \rangle} \sigma_j \; , \qquad \pi = \exp\{-2\beta\sigma_i b\} \; .$$

2. Flip $\sigma_i \to -\sigma_i$ with probability $p = \min(1, \pi)$.

Here we have indicated with $j \in \langle ij \rangle$ a site j that is nearest neighbour of i. Note that we implement step 2 by drawing a uniformly distributed random number $u \in [0, 1]$ and flipping σ_i if $p \leq u$. If σ_i is flipped, this value will replace the old σ_i in the configuration, giving rise to a new configuration in the chain. If σ_i is not flipped, the previous configuration is repeated as a further configuration. This step is very important: neglecting to repeat configurations whereby the spin flip has not been accepted is quite a common beginner mistake that spoils the Markov dynamics leading up to incorrect results.

Similarly to the Metropolis case, step 2 is implemented by drawing a uniformly random number $u \in [0, 1]$ and setting $\sigma_i = 1$ if $p \leq u$ and $\sigma_i = -1$ otherwise. An implementation of the Heat Bath consists of the following steps:

For a site i and the spin σ_i

1. Calculate

$$b = \sum_{j \in \langle ij \rangle} \sigma_j \,, \qquad p = \frac{\exp\{-\beta b\}}{\exp\{\beta b\} + \exp\{-\beta b\}} \,.$$

2. Set $\sigma_i = 1$ with probability p and $\sigma_i = -1$ with probability $1 - p$.

In both cases, we call a sweep the update of $L_x \times L_y$ spins. In order to complete a sweep, one could move in a predefined manner along the grid. For instance, one can start by visiting the point $(i = 0, j = 0)$, then $(i = 1, j = 0)$ and keeping incrementing the i coordinate until $i = L_x - 1$; at that point, one repeats the sequence of i coordinates at $j = 1$, incrementing the j coordinate each time $i = L_x - 1$; the sweeps is complete once $i = L_x - 1$ and $j = L_y - 1$. An alternative way of proceeding consists in visiting $L_x \times L_y$ spins by choosing each one randomly. This procedure (which could give rise to some spins being visited multiple times in a sweep and others being never visited during the same sweeps) is known to produce a more decorrelated Markov sequence, and hence a more efficient sampling. Likewise, the Heat Bath algorithm is generally better than the Metropolis one at decorrelating the system and hence should be preferred[11].

An example set of data obtained with an implementation of the Metropolis algorithm is provided in Figure 4.20. In the case displayed, a simulation has been set up for 45 β values in the interval $[0.025, 0.85]$, which contains the value $\beta_c = 0.5 \log(1 + \sqrt{2})$. The choice of the interval has relied on the knowledge of the value of β_c and is done in such a way that the simulation shows the high-temperature (low-β) regime, the low-temperature (high-β) regime and the critical region. Note that the β are not equally spaced, but more densely packed around the critical value. This reflects our objective of exploring in

---

[11]Note however that, while a Metropolis algorithm can be implemented with any Hamiltonian, not for every system is it known how to derive a Heat Bath algorithm.

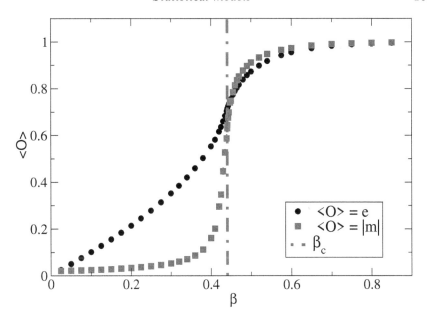

FIGURE 4.20: Simulation results for the Ising model in two dimensions on a square grid with number of sites 40^2. The data have been obtained with 100000 measurements preceded by 10000 thermalisation steps. Error bars are smaller than the symbols.

more details the physics of the critical region, where the behaviour of the observables that are sensitive to the phase transition changes more quickly. In the figure, which has been obtains considering an $L_x = L_y = 40$ lattice, the two observables that are plotted are the energy per unit link,

$$e = \frac{1}{2L^2} \langle H \rangle \qquad (4.167)$$

and the absolute value of the magnetisation per unit site $|m|$. The vertical line indicates the expected critical β. Error bars (invisible on the scale of the plot) have been computed accounting for the fact that in general configurations generated with Monte Carlo methods have a non-zero autocorrelation time; in particular, the Jackknife method (see subsection 3.6.2) has been used to determine the statistical errors. At low β, both e and m approach zero, while at high β they go to one. This behaviour is compatible with our analytic expectation. The transition between the low-β and the high-β phase happens in the critical region, where the change of behaviour is dramatic, in particular for m. The figure does not suggest the presence of any singularity in the simulated system, which has to be expected at finite size. Likewise, the fact, while visibly smaller than at high β, $|m|$ is non-zero at lower β is also ascribed to finite size.

The reader may wonder why we have studied $|m|$ rather than m. The reason is that at finite size

$$\langle m \rangle = 0 \tag{4.168}$$

regardless of the phase the system in. In the would-be symmetry broken phase, this is a consequence of the fact that the tunneling probability between the two non-degenerate equilibrium states is finite, albeit exponentially decreasing with the volume.

The result of our simulation is qualitatively compatible with expectations. To verify quantitatively that the behaviour is correct, we can use the framework of finite size scaling [32]. This will provide asymptotic formulae for how observables depend on the size at large sizes. For instance, for the position $\beta_c(L)$ of the peak χ_M of the magnetic susceptibility at finite lattice size L^2, one find, as a function of the linear size L,

$$\beta_c - \beta_c(L) \propto L^{-\frac{1}{\nu}} . \tag{4.169}$$

Therefore, not only is

$$\lim_{L \to \infty} \beta_c(L) = \beta_c , \tag{4.170}$$

but an asymptotic fit of this behaviour enables us to extract the critical exponent ν. Similarly, one finds

$$\chi_M(L) \propto L^{\frac{\gamma}{\nu}} , \tag{4.171}$$

which, after we have obtained ν from Eq. (4.169), enables us to find γ. Through the relations (4.166), γ and ν give access to all other critical exponents. Therefore, a strategy for extracting critical properties consists in simulating progressively larger sizes and using the asymptotic formulae to obtain β_c and the critical exponents through a fit.

While in this section we have applied Monte Carlo methods to verify known analytic expectations for the Ising model, Monte Carlo methods can be used also in cases in which exact results are not known, providing valuable insights on the physics of the system. In applications of the latter type, one will use some minimal analytic guidance such as the expected behaviour of the Hamiltonian and of the order parameter at low and high β to locate (e.g. through bisection) the critical region and then perform an accurate analysis to determine the critical β and the critical exponents.

4.4.7 Upper and lower critical dimension

We have discussed the Ising model in one dimension, finding no transition, and in two dimensions, finding a second-order phase transition. What happens at higher dimension?

In three dimension, like in the two-dimensional case, there is a second-order phase transition. This time, the model is not solvable, therefore, to find the critical exponents, we need to rely on other methods. Analytical computations make use of the concept of universality, to determine the critical exponents in models in the same universality class in which they can be computed, e.g. with some controlled expansion. Numerical models are based on a Monte Carlo analysis, which does not present technical complications with respect to the two-dimensional case. Both methods are by now established, and produce robust results that are reassuringly in agreement. For details, we refer to the review [33].

In four dimensions, there is still a phase transition, and the exponents in this case are the mean-field exponents that we computed in subsection 4.4.3. This is the consequence of a general fact. In fact, when we have compared the mean-field solution with the exact solution, we said that the discrepancy is due to fluctuations, which the mean-field approach disregards. The relative importance between fluctuations and mean value is given by the ratio

$$R = \langle (\delta m) \rangle / \langle m \rangle \ . \tag{4.172}$$

Replacing these quantities with their asymptotic values near the critical point, at criticality we have $R \to \infty$ for dimension $d < 4$ and $R \to 0$ for $d > 4$. In the former case, fluctuations dominate and we have a critical behaviour different from the mean field prediction. In the latter case instead fluctuations become irrelevant, and mean field holds. The limiting case is special, since fluctuations and ensamble average count equally. This case approaches criticality with logarithmic corrections to mean field. As above this dimension, the transition is of the mean-field type, $d = 4$ is called the upper critical dimension, already introduced in subsection 4.4.5. The heuristic argument we have presented for the determination of the upper critical dimension is called the *Ginzburg criterion*, after its proposer, the renowned physicist Vitaly Ginzburg, also coauthor of the Ginzburg-Landau theory of phase transitions, for which he received the Nobel Prize in 2003.

The adjective *upper* that qualifies the critical dimension $d = 4$ suggests that there may be also a lower critical dimension. We have seen that in the Ising model, there is no phase transition for $d = 1$. This fact is more general and applies to all spin systems with a discrete symmetry group: for $d = 1$ no spontaneous symmetry breaking can occur. For those systems then, $d = 1$ is the lower critical dimension. Things are different for systems with a continuous symmetry group, some of which will be discussed in the next section. For them, one can prove that even in $d = 2$ any non-invariant *local* operator has a null ensemble average. Therefore, no *local* symmetry can be broken in $d = 2$, which is the lower critical dimension in this case. However, there is a different class of symmetries, called *topological symmetries*, that descend from the conservation laws for extended objects of a topological nature. No general theorem prevents those symmetries from being spontaneously broken for certain values of the control parameters and restored for others. In two dimensions, phase transitions for systems with a local continuous global symmetry are related to the spontaneous breaking of symmetries of topological nature or of discrete subgroups of the continuous symmetry.

4.5   An overview of other models

Due to its simplicity, the Ising model is the archetype of several classical spin systems. There are various extensions and generalisations of the Ising spin model. For instance, one can add next-to-neighbour interactions or interactions that extend further in the lattice, consider interactions with four or a higher number of spins, change the degrees of freedom to a larger ensemble (e.g. considering the integers $0, \pm 1$), etc. One can also act on the coupling, making it negative (going in this way from the ferromagnetic to the antiferromagnetic case) or/and link-dependent, to account for inhomogeneities of the material. All these systems can be used to model different types of physical interactions. For instance, a simple model of a spin glass is an Ising model in which the sign and magnitude of the coupling depends on the link.

The study of classical spin systems is still an active field of investigation, for a variety of reasons, including connections with machine learning that we will explore in Chapter 8. In two dimensions, they are rather well-understood prototype models for testing new tools and ideas. In addition, they define universality classes that are relevant for the investigation of more complex systems based on the same Hamiltonian symmetries. Therefore, especially in three dimensions, they are the subject of precision investigations that focus on their critical properties.

In this section, we will provide an overview of a selection of classical spin models based on larger symmetry groups with a constant ferromagnetic coupling ($J > 0$). We start with **clock models** as generalisation of the Ising model. At zero external field, the N-state clock model is described by the Hamiltonian

$$H = -J \sum_{\langle ij \rangle} \cos\left(\frac{2\pi}{N} (l_i - l_j) \right) , \tag{4.173}$$

where N, l and m are integers, with N fixed and $0 \leq l_i, l_j < N$. The degrees of freedom of the system are the set of integers $\{l_i\}$ living on sites i, and the interaction is nearest-neighbour. Alternatively, we can consider complex spin variables

$$\sigma_i = e^{i \frac{2\pi}{N} l_i} , \tag{4.174}$$

and rewrite the Hamiltonian as

$$H = -J \sum_{\langle ij \rangle} \mathcal{R}e \left(\sigma_i \sigma_j^* \right) , \tag{4.175}$$

where the asterisk $*$ indicates complex conjugation and $\mathcal{R}e$ selects the real part of the product. In this rewriting, H is invariant under the transformation

$$\sigma_i \rightarrow e^{i\frac{2\pi}{N}n}\sigma_i = e^{i\frac{2\pi}{N}(l_i+n)\bmod N} = e^{i\frac{2\pi}{N}l_i'} , \tag{4.176}$$

where the $\bmod N$ function ensures that the argument of the exponential after the transformation still satisfies the constraint $0 \leq l' < N$. The complexification of the degrees of freedom given by Eq. (4.174) shows that the N-state clock model is invariant under transformations of the group

$$\mathbb{Z}_N = \{e^{i\frac{2\pi}{N}n} , \ 0 \leq n < N\} . \tag{4.177}$$

Phase transitions in these models are related to the spontaneous breaking of the $\mathbb{Z}_N$ symmetry at low temperatures. An order parameter for this transition is the magnetisation

$$m = \sum_i \sigma_i . \tag{4.178}$$

If all the σ_i have the same value, $|m| = 1$. Conversely, if the σ_i are randomly distributed, in the thermodynamic limit $m = 0$.

Note that for $N = 2$ this is the $\mathbb{Z}_2$ symmetry of the Ising model. Indeed it is possible to recast the Ising model into a clock model.

We now go back to Eq. (4.173). The nearest-neighbour interaction term gives its largest contribution when $l_i = l_j$, which results in the maximum value of one for the cosine function. In the ordered phase, this encourages alignment of the system along one of the possible N values of the spins. We can further reinforce this tendency to align replacing the cosine with a Kronecker delta function:

$$H = -J\sum_{\langle ij \rangle} \delta_{l_i,l_j} , \tag{4.179}$$

where

$$\delta_{l_i,l_j} = \begin{cases} 1 \text{ if } l_i = l_j \\ 0 \text{ if } l_i \neq l_j \end{cases} . \tag{4.180}$$

Equation (4.179) defines the *N-state Potts model*. The system is symmetric under the group of permutations of N elements. Note that this symmetry contains the $\mathbb{Z}_N$ group. In two dimensions, Potts models are found to have a phase transition at

$$\beta_c = \log\left(1 + \sqrt{N}\right) , \tag{4.181}$$

with the transition being of second order for $N \leq 4$ and first order for $N > 4$. A detailed tractation of spin models with emphasis on exact results can be found, e.g. in [34].

Starting again from the Hamiltonian (4.173) and taking the limit $N \rightarrow \infty$, we obtain the XY model

$$H = -J\sum_{\langle ij \rangle} \cos\left(\theta_i - \theta_j\right) , \qquad 0 \leq \theta_i, \theta_j < 2\pi , \tag{4.182}$$

whose degrees of freedom are angular variables. This Hamiltonian is symmetric under a constant shift of the angles

$$\theta_i \to (\theta_i + q) \mod(2\pi) , \qquad 0 \le q < 2\pi . \tag{4.183}$$

Using complex variables

$$\sigma_i = e^{i\theta_i} , \tag{4.184}$$

exposes the fact that a constant shift q of the angle is equivalent to a $U(1)$ transformation. Since $U(1)$ is isomorphous to $O(2)$, the symmetry group can be equivalently thought of as $O(2)$. This introduces us to the $O(N)$ models, which are a generalisation of the XY model. For these models, the Hamiltonian takes the form

$$H = -J \sum_{\langle ij \rangle} \vec{s}_i \vec{s}_j , \tag{4.185}$$

where $\vec{s}_i$ is a unit vector in a N-dimensional space. For $N = 1$, the model is once again the Ising model. For $N = 2$, as anticipated, we obtain the $O(2)$ model, or, equivalently, the XY model, which will be discussed more in detail in section 6.1. For $N = 3$, the model is called the Heisenberg ferromagnet, since it was introduced and used by Heisenberg as a description of a ferromagnet.

4.6   Exercises

Exercise 4.1 *Knowing that in the differential form the Gibbs free energy G for a magnetic system is given by*

$$dG(T, h) = -S \, dT - M \, dh ,$$

where T is the temperature, S the entropy, M the magnetisation and h is the external field, prove the Maxwell's identity

$$\left(\frac{\partial S}{\partial h} \right)_T = \left(\frac{\partial M}{\partial T} \right)_h .$$

Exercise 4.2 *Prove that the partition function Z in a canonical system takes the form*

$$Z = e^{-\beta F} ,$$

with F the Helmholtz free energy.

Exercise 4.3 *A system is made of N non-interacting particles and each particle can only occupy either the energy level $E_1 = -\epsilon$ or $E_2 = \epsilon$, with $\epsilon > 0$. Derive the average energy $\overline{E}(T)$ using*

a. the microcanonical approach

b. the canonical approach

and prove the equality of the canonical and microcanonical results in the thermodynamic limit, $N \to \infty$.

Exercise 4.4 *A quantum harmonic oscillator has energy levels*

$$E_n = \hbar\omega \left(n + \frac{1}{2} \right) . \tag{4.186}$$

Write the partition function, the internal energy and the heat capacity of a gas of N non-interacting harmonic oscillators.

Exercise 4.5 *A simplified model of a polymer represents this system as a linear sequence of N independent monomers. Each monomer can have length $l = 0$ or $l = 1$ (in appropriate units). A monomer in a state $l = 0$ contributes 0 to the total energy, while a monomer with $l = 1$ contributes ϵ, with $\epsilon > 0$. We define N_0 as the number of monomers of length $l = 0$ and N_1 as the number of monomers of length $l = 1$. We call N_E the number of states at given energy E.*

a. *Write the total energy E of the system as a function of N_0 and N_1 and derive the expression for N_E as a function of the same two variables.*

b. *Show that the partition function Z of the system at given temperature T is given by*

$$Z = \left(1 + e^{-\beta\epsilon} \right)^N .$$

c. *Derive the expressions for F, U and C_V.*

Exercise 4.6 *The* Gibbs paradox *exposes the necessity of including the quantum mechanics concept of indistinguishable particles in a classic tractation of Statistical Mechanics in order to get correct results. The paradox consists in the impossibility of giving a coherent definition of entropy in the case of distinguishable particles, since the entropy of a system of non-interacting identical but distinguishable particles is different from the sum of the entropies of its subcomponents. For pedagogical reasons, in this exercise, we will walk through the argument using the canonical ensemble and use the micro- canonical ensemble only at the end. However, note that historically the argument has been formulated in terms of a mixing entropy (introduced below) using only micro-canonical ensemble considerations.*

a. *Show that for an ideal gas of N identical particles of mass m in three-dimensional space, with Hamiltonian given by*

$$H = \sum_{i=1}^{3N} \frac{p_i^2}{2m} ,$$

the partition function at temperature T is given by

$$Z = \frac{V^N}{N! h^{3N}} \left(\frac{2\pi m}{\beta} \right)^{\frac{3}{2}N} . \tag{4.187}$$

b. *Consider a container of volume V with a wall dividing it into two regions of volume V_1 and V_2, respectively, containing respectively N_1 and N_2 particles of the same ideal gas in equilibrium at temperature T. Show that the free energy of the system of $N = N_1 + N_2$ particles obtained removing the wall without disturbing equilibrium is the sum of the free energy of the subsystem in V_1 and of that in V_2.*

c. *Now repeat the calculation disregarding the factor $N!$ in Eq. (4.187) and show that in this case the additivity of the free energy does not hold any longer. By explicit calculation, show that the difference in free energy before and after the removal of the wall can be explained in terms of a difference in entropy (known as the* mixing entropy) *in the two cases.*

d. *Calculate the mixing entropy using the micro canonical ensemble and show that also in this case the mixing entropy is zero if the $N!$ term is correctly considered.*

Exercise 4.7 *Find the expressions for the critical parameters P_c, v_c and T_c for the Van der Waals gas and show that, when expressed in terms of the rescaled quantities*

$$\tilde{P} = \frac{P}{P_c} \quad , \quad \tilde{T} = \frac{T}{T_c} \quad , \quad \tilde{v} = \frac{v}{v_c}$$

the Van der Waals equation takes the universal form

$$\tilde{P} = \frac{8\tilde{T}}{3\tilde{v} - 1} - \frac{3}{\tilde{v}^2} \; ,$$

where the parameters a and b do not appear explicitly.

Exercise 4.8 *Show that the isothermal compressibility diverges at the critical point of the Van der Waals gas.*

Exercise 4.9 *In one dimension, the Ising model with next-to-nearest-neighbour interaction has Hamiltonian*

$$\mathcal{H} = -\sum_i \sigma_i \left(J_1 \sigma_{i+1} + J_2 \sigma_{i+2} \right) \; .$$

Show that the model is equivalent to the one-dimensional Ising model with only nearest-neighbour interaction (with coupling J_2) in an external magnetic field J_1 and using this result argue that in the original model there is no phase transition at any non-zero value of the temperature T.

Exercise 4.10 *Consider a system consisting of four spin variables, each residing on a corner of a square. Each corner is occupied only by one spin, and each spin can take the values ± 1. The Hamiltonian of the system is*

$$\mathcal{H}(\sigma) = -\left(\sigma_1 \sigma_2 + \sigma_2 \sigma_3 + \sigma_3 \sigma_4 + \sigma_4 \sigma_1 \right) \; ,$$

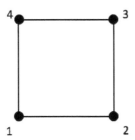

FIGURE 4.21: Ising model with sites on the four corners of a square.

where the indices 1, 2, 3, 4 above label the corners of the square (see Figure 4.21).

a. *List all possible configurations of the system and show that the only non-zero contributions to the Hamiltonian come from configurations with alternate opposite spins or configurations in which all spins are equal.*

b. *Show that the partition function Z at temperature T is given by $Z = 4\left(\cosh(4\beta) + 3\right)$ and compute the specific heat at constant volume.*

c. *Show that the partition function can be recast into the expression*

$$Z = 4 \sum_{\sigma_1, \sigma_3} \left(\cosh^4 \beta + 2\sinh^2 \beta \cosh^2 \beta \left(\sigma_1 \sigma_3\right) + \sinh^4 \beta\right) \ .$$

Exercise 4.11 *Writing a Monte Carlo code for the Ising model in two dimensions is a classic propaedeutic exercise for numerical studies in statistical and lattice gauge systems. This exercise will guide through the steps that will help the reader to build their first Monte Carlo code.*

a. *Consider a grid in two dimensions. How would you label its points? How would you move from a given point to one of its nearest neighbour? How would you implement periodic boundary conditions? Translate your considerations into a computer programme and test that your expectations are verified.*

b. *Using the geometry obtained in the previous point, write a computer program implementing the Metropolis and the Heat Bath update algorithms for the Ising model in two dimensions keeping the lattice dimensions L_x and L_y, β and the number of sweeps nsweeps as general parameters. This part of the exercise includes assigning spin variables to the lattice points, defining the product of nearest-neighbour spins, writing the update routines and sweeping through the lattice systematically.*

c. *Write the measurement routines for the energy and for the magnetisation, which can be called after a number of updates of the whole lattice. It is better to write routines for the single measures, store the latter in an array and then post-process the results keeping into considerations the fact that updates are correlated.*

d. *Run the code for $L_x = L_y = 20$ at $\beta = 0.43$ (symmetric phase) and $\beta = 0.45$ (broken phase), and compare the magnetisation in the two cases.*

e. *Now run the code for $\beta \in [0.4, 0.5]$, taking 11 values spaced 0.1 apart. Give appropriate consideration to thermalisation and size of the measure sample. Does the magnetisation display the expected behaviour?*

f. *For the same lattice size, measure the energy, the magnetisation and the magnetic susceptibility for $\beta \in [0.43, 0.45]$, showing that there is a peak in the magnetic susceptibility in this interval. You may start with 11 β spaced 0.002 apart.*

g. *Repeat the previous point for $L = 30, 40, 50$, and show that the peak position of the magnetic susceptibility moves towards the critical value $\beta_c = \frac{1}{2} \log \left(1 + \sqrt{2}\right)$.*

Exercise 4.12 *Adapting the strategy proposed in the previous exercise, write a Monte Carlo code with Metropolis updates for*

a. *clock models;*

b. *Potts models;*

c. *$O(N)$ models.*

Perform an appropriate scan in β values searching for phase transitions and compare your findings with analytic results when possible.

Chapter 5

Advanced Monte Carlo simulation techniques

Numerical methods introduced in Chapter 3, essentially based on Markov Chain Monte Carlo, have been implemented in Chapter 4 for the numerical simulation of the Ising model, which is the prototype of statistical systems made up of many interacting degrees of freedom and with a non-trivial phase structure.

Most of these models are characterized by a *local* interaction. They are described many stochastic variables which are not independent of each other, i.e. their distribution does not factorize in the product of distributions over the single variables. However, the probability distribution is still written as the product of terms involving a few variables at a time (typically pair by pair), and each stochastic variable is coupled only to a few other variables, typically based on a relation of spatial proximity. For example, in the Ising model, the interaction couples only spin pairs living on nearest neighbours lattice sites. In those cases, local algorithms such as the one proposed in Chapter 4, for which the elementary Markov step changes a limited number of variables, typically just one, represent the simplest and usually most efficient choice: because of locality, large part of the probability distribution factorizes out unchanged at each step, so that computations are cheap and convenient.

The need for more advanced algorithms may emerge essentially for two reasons, in some unlucky cases the two reasons occur at the same time. The first is when interactions are non-local: they may still couple the stochastic variables pair by pair, however the coupling is all-to-all: we will briefly discuss a case like that in Chapter 6, when considering a particular kind of Quantum Field Theories. In this situation, a local move changing a single stochastic variable will still require the evaluation of a large number of factors in the probability distribution, implying a considerable amount of numerical work for just a little step in the configuration space. This is why a non-local algorithm such as the Hamiltonian Monte Carlo discussed in section 5.1, where a large number or even all the stochastic variables are evolved at each Markov Chain step, may represent a much more convenient approach.

The second case is when, even in the presence of a local interaction, the system develops collective modes, for which a large number of stochastic variables are strongly correlated among each other on macroscopic (i.e. non-local) scales. This is a situation found for instance in statistical systems close to a phase transition, where a critical behavior takes place or different phases coexist, or also in other cases, such as in the presence of topological excitations. When this occurs, simple local algorithms may fail in exploring efficiently the configuration space, leading to a *critical slowing down* in the Markov Chain dynamics and to the need for more advanced algorithms.

5.1 Hamiltonian (Hybrid) Monte Carlo (HMC) simulations

HMC represents a whole class of Markov Chain Monte Carlo algorithms which have been originally introduced in the context of Lattice QCD simulations in 1987 [35] under the name of Hybrid Monte Carlo; it has then found application in a wide variety of stochastic problems. Nowadays it is mostly known as Hamiltonian Monte Carlo, leaving however the original acronym HMC unchanged.

The main idea is to extend the original set of stochastic variables by doubling it, after the introduction of an equal number of auxiliary stochastic variables. Each original/auxiliary variable pair is given the interpretation of a coordinate/momentum pair, in the sense of Hamiltonian dynamics, and the integration space is interpreted as a phase space. The new set of stochastic variables is given a canonical probability distribution over phase space such that, after integration over the momenta, the coordinates are distributed according to the original distribution.

This idea reveals extremely powerful, since one can develop a class of algorithms which is based on the integration of Hamilton equations and is capable of exploiting many peculiar properties of Hamiltonian dynamics. In the following, we will first expose the main lines of the algorithm and discuss the reasons underlying its success in many applications. We will then briefly sketch some recent developements and refer to the specialized literature for more details.

5.1.1   Molecular dynamics simulations and HMC

The starting problem is completely analogous to the one considered above to illustrate Langevin simulations, i.e. a set of n real stochastic variables $x_1, \ldots, x_n$ distributed according to

$$[dx]\, P(x_1 \ldots x_n) = \exp\left(-S(x_1 \ldots x_n)\right) \prod_{k=1}^{n} dx_k \,. \tag{5.1}$$

The introduction of $S = -\log P$ is convenient for the description of the HMC algorithm. On the other hand S is the primary quantity entering many stochastic problems in statistical mechanics and quantum field theory, where it represents either the energy or the action of the system. It is understood that a proper additive constant enters S (which in the language of statistical mechanics is the logarithm of the partition function), such that p is properly normalized.

At this point we introduce n auxiliary independent stochastic variables, $p_1, \ldots, p_n$ and a joint probability distribution

$$[dx]\, [dp]\, \tilde{P}(x_1 \ldots x_n, p_1 \ldots p_n) \tag{5.2}$$

whose only requirement is to reproduce the original distribution after integration over the auxiliary variables, i.e.:

$$\int [dp]\, \tilde{P}(x_1 \ldots x_n, p_1 \ldots p_n) = P(x_1 \ldots x_n)\,. \tag{5.3}$$

Under these very general constraints, the new distribution can always be rewritten as

$$[dx]\, [dp]\, \exp\left(-H(p_1 \ldots p_n, x_1 \ldots x_n)\right) \tag{5.4}$$

where

$$H = K(p_1 \ldots p_n, x_1 \ldots x_n) + S(x_1 \ldots x_n) \qquad (5.5)$$

and the function K is defined in terms of the conditional probability for p given x, i.e.

$$K = -\log \pi(p_1 \ldots p_n | x_1 \ldots x_n) \qquad (5.6)$$

since, given the constraint in Eq. (5.3), the joint probability distribution can always be written as

$$P(x_1 \ldots x_n, p_1 \ldots p_n) = \pi(p_1 \ldots p_n | x_1 \ldots x_n) \, P(x_1 \ldots x_n). \qquad (5.7)$$

In fact, in the following the very definition of the joint distribution $\tilde{P}$ is based on the definition of a proper function K. The only constraint on K, apart from the standard regularity ones, is that it must be the logarithm of a probability, hence $\int [dp] \exp(-K(x,p)) = 1$ independently of the values of the x variables, meaning that, given a generic function of the x and of the p variables, one should consider the addition of an x-dependent function to satisfy such constraint.

The Hamiltonian interpretation comes in at this point. Each couple of stochastic variables (x_k, p_k) is interpreted as conjugate coordinate/momentum pair, and the joint distribution is interpreted as the canonical distribution, over the $2n$-dimensional phase space, associated with the Hamiltonian $H = K + S$. The function S is interpreted as an interaction potential acting among the coordinates, while the function K is interpreted as the kinetic terms of the Hamiltonian, involving both momenta and coordinates[1]. In the following, in order to fix ideas for a description of the algorithm, we will consider a particular simplified form of the kinetic term, such that the momenta variables are stochastically independent of the x and normally distributed, i.e.

$$K(x,p) = K(p) = \frac{1}{2} \sum_{k,h} (M^{-1})_{k,h} \, p_k p_h + \text{const.} \qquad (5.8)$$

where the matrix M is positive definite and usually known as the mass matrix, in analogy with the Hamiltonian for a system of non-relativistic particles, and the constant term serves for normalisation and, playing no role in the description of the algorithm, will be dropped in the following. We will further assume that M is the identity matrix, so that, finally

$$K(x,p) = \frac{1}{2} \sum_k p_k^2. \qquad (5.9)$$

---

[1] Is not unusual to find examples in Classical Mechanics where coordinates and momenta are mixed in the kinetic term, think for instance of the Hamiltonian for a particle constrained on a spherical surface.

While useful for a clear description of the algorithm, this simple choice is also quite common in many practical applications. It should be stressed, however, that the choice of K is, apart from the minimal constraints exposed above, completely arbitrary, and in fact an important point when one considers the efficiency of the HMC algorithm. This issue will be discussed in more detail later on.

The microcanonical step

The purpose of the HMC algorithm is to sample the joint probability distribution $\tilde{P}$, with the idea that, when this is marginalized to the x variables, one obtains the original distribution P. Since $\tilde{P}$ is a canonical distribution, various properties of Hamiltonian dynamics can be considered to construct an efficient Monte Carlo algorithm.

In particular, one can introduce a fictitious variable t, which is considered as the evolution time of the Hamiltonian system, and then devise a step corresponding to the integration of Hamilton equations

$$\frac{dx_k}{dt} = \frac{\partial H}{\partial p_k} \quad ; \quad \frac{dp_k}{dt} = -\frac{\partial H}{\partial x_k} \tag{5.10}$$

over a finite time interval T. In this way, the stochastic variables will perform a global move along a trajectory corresponding to the solution of the fictitious equations of motion. Such a step is usually called *microcanonical*, in the statistical mechanics sense, since, for a time-independent H, Hamilton equations leave the value of the Hamiltonian unchanged, hence the system moves along a surface of constant "energy" H, as if it were exploring the micro canonical ensemble. Hamilton equations are a set of first order differential equations: given the starting point (i.e. for fixed initial conditions), the solution is unique, therefore this kind of step in the space of stochastic variables is deterministic. The step is sketched in Figure 5.1, where the phase space is represented for simplicity as a two-dimensional space. The starting point is (x, p), the trajectory brings it to the final point (x', p'), which satisfies $H(x', p') = H(x, p)$. We would like to check if this kind of step satisfies the detailed balance condition: since this is a relation between probabilities, and given that the space of stochastic variables is continuous, it is necessary to discuss in terms of portions of phase space instead of single points. For this reason, we consider a neighborhood of the starting point (x, p) of volume $dx\, dp$, as depicted in Figure 5.1, which is mapped by the Hamiltonian evolution onto a neighborhood of the final point (x', p') of volume $dx'dp'$.

Since the evolution is deterministic, the transition matrix of this microcanonical step connects the state around (x, p) to the state around (x', p') with probability equal to one, and with zero probability to any other region of phase space. It is clear that, in this form, the step cannot satisfy detailed balance, since starting from point (x', p') the time evolution will not move the stochastic variables back to (x, p), but instead farther away from that point.

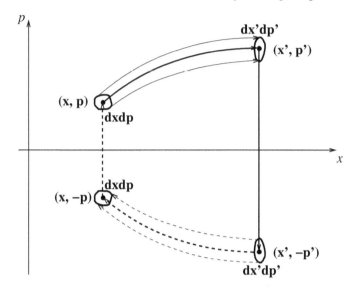

FIGURE 5.1: Sketch of the deterministic step of the HMC algorithm. The continuous line represents the foward step connecting the region of phase space around (x, p) to the region around $(x', -p')$. The dashed line represents the backward step.

In order to get back, on should reverse the evolution, i.e. the sign of time, exploiting the reversibility of Hamilton equations. If the Hamiltonian is an even function of moment, like for the choice done in Eq. (5.8), reversing the sign of time is equivalent to a change of sign for all momenta. Therefore, as a simple adjustment of the microcanonical step, we add a change of sign of all momenta at the end of the evolution trajectory, as depicted in Figure 5.1. In this way the new microcanonical step will bring the region around (x, p) with probability 1 to the region around $(x, -p)$, and the latter back to the original region, again with probability 1.

Now, given that the element of the transition matrix connecting the two regions is one in both directions, the verification of detailed balance, i.e. Eq. (3.29), consists simply in checking that the two regions have equal probabilities according to the canonical distribution that we want to sample, i.e. that:

$$dx\, dp\, e^{-H(x,p)} = dx'\, dp'\, e^{-H(x',-p')} \,. \tag{5.11}$$

The condition $H(x, p) = H(x', -p')$ is guaranteed by the conservation of the Hamiltonian under the time evolution and by the fact that H is an even function of the momenta. The condition $dx\, dp = dx'\, dp'$ is instead guaranteed by another important property of Hamilton equations, usually known under the name of *Liouville theorem*: Hamiltonian time evolution leaves the volume of phase space unchanged.

We would like to stress that the sign flip of momenta at the end of the trajectory is just a trick, which is irrelevant to the evolution of the x variables and has been introduced just to ensure detailed balance. An alternative trick could have been to allow with equal probability for a forward or backward evolution at the beginning of each step. However, the algorithm would have been fine even without any adjustment, satisfying the general balance condition in Eq. (3.28) rather than the detailed one. In fact, it is well known that the canonical distribution is an equilibrium distribution for Hamilton equations.

The Heat-bath step

In the present form, the algorithm is useless, since it is manifestly non-ergodic. The process illustrated in Figure 5.1 would bring the system back to the original position every other step: that could be cured by allowing for a random time evolution T at each step, possibly also changing stochastically the direction of evolution, forward or backward in time. However, the main problem is that the process is microcanonical, i.e. the best one could obtain is a random walk uniformly covering a $(2n-1)$-dimensional hypersurface of the phase space at fixed value of the Hamiltonian (isoenergetic).

For this reason, the microcanonical move must be alternated with another step, capable of moving the system stochastically across different isoenergetic hypersurfaces. When the kinetic term takes the simple form in Eq. (5.8), one can devise an efficient and simple heat-bath step which changes all momenta variables at the same time while leaving the x variables unchanged. Indeed, the p variables are normally distributed, and we already know how to obtain independent draws out of a Gaussian distribution with relative ease. When K takes the form in Eq. (5.9), each momentum p_k is distributed according to $dp_k \exp(-p_k^2/2)$, i.e. as a normal variate with unit variance, and can be drawn using, e.g. the Box-Müller algorithm. However, even when the mass matrix M is non-trivial, one can draw a vector of unit normal variates η_k and then set $p_k = \sqrt{M}_{k,h}\, \eta_h$ (the square root is well defined since M is positive definite).

The combination of the heat-bath step with the microcanonical one leads, at least in principle, to an efficient algorithm, which is capable on one hand to rapidly explore an isoenergetic surface by global and coherent moves of all coordinates and momenta (the duration T of the time evolution can be tuned to optimize this step), and on the other hand to efficiently move from one isoenergetic surface to the other by a global heat-bath algorithm.

There is, however one question we have not considered yet: are we able to exactly integrate Hamilton equations (5.10) over a finite time interval T? That of course depends on the interaction function $S(x_1 \ldots x_n)$, however the answer is that, for the cases in which an exact solution is known, one would not resort to such a complex weapon like the HMC. For instance, the best known exactly solvable system is a collection of harmonic oscillators; however, in that case S would be a quadratic function of the x variables, in which case the original distribution P would be Gaussian and could be sampled by much simpler means.

Therefore, in practical applications of the HMC algorithm, one is compelled to resort to a numerical integration of Hamilton equations. The numerical integration of differential equations is almost exact, meaning that small violations of energy conservation and of phase space volume conservation are typically present, leading to a violation of detailed balance. As we are going to show in the following, both problems can be cured effectively, turning the microcanonical step into an efficient Metropolis step.

5.1.2   Practical implementation of the HMC algorithm

The simplest method to integrate numerically the first order differential equations (5.10) was devised by Euler a few centuries ago. One divides the total evolution time T in N finite intervals $\Delta t = T/N$, defining an elementary discrete evolution step as follows

$$x_k(t + \Delta t) = x_k(t) + \Delta t \, \frac{\partial H}{\partial p_k}(x(t), p(t))$$

$$p_k(t + \Delta t) = p_k(t) - \Delta t \, \frac{\partial H}{\partial x_k}(x(t), p(t)) \tag{5.12}$$

which is then iterated N times. The problem is that now, for each time step, both energy conservation and phase space volume conservation are only satisfied at the first order in Δt. Indeed, by Taylor expansion of the Hamiltonian computed as a function of the variables at time $t + \Delta t$, one finds

$$
\begin{aligned}
H(t + \Delta t) &= H(x(t), p(t)) + \Delta t \sum_k \left(\frac{\partial H}{\partial x_k} \frac{\partial H}{\partial p_k} - \frac{\partial H}{\partial p_k} \frac{\partial H}{\partial x_k} \right) + O(\Delta t^2) \\
&= H(x(t), p(t)) + O(\Delta t^2)
\end{aligned}
\tag{5.13}
$$

where $O(\Delta t^2)$ corrections are in general non-zero. In the same way, for the determinant of the Jacobian matrix, which fixes how volumes transform, one can write, exploiting the fact that at the first order in Δt only diagonal elements contribute,

$$
\begin{vmatrix}
\frac{\partial x_k(t+\Delta t)}{\partial x_h(t)} & \frac{\partial x_k(t+\Delta t)}{\partial p_h(t)} \\
\frac{\partial p_k(t+\Delta t)}{\partial x_h(t)} & \frac{\partial p_k(t+\Delta t)}{\partial p_h(t)}
\end{vmatrix}
= 1 + \Delta t \sum_k \left(\frac{\partial^2 H}{\partial x_k \partial p_k} - \frac{\partial^2 H}{\partial p_k \partial x_k} \right) + O(\Delta t^2)
$$

$$
= 1 + O(\Delta t^2)
\tag{5.14}
$$

so that the phase space measure remains invariant at order Δt, but $O(\Delta t^2)$ are in general non-zero. Along a trajectory made up of N steps3, both the energy and the phase space volume will undergo a change of the order of $N\Delta t^2 \propto \Delta t$. The adoption of higher order integrators, instead of the Euler one, could make the violation smaller and of higher order in Δt, but in general it will not vanish.

Another drawback of the Euler integrator is that, contrary to the continuous time evolution, it is not reversible. This is clearly visible from the elementary time step in Eq. (5.12): if we start from $x_k(t+\Delta t)$ and $p_k(t+\Delta t)$ and change the sign of Δt, we will not end up exactly in $x_k(t)$ and $p_k(t)$, since now the derivatives of H will be computed at time $t+\Delta t$.

To summarize, let us reconsider the microcanonical step depicted in Figure 5.1 and see what are the effects of adopting a finite step numerical integrator to evolve the system from (x,p) to (x',p'):

1. $H(x',p') \neq H(x,p)$;

2. $dx'\,dp' \neq dx\,dp$;

3. the evolution is still deterministic; however, in general it is not reversible any more.

Each of the three modifications above ruins the proof of detailed balance: Eq. (5.11) is violated because of points 1) and 2); moreover, while the region around (x,p) goes for sure to the region around some $(x',-p')$, we are not sure that the latter goes back to the former. Let us see how the problems above can be cured.

Let us suppose we have been able to devise some improved integrator which solves the reversibility problem, we will come back to this issue later one. Then, the fact that $dx\,dp\,e^{-H(x,p)} \neq dx'\,dp'\,e^{-H(x',-p')}$ could be taken into account by considering the discrete time evolution trajectory simply as the proposal move for a Metropolis algorithm, accepting the step with probability

$$r = \frac{dx'\,dp'\,e^{-H(x',-p')}}{dx\,dp\,e^{-H(x,-p)}} = \frac{dx'\,dp'}{dx\,dp}\,\exp(-\Delta H)\,. \tag{5.15}$$

However, while the energy difference ΔH between the final and the initial configuration can be easily computed, the information about the amount of volume change, $dx'\,dp'/(dx\,dp)$, is in general not easily accessible. For that reason, it is essential that one can devise numerical algorithms for which at least the conservation of phase space volume is guaranteed for any finite time step Δt. Luckily enough, the mathematical structure of phase space in Hamiltonian mechanics permits a precise strategy to built a class algorithm sharing this property, which are known under the name of *symplectic integrators*.

The idea is that phase space volume conservation is a property that is not only shared by continuous time evolution, but by the more general class of finite transformations known as *canonical* (or *symplectic*) transformations. The Jacobian matrix of a canonical transformation is symplectic, hence it has unit determinant.

The discrete transformation in Eq. (5.12), which leads from $x_k(t), p_k(t)$ to $x_k(t+\Delta t), p_k(t+\Delta t)$, is not a canonical transformation[2], as can be checked

---

[2]Let us consider for instance the Poisson bracket $\{x_k(t+\Delta t), p_h(t+\Delta t)\}$, which should

by computing the Poisson brackets among the new coordinates and momenta after the transformation, which are canonical up to corrections of $O(\Delta t^2)$. However, the Euler integrator can be turned into a canonical transformation by a simple modification, which consists of computing derivatives not exactly at time t, but as a function of the coordinates at time t and momenta at time $t + \Delta t$, or viceversa, i.e. as follows:

$$
\begin{aligned}
x_k(t + \Delta t) &= x_k(t) + \Delta t \, \frac{\partial H(x(t), p(t + \Delta t))}{\partial p_k(t + \Delta t)} \\
p_k(t + \Delta t) &= p_k(t) - \Delta t \, \frac{\partial H(x(t), p(t + \Delta t))}{\partial x_k(t)} .
\end{aligned}
\tag{5.17}
$$

or alternatively

$$
\begin{aligned}
x_k(t + \Delta t) &= x_k(t) + \Delta t \, \frac{\partial H(x(t + \Delta t), p(t))}{\partial p_k(t)} \\
p_k(t + \Delta t) &= p_k(t) - \Delta t \, \frac{\partial H(x(t + \Delta t), p(t))}{\partial x_k(t + \Delta t)} .
\end{aligned}
\tag{5.18}
$$

The simplest way to prove that the transformations above are indeed canonical is by checking that there is a generating function associated with them[3].

The above prescription seems to be not easily applicable in practice: the dependence of the new coordinates and momenta is given in an implicit way, i.e. one needs to already know some of their values at time $t + \Delta t$ to really

---

be δ_{kh} were the transformation canonical. From Eq. (5.12), after a few easy computations, we obtain

$$
\begin{aligned}
\{x_k(t + \Delta t), p_h(t + \Delta t)\} &\equiv \sum_l \left(\frac{\partial x_k(t + \Delta t)}{\partial x_l(t)} \frac{\partial p_h(t + \Delta t)}{\partial p_l(t)} - \frac{\partial x_k(t + \Delta t)}{\partial p_l(t)} \frac{\partial p_h(t + \Delta t)}{\partial x_l(t)} \right) \\
&= \delta_{kh} + \Delta t^2 \left(\frac{\partial^2 H}{\partial x_l \partial p_k} \frac{\partial^2 H}{\partial p_l \partial x_h} - \frac{\partial^2 H}{\partial p_l \partial p_k} \frac{\partial^2 H}{\partial x_l \partial x_h} \right)
\end{aligned}
\tag{5.16}
$$

where the Δt^2 term is computed at time t and in general is non-zero. Similar considerations apply to $\{x_k(t + \Delta t), x_h(t + \Delta t)\}$ and $\{p_k(t + \Delta t), p_h(t + \Delta t)\}$, which should vanish but are instead proportional to Δt^2.

[3] For instance, for the transformation in Eq. (5.17), one finds a type II generating function (i.e. function of the old coordinates and of the new momenta) as follows:

$$
F_2(x(t), p(t + \Delta t)) = \sum_k x_k(t) p_k(t + \Delta t) + \Delta t \, H(x(t), p(t + \Delta t))
$$

which indeed gives

$$
\begin{aligned}
x_k(t + \Delta t) &= \frac{\partial F_2}{\partial p_k(t + \Delta t)} = x_k(t) + \Delta t \, \frac{\partial H(x(t), p(t + \Delta t))}{\partial p_k(t + \Delta t)} \\
p_k(t) &= \frac{\partial F_2}{\partial x_k(t)} = p_k(t + \Delta t) + \Delta t \, \frac{\partial H(x(t), p(t + \Delta t))}{\partial x_k(t)} .
\end{aligned}
$$

Similarly, for the transformation in Eq. (5.18), one finds a type IV generating function $F_4(x(t + \Delta t), p(t))$.

implement the step. However, the integrator can be made explicit quite easily if the Hamitonian has a separate dependence on coordinates and momenta

$$H(x, p) = K(p) + S(x) \tag{5.19}$$

like it happens for the choice of the kinetic term done in Eq.(5.8). Indeed, in this case, Eqs. (5.17) and (5.18) can be rewritten as follows:

$$p_k(t + \Delta t) = p_k(t) - \Delta t \frac{\partial S(x(t))}{\partial x_k(t)}$$
$$x_k(t + \Delta t) = x_k(t) + \Delta t \frac{\partial K(p(t + \Delta t))}{\partial p_k(t + \Delta t)} \tag{5.20}$$

$$x_k(t + \Delta t) = x_k(t) + \Delta t \frac{\partial K(p(t))}{\partial p_k(t)}$$
$$p_k(t + \Delta t) = p_k(t) - \Delta t \frac{\partial S(x(t + \Delta t))}{\partial x_k(t + \Delta t)} \tag{5.21}$$

and the two updating steps for the p's and for the x's can be performed one after the other without any interference, provided they are written in the code exactly in the same order in which they are written in the text[4]. Notice in particular that, in the second case, we had to exchange the order, so that each step makes use of already computed quantities.

The only missing point seems to be reversibility: indeed, none of the two symplectic integrators brings back to the starting point when $\Delta t \rightarrow -\Delta t$. However, if we solve Eq. (5.20) for the original variables, we obtain

$$x_k(t) = x_k(t + \Delta t) - \Delta t \frac{\partial K(p(t + \Delta t))}{\partial p_k(t + \Delta t)} \tag{5.22}$$
$$p_k(t) = p_k(t + \Delta t) + \Delta t \frac{\partial S(x(t))}{\partial x_k(t)}$$

which is nothing but the elementary step (5.21) after sending $\Delta t \rightarrow -\Delta t$, i.e. the two steps (5.20) and (5.21) are, modulo an inversion of time evolution, the reverse of each other. That gives the opportunity to devise an elementary and reversible step of size Δt by combining in sequence the two different steps,

---

[4]It is easy to prove that the elementary steps (5.20) and (5.21) preserve phase space volume even without making explicit reference to canonical transformations. Indeed, each of them is made up of two sub-steps which are separately volume preserving. For instance, the first sub-step of (5.20) leaves the x_k unchanged and just changes the p_k by x-dependent functions, from that one easily proves that the associated Jacobian matrix is triangular with unit determinant.

each taken of size $\Delta t/2$. That leads to the so-called *leapfrog* integrator:

$$p_k(t + \Delta t/2) = p_k(t) - \frac{\Delta t}{2} \frac{\partial S(x(t))}{\partial x_k(t)}$$

$$x_k(t + \Delta t) = x_k(t) + \Delta t \frac{\partial K(p(t + \Delta t/2))}{\partial p_k(t + \Delta t/2)} \tag{5.23}$$

$$p_k(t + \Delta t) = p_k(t + \Delta t/2) - \frac{\Delta t}{2} \frac{\partial S(x(t + \Delta t))}{\partial x_k(t + \Delta t)}$$

which is the most elementary symplectic and reversible integrator for a set of canonical evolution equations. It should be stressed that the theory underlying the numerical integration of the equations of motion of complex systems, which we are touching here just to adjust one part of our Monte Carlo algorithm, is of interest for a vast amount of research fields, going from molecular dynamics to celestial mechanics. In those contexts, Eq. (5.24) is also known under the name of *velocity Verlet* algorithm. The name *leapfrog* instead derives from the fact that, when Eq. (5.24) is iterated several times, the x and the p variables stay alternatively half a step ahead of each other in the evolution time, like in the leapfrog game.

It is easy to check that the leapfrog step is indeed reversible, i.e. that starting from the variables at time $t + \Delta t$ and iterating it for a time step $-\Delta t$ one goes back to the original variables[5]. Apart from reversibility, which is the major gain, there is another advantage obtained by the leapfrog algorithm: energy violation after one step is not of order Δt^2, like for the Euler integrator or for the non-reversible symplectic integrators (5.17) and (5.18), but instead of order Δt^3. One could check that explicitly after a few tedious computations, however it is easier and better to realize that this is actually a gift of reversibility. Indeed, by reversibility, the Hamiltonian must go back to its original value by re-doing the step after $\Delta t \to -\Delta t$, hence the lowest order in the integration error must be odd in Δt, and since it is not $O(\Delta t)$ for sure, it must be at least of order $O(\Delta t^3)$. The leapfrog is in fact of order $O(\Delta t^3)$, as can be checked by doing the tedious computations mentioned above[6]. That means that, when

---

[5]In actual implementations of the algorithm, the computation of derivatives of S and K may require approximations or involve rounding errors. One should make sure that such approximations do not imply any bias towards forward or backward propagation, so that reversibility is preserved.

[6]One way to simplify computations is by the so-called *splitting* approach. Time evolution can be viewed as the action of differential operators written in terms of Poisson brackets:

$$x_k(t + dt) = x_k(t) + dt\{x_k, H\} = x_k(t) + dt\{x_k, K + S\} = (1 + dt\{., K + S\}) x_k(t)$$
$$p_k(t + dt) = p_k(t) + dt\{p_k, H\} = p_k(t) + dt\{p_k, K + S\} = (1 + dt\{., K + S\}) p_k(t).$$

The exact integration for a finite time interval can be exponentiated, so that the classical evolution operator looks very similar to the evolution operator in quantum mechanics: $U(\Delta t) \equiv \exp(\Delta t\{., K + S\})$. It is formally defined by the Taylor expansion of the exponential and it is usually difficult to explicit it in a finite form. However, when $K = K(p)$ and $S = S(x)$ it can be split in the product of single exponentials which evolve separately the x's or the p's, plus an error which can be computed by the standard Baker-Campbell-Hausdorff

considering the whole trajectory, which is made up of $N = T/\Delta t$ steps, the total expected integration error is, apart from fortuitous cancellations, of the order of $N\Delta t^3 \propto \Delta t^2$; for this reason one says that the leapfrog is a second order integrator.

At this point, let us summarize the main steps of the HMC algorithm:

1. Conjugate momenta p_k are refreshed completely by a global heat-bath step, while coordinates are kept fixed;

2. Coordinates and momenta are evolved simultaneously, $(x, p) \rightarrow (x', p')$, for a fictitious time interval T by integrating the canonical equations, e.g. by performing $N = T/\Delta t$ steps of the leapfrog algorithm in Eq. (5.24). This part, because of its similarity with the numerical integration performed for many body systems, is usually called the Molecular Dynamics (MD) evolution;

3. A Metropolis test is performed according to the probability ratio $\exp(H(x, p) - H(x', p'))$ and, if successful, the new state (x', p') is accepted.

We have skipped here the momenta inversion $p'_k \rightarrow -p'_k$ which is only instrumental to prove detailed balance and completely irrelevant otherwise.

5.1.3 Effectiveness of the HMC algorithm and further improvements

The HMC algorithm is in general indicated for systems with many degrees of freedom, where the probability density is strongly localized in restricted regions of the stochastic variables space, and when strong non-local interactions (correlations) are present between the stochastic variables, which prevent the optimal use of other standard algorithms, like the local ones. Indeed, the algorithm has been invented for the Monte Carlo simulation of Lattice QCD in the presence of dynamical fermions, where a typical non-local object like the determinant of a very large and sparse matrix appears in the probability distribution of gauge fields.

The success of the algorithm is mostly due to the fact that it permits a global move of the stochastic variables along directions which keep the system in a region relevant to the original distribution $S(x)$. Parameters like the total length of the trajectory, T, and the integration step, Δt, can be optimized so

---

formula, very similarly to what happens for the transfer matrix in quantum mechanics. In particular, the leapfrog integrator corresponds to the following splitting approximation:

$$U(\Delta t) = \exp(\Delta t/2\{., S(x)\}) \exp(\Delta t\{., K(p)\}) \exp(\Delta t/2\{., S(x)\}) + O(\Delta t^3).$$

where the single exponentials, being dependent either on the x's or on the p's, give exact linear evolutions, i.e. higher order terms in the exponential expansions vanish exactly.

that the move is as long as possible while keeping a reasonable high acceptance and a contained computational effort.

An important point is also that the HMC is a beautiful and complex object, which is amenable to many further improvements, some of which can be potentially dramatic. This is certainly not the right place to make a detailed treatment of the most recent advancements, some of which are still the subject of ongoing research. However we would like to shortly mention a few lines of development with respect to the basic treatment given above.

1. One can make and improved choice, with respect to the leapfrog algorithm, for the MD integrator. Indeed, a careful sub-step division of the whole time step Δt can lead to the elimination of $O(\Delta t^3)$ (or even higher order) corrections to energy conservation, leading to improved integrators for which a reduced number of MD steps (i.e. a larger Δt) is needed to keep a good level of acceptance during the Metropolis test, thus saving, at least in principle, computational time. Of course that could make the elementary Δt step significantly more complex from the computational point of view, so that an optimal balance should be found. A good choice is for instance the one proposed by Omelyan *et al* in 2002 [36, 37], where a significant improvement is obtained, with a relatively simple modification of the algorithm, even by just reducing the prefactor in front of the Δt^3 term.

2. The actual Hamiltonian used to define the equations of motion does not need to coincide with the one used in the Metropolis step: the latter is constrained, while the former can be changed, if this is convenient, without affecting detailed balance. This simple observation is at the basis of the following possible improvement.

   Since symplectic integrators are exact canonical transformations, one can always find, for any given value of Δt, a modified Hamiltonian $\tilde{H}_{\Delta t}$, usually known as the *shadow Hamiltonian*, which is exactly conserved along the curves defined by the chosen integrator. Using this fact, one could try to modify the Hamiltonian used to define the equations of motion, so that the shadow Hamiltonian associated with it is closer to the actual Hamiltonian used in the Metropolis test [38].

3. As we have stressed at the beginning, the choice of the kinetic term is completely arbitrary, apart from the normalisation condition $\int [dp] \exp(-K(x,p)) = 1$. One can try to exploit this freedom to improve the way the equations of motion explore the probability distribution.

   One successful modification is the one known as Riemann Manifold HMC (RMHMC) [39], which consists simply in allowing for a dependence on the x coordinates of the mass matrix M which defines K in Eq. (5.8), like it happens for the kinetic term of a mechanical system in curvilinear coordinates.

In principle, a careful choice of the mass matrix can be very effective to define equations of motion which better follow the particular geometry of the given probability distribution. There are a few drawbacks, however, which should be taken into account: *(i)* the mixed dependence on the x's and the p's of the kinetic term does not allow for explicit symplectic integrators, like the leapfrog, so that one has to resort to more complex *implicit* integrators; *(ii)* a generic x-dependent mass matrix will violate the normalisation condition of the kinetic term, since a determinant of M will appear after over the integration over the p's, which must be properly compensated by an *ad hoc* modification of the S function.

A particularly appealing application of this scheme, which makes worth the effort when one has to update an extended system with long-range and slow modes (like a field theory or a statistical system close to a critical point), is the tentative to differentiate the evolution of different Fourier modes, by making the effective mass in the kinetic term mode-dependent, in order to speed up the slowest ones. While the general idea goes back to a few decades ago [40], an application to the HMC algorithm within the general scheme of RMHMC is only recent, with promising results [41, 41].

4. Finally, one can think of modifications of the accept/reject algorithm which might violate detailed balance while preserving the general balance condition. This is the case of the so-called Look Ahead HMC algorithm (LAHMC) [42], which allows for the possibility of continuing the evolution along a trajectory which would have been rejected otherwise.

5.1.4 An example of the HMC implementation

We would like to close this brief description of the HMC algorithm by discussing a simple example. Let us consider a probability distribution for two stochastic variables x, y, defined by

$$P(x, y)\, dx\, dy = \exp(-S(x))\, dx\, dy \tag{5.24}$$

where

$$S(x, y) = A \left(\sqrt{x^2 + y^2} - R \right)^2 \tag{5.25}$$

where A and R are given constants. In practice, this is a two-dimensional distribution which is peaked, with width $\sim 1/\sqrt{A}$, around a circumference of radius R.

We make the following choice of the Hamiltonian

$$H(x, y, p_x, p_y) = \frac{p_x^2}{2} + \frac{p_y^2}{2} + S(x, y) \tag{5.26}$$

and define the *force* by

$$f_x = -\frac{\partial S}{\partial x} = -2A\left(1 - \frac{R}{\sqrt{x^2 + y^2}}\right)x \,,$$

$$f_y = -\frac{\partial S}{\partial y} = -2A\left(1 - \frac{R}{\sqrt{x^2 + y^2}}\right)y \,.$$

To shorten the notation, it is convenient to introduce vector notation:

$$\vec{p} = \begin{pmatrix} p_x \\ p_y \end{pmatrix} \equiv \begin{pmatrix} p_1 \\ p_2 \end{pmatrix}, \qquad \vec{x} = \begin{pmatrix} x \\ y \end{pmatrix}, \qquad \vec{f} = \begin{pmatrix} f_x \\ f_y \end{pmatrix}.$$

Starting from an ensemble member $\vec{x}_0$, we now need to generate a new *relevant* random vector $\vec{x}$. The HMC process for this task is as follows:

1. Choose a random momentum vector $\vec{p}_0$ with probability distribution $\propto \exp\{-p_x^2/2 - p_y^2/2\}$.

2. Evolve the set $\vec{p}_0, \vec{x}_0$ to $\vec{p}, \vec{x}$ at some HMC time t using the equations of motion:

$$\frac{d}{dt}\vec{p} = \vec{f} \,, \qquad \frac{d}{dt}\vec{x} = \vec{p} \,.$$

The latter equation is 1st order Ordinary Differential Equation (ODE). We have an initial value problem at hand:

$$\vec{p}(0) = \vec{p}_0 \,, \qquad \vec{x}(0) = \vec{x}_0 \,.$$

3. Stop after some time t and consider $\vec{x}(t)$ as possible new ensemble members. Finally, accept proposal with probability

$$\min\left(\exp\{-H(\vec{p}(t), \vec{x}(t)) + H(\vec{p}(0), \vec{x}(0))\}, 1\right) \,.$$

If we could carry out step 2 exactly, we would not need step 3 since the Hamilton function is a constant of motion, i.e. does not depend on t. This is a general result already discussed above; however, it is instructive to check it explicitly for this particular example, by calculating:

$$\frac{d}{dt}H\left(\vec{p}(t), \vec{x}(t)\right) = \sum_{i=1}^{2}\left[\frac{\partial H}{\partial p_i}\frac{dp_i}{dt} + \frac{\partial H}{\partial x_i}\frac{dx_i}{dt}\right] = \sum_{i=1}^{2}\left[p_i\frac{dp_i}{dt} - f_i\frac{dx_i}{dt}\right]$$

$$= \sum_{i=1}^{2}[p_i f_i - f_i p_i] = 0 \,,$$

where we have used

$$\frac{\partial H}{\partial x_i} = \frac{\partial S}{\partial x_i} = -f_i$$

and the equation of motion.

Only in rare cases, it is possible to integrate the ODE in step 2 exactly. In all other cases, as discussed above, we need to resort to numerical integration. To this aim, we decompose the HMC time t into steps of size Δt: $t = n_h \Delta t$ and use a difference scheme to estimate the new ensemble member $\vec{x}(t)$. We then only obtain an approximation afflicted by discretisation errors, which would vanish with $\Delta t \to 0$. The beauty of the HMC approach is that the approximation for $\vec{x}(t)$ is still an ensemble member thanks to the accept-reject step 3 above. For the approach to be exact, we only need that the numerical integrator preserves phase space volumes exactly, and time reversal symmetry

$$\vec{p}_0, \vec{x}_0 \xrightarrow{\Delta t} \vec{p}, \vec{x} \xrightarrow{-\Delta t} \vec{p}_0, \vec{x}_0.$$

The price to pay is that the Hamilton Function is no longer strictly independent and the probability to accept (see step 3 above) can be small if we choose Δt too large. Numerical schemes that preserve volumes can be found within the class of *symplectic integrators* discussed above, which are generally also good in minimizing violations to energy conservation. An easy to implement algorithm is the leapfrog integration discussed previosuly, for which the elementary step from t to $t + \Delta t$ works as follows:

$$\vec{p}(t + \Delta t/2) = \vec{p}(t) + \frac{\Delta t}{2} \vec{f}\big(\vec{x}(t)\big),$$
$$\vec{x}(t + \Delta t) = \vec{x}(t) + \Delta t\, \vec{p}(t + \Delta t/2),$$
$$\vec{p}(t + \Delta t) = \vec{p}(t + \Delta t/2) + \frac{\Delta t}{2} \vec{f}\big(\vec{x}(t + \Delta t)\big).$$

Observe that these elementary steps can be efficiently stack together since:

$$\vec{p}(t + 3\Delta t/2) = \vec{p}(t + \Delta t) + \frac{\Delta t}{2} \vec{f}\big(\vec{x}(t + \Delta t)\big)$$
$$= \vec{p}(t + \Delta t/2) + \frac{\Delta t}{2} \vec{f}\big(\vec{x}(t) + \Delta t\big) + \frac{\Delta t}{2} \vec{f}\big(\vec{x}(t + \Delta t)\big)$$
$$= \vec{p}(t + \Delta t/2) + \Delta t\, \vec{f}\big(\vec{x}(t + \Delta t)\big).$$

Let us study the HMC approach for the example at hand. We choose the parameters of the action as

$$A = 10 \qquad R = 2.$$

The corresponding action is shown in Figure 5.2. It is rotational symmetric and has a global minimum for a circle with radius $R = 2$. Any Monte Carlo configuration (x, y) is a point in the xy-plane. The challenge is that the MC ensemble needs to spread out across the circle, which is the area of high probability density. It is expected that small-step MC experiences sizeable autocorrelations.

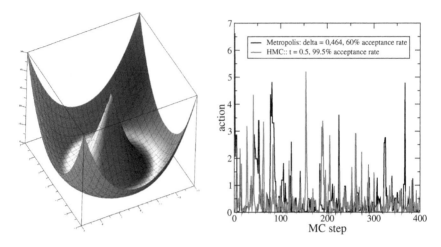

FIGURE 5.2: Left panel: The action $S(x, y)$. Right panel: MC time series for the action S for the Metropolis algorithm (black) and the 10-step HMC simulation (red).

We will benchmark our findings against those from a standard Metropolis-Hastings approach. Starting at the point (x_0, y_0), which can be viewed as a MC configuration, a new configuration (x, y) is proposed with probability:

$$(P(x, y) \propto \exp \left\{ -\frac{(x - x_0)^2}{\delta_M^2} - \frac{(y - y_0)^2}{\delta_M^2} \right\} .$$

The new configuration is then accepted with probability

$$p = \min \left(\exp\{-S((x, y) + S(x_0, y_0)\}, 1 \right) .$$

We fine-tune the "step size" δ_M such that we achieve an acceptance rate of approximately 60% and find $\delta_m = 0.464$. Starting point is $(0.1, 0)$ and we perform $1,000$ MC steps for thermalisation. The time series for the action S as a function of the MC step number is also shown in Figure 5.2 (right panel).

We compare results with a 10-step HMC simulation, $n_h = 10$ and vary Δt to achieve a spread of the HMC time t via $t = n_h \, \Delta t$. The result for the action time series for $t = 0.5$ is shown in Figure 5.2 (right panel, red). Note that the HMC approach has an acceptance rate of 99.5% (compared to the 60% of the Metropolis algorithm). Figure 5.3 shows the acceptance rate of the 10-step HMC simulation as a function of the HMC time t. The first observation is that it is generically high thanks to the leapfrog integrator. HMC simulations can show a periodic behaviour with HMC time, which implies that large t do not necessarily mean more uncorrelated MC configurations. This can be seen

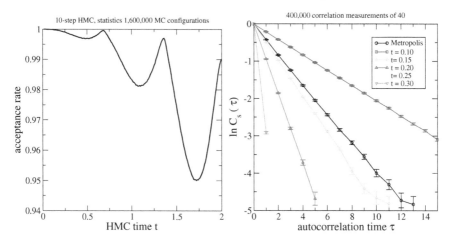

FIGURE 5.3: Left panel: The acceptance rate for the 10-step HMC simulation as function of the HMC time t. Right panel: Autocorrelation functions $C_s(\tau)$ for the action S as a function of the autocorrelation time τ for the Metropolis algorithm (black) and several 10-step HMC simulations with different lengths τ of the HMC trajectory.

in the rate function, when the acceptance rate returns to very high values after a decline. As a rule of thumb, we try to choose t small enough to avoid this periodic behaviour, which only costs computational resources. On the other hand, we would like to choose t large enough to generate configuration that are "enough" statistically independent. The latter can be monitored by estimating autocorrelation functions.

To quantify the difference in performance of Metropolis and HMC simulations, we calculate the autocorrelation function of the action. If $S(t)$ denotes the action evaluated with configurations $(x(t), y(t)$ at HMC time t, the auto-correlation function is defined as

$$C_s(\tau) = \frac{\left\langle \Big(S(t) - s\Big)\Big(S(t + \tau) - s\Big)\right\rangle}{\left\langle \Big(S(t) - s\Big)^2\right\rangle} . \qquad (5.27)$$

Note that after thermalisation, the autocorrelation time does not depend on t. Thereby, we call the *relative* HMC time τ " autocorrelation time". Also note that s is the action expectation value:

$$s := \langle S(t) \rangle , \qquad (5.28)$$

which also does not depend on t. Our numerical findings for several autocorrelation functions are summarised in Figure 5.3, right panel. The black symbols show the result for the Metropolis simulation with 60% acceptance rate. Note that if the HMC trajectory length t is as small as 0.1, the Metropolis algorithm outperforms HMC. On the other hand, the slope of the autocorrelation functions rapidly rises with an increase and already for $t = 0.3$, which still enjoys an excellent acceptance rate of 99.86%, any correlations between MC samples are eliminated.

Actually, the present comparison between Metropolis and HMC could be to a large extent unfair. First of all, we have not tried any optimisation on the Metropolis; moreover, the comparison does not take into account in any way the different "computational cost" involved in each step of the two algorithms. A more careful comparison is left as an exercise for the reader (see, in particular, exercise 5.2).

5.2 Non-local Monte Carlo update

The update of single degrees of freedom (local update) is often dictated by the requirement for a significant acceptance rate (importance sampling). An approach that allows updating a group of fundamental degrees of freedom (cluster update) in one Monte Carlo step usually leads to an unparalleled performance. Such an approach is not always possible. In this section, we will introduce the concept of cluster update algorithms using examples where this approach is feasible.

5.2.1 Critical slowing down

Although in an ideal MC simulation, the sets of configurations, which are used to estimate the observables of interest, are *statistically independent*. If M_i is the observable evaluated with set i, independence means that

$$\langle M_i\, M_{i+k}\rangle \; - \; \langle M_i\rangle\,\langle M_k\rangle \; = \; \delta_{ik}\,\sigma_M^2\,. \tag{5.29}$$

We have assumed that the system has been sufficiently thermalised such that we can assume translational invariance, i.e. independence of i. In practice, the configurations are afflicted by memory effects, and, instead of (5.29), we find:

$$\langle M_i\, M_{i+k}\rangle \; - \; \langle M_i\rangle\,\langle M_k\rangle \; = \; C(i-k)\,. \tag{5.30}$$

Generically, the autocorrelation function $C(x)$ decreases monotonically with $|x|$. The amount of autocorrelations can be quantified by the so-called *integrated autocorrelation time* (see section 3.5 and in particular (3.48)):

$$\tau_{\text{int}} \; = \; \frac{1}{\sigma_M^2}\sum_{k=1}^{\infty} C(k)\,. \tag{5.31}$$

The impact of autocorrelations is to reduce the signal-to-noise ratio of the estimators. If we estimate the observable $\langle M \rangle$ by N MC measurements, i.e.

$$\langle M \rangle \approx \frac{1}{N} \sum_{i=1}^{N} M_i \, ,$$

the variance for the estimator is given by (see section 3.5 for a derivation)

$$\sigma^2 = \frac{\sigma_M^2}{N} \left(1 + 2\,\tau_{\text{int}} \right) . \tag{5.32}$$

This can be intuitively understood for a simple example: assume that we have devised an MC update that copies a given configuration m times and then generate a completely independent configuration. In this simulation, the sequence of observables breaks down into chunks of m equals (here for $m = 6$):

$$M_1 \; M_1 \; M_1 \; M_1 \; M_1 \; M_1 \; M_2 \; M_2 \; M_2 \; M_2 \; M_2 \; M_3 \; M_3 \; M_3 \; M_3 \; M_3 \; M_3 \; M_4 \; M_4 \; \ldots \; .$$

Let us calculate the correlation function: assume i is selects the i-th position of one chunk. We would then find that $M_i = M_{i+k}$ for $k = 1 \ldots (m - i)$. For $k > m - i$, the M_{i+k} values are independent of M_i. If we would always choose the i-th position within one chunk, we would find:

$$\langle M_i \, M_{i+k} \rangle - \langle M_i \rangle \, \langle M_k \rangle_i == \sigma_M^2 \, \theta\Big(k - (m - i)\Big) \, ,$$

where $\theta(x) = 1$ for $x > 0$, and $\theta(x) = 0$ else. In the simulation, we would choose any position within one chunk with equal probability. Hence, we find:

$$C(i - k) = \sigma_M^2 \, \frac{1}{m} \sum_{i=1}^{m} \theta\Big(k - (m - i)\Big) = \sigma_M^2 \, \frac{k}{m} \, \theta\big(m - k\big) .$$

Hence, the integrated autocorrelation time for our example is given by

$$\tau_{\text{int}} = \frac{1}{m} \sum_{k=1}^{m-1} k = \frac{m - 1}{2} .$$

According to (5.32), the variance for an observable is thus given by

$$\sigma^2 = \frac{\sigma_M^2}{N} \, m = \frac{\sigma_M^2}{(N/m)} . \tag{5.33}$$

This is a very intuitive result: if our simulation contains $N_c = N/m$ chunks of same values M_i, we find for the estimator:

$$\langle M \rangle \approx \frac{1}{N} \sum_{i=1}^{N} M_i = \frac{1}{N_c \, m} \sum_{k=1}^{N_c} m \, M_k = \frac{1}{N_c} \sum_{k=1}^{N_c} M_k .$$

Hence, N_c independent measurements M_k contribute and, consequently, the variance for the observable is given by

$$\sigma^2 = \frac{\sigma_M^2}{N_c} = \frac{\sigma_M^2}{(N/m)} \, ,$$

which coincides with our finding in (5.33).

The value of the autocorrelation time depends on the observable and on the MC algorithm. For efficient MC simulations, it is key the algorithm under consideration can keep τ_{int} at bay. As we will explain in this subsection, *local update* algorithm combined with *importance sampling* necessarily fail for certain statistical systems for physical reasons: We will find that the autocorrelation time strongly increases for a certain parameter range (we say "near criticality") to an extent that those algorithms cannot be used. This is called *critical slowing down*.

To illustrate critical slowing down, we come back to the Ising model in two dimensions, which we have discussed in great detail in section 4.4. The spins $\sigma_x \in \{-1, 1\}$ are associated with the sites of the $N \times N$ lattice. The partition function is given by

$$Z = \sum_{\sigma=\pm 1} \exp\left\{\beta \sum_{\langle xy \rangle} \sigma_x \sigma_y\right\} , \tag{5.34}$$

where $\langle xy \rangle$ next to nearest-neighbours on the lattice. We will work below with periodic boundary conditions. The observable of interest is the magnetisation:

$$\widetilde{M} = \frac{1}{N^2} \sum_x \sigma_x . \tag{5.35}$$

Since measure and Gibbs factor of the partition function (5.34) are invariant under the global reflection $\sigma_x \to -\sigma_x$, $\forall x$, the expectation value of the magnetisation necessarily vanishes on finite lattices:

$$\langle M \rangle = \frac{1}{Z} \sum_{\sigma=\pm 1} \left(\frac{1}{N^2} \sum_x \sigma_x\right) \exp\left\{\beta \sum_{\langle xy \rangle} \sigma_x \sigma_y\right\} .$$

We instead consider as observable

$$M = \sigma_{x_0} \frac{1}{N^2} \sum_x \sigma_x , \tag{5.36}$$

which can be viewed as the magnetisation for which the spin at x_0 serves as reference. For a study of the autocorrelations, we use a standard heat bath algorithm, as described in Chapter 4.

If we $\{\sigma_x^i, \forall x\}$ is the configuration of the i-th MC sweep, the corresponding magnetisation is given by

$$M_i = \sigma_{x_0}^i \frac{1}{N^2} \sum_x \sigma_x^i .$$

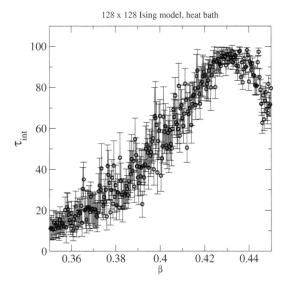

FIGURE 5.4: The integrated autocorrelation time for the 128×128 Ising model as a function of the inverse temperature β for the local update Heat Bath alogorithm.

We are now ready to estimate the autocorrelation function C is (5.30) as a function of the inverse temperature β. To this aim, we have generated, for each β, $10\,000$ configurations. To estimate the integrated autocorrelation time τ_{int}, we randomly choose $i \in [1,\ 10\,000 - 100]$ and calculate

$$\sum_{k=1}^{100} M_i \, M_{i+k} \ .$$

Note that we are using a cutoff of 100 for the summing the correlation function. Repeating this several times generates a sequence of values that can be used as estimator for expectation values such as $\langle M_i M_{i+k} \rangle$. Error bars can be obtained by the boost trap method.

The result of this approach is shown in Figure 5.4. The critical inverse temperature for the 2D Ising model is

$$\beta_c = \frac{1}{2} \ \ln(1 + \sqrt{2}) \ = \ 0.4406867 \ldots \ .$$

Our central observation is that for small β, autocorrelations are well controlled by choosing the number of sweeps between measurements. However, if we approach the critical value $\beta \to \beta_c$, the autocorrelation time rapidly increases. If we were to estimate the statistical error for $\beta = 0.44$ by $\sigma_M / \sqrt{N_{MC}}$, we underestimate the error bars by roughly one order of magnitude:

$$\sqrt{1 + 2\tau_{int}} \approx \sqrt{1 + 2 \times 100} \approx 14.2 \ .$$

The reason for this critical slowing down becomes apparent by an inspection of a sample Ising configuration for $\beta = 0.44$ (see Figure 5.5, right panel). A black symbol means at position x indicates $\sigma_x = 1$ (white symbol for $\sigma_x = -1$. The Ising spins tend to form rather large clusters of spins with the same orientation. This is not too surprising since for large β all spins would have roughly the same orientation. Figure 5.5, left panel, shows the spin configuration after the symmetry operation $\sigma_x \to -\sigma_x$. Hence, both configurations are equally relevant.

How much sweeps would be needed to significantly change, let us say, the left configuration?

We would need to resolve and revert whole clusters. However, this is not easily done by the local update algorithm: if our update procedure has chosen a spin *inside* a $\sigma = 1$ cluster, the probability p that this spin remains a $\sigma = 1$ spin is very high:

$$p = \frac{\exp\{4\beta\}}{\exp\{4\beta\} + \exp\{-4\beta\}} \approx 97.1\% \ .$$

We conclude that only spins at the interface between clusters are updated efficiently. The interface is one dimensional. For a $N \times N$ lattice, let αN denote the number of the spins at the interfaces. This leaves $N^2 - \alpha N$ spins inside clusters. If we choose to do $\nu \times N$ local updates (ν fixed if we change the size N), the probability that nothing changes is therefore estimated by

$$q := \left(\frac{N^2 - \alpha N}{N^2}\right)^{\nu N} = \left(1 - \frac{\alpha}{N}\right)^{\nu N} \to \exp\{-\nu\alpha\}.$$

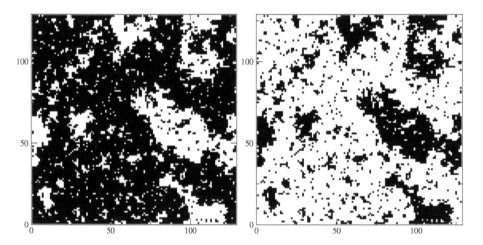

FIGURE 5.5: . Two sample configurations for $\beta = 0.44$.

By adjusting ν, we can achieve a sizeable probability, i.e. $1 - q$, that at least one spin has been changed. To completely change the configuration, we have to do this at least for every spin of the lattice once. Hence, the total amount of local updates for changing a configuration approximately scales like $\nu N \times N^2$ (for large N). We hence expect that the autocorrelation time *diverges with the system volume* near criticality ($\beta \approx \beta_c$) as

$$\tau_{\text{int}} \propto V^z , \qquad V = N^2 . \qquad (5.37)$$

The parameter z is called *dynamical critical index*. Our crude estimate above would give $z = 1.5$.

5.2.2 Cluster update algorithms

If we want to simulate statistical physics models near criticality, we need to devise a new Monte Carlo update strategy since, as we explained in the previous subsection, local update algorithms fail due to critical slowing down. Key to the resolution of this issue are so-called *non-local update* algorithms. For those, a single MC step updates a number of spins at a time. If we succeed to update most of the spins of a cluster of spins with the same sign, we would be able to significantly change the configuration within a single MC step, which is a key ingredient to keep autocorrelation times low. We call such a MC procedure *cluster update algorithm*. Any such cluster algorithm needs to "grow" the clusters according to the parameter of the model. An algorithm would fail if the proposed clusters are largely different from the actual Ising clusters near criticality.

Let us reconsider the partition function of the Ising model (5.34) with the Boltzmann-Gibbs factor

$$\exp\left\{ \beta \sum_{\langle xy \rangle} \sigma_x \sigma_y \right\} .$$

If two neighbouring spins σ_x and σ_y are equal, the product is given by $\sigma_x \sigma_y = 1$. In any other case, the product equals -1. Let us exploit this observation for the factor $\exp\{\beta \sigma_x \sigma_y\}$:

$$
\begin{aligned}
\exp\{\beta \sigma_x \sigma_y\} &= e^{-\beta} + \left(e^{\beta} - e^{-\beta} \right) \delta(\sigma_x, \sigma_y) \\
&= e^{\beta} \left[(1 - p) + p\, \delta(\sigma_x, \sigma_y) \right] , \\
p &= 1 - e^{-2\beta} .
\end{aligned}
$$

where $\delta(i, k) = 1$ for $i = k$ and $\delta(i, k) = 0$ else (Kronecker δ symbol). We therefore find the Boltzmann-Gibbs factor:

$$\prod_{\langle xy \rangle} \exp\{\beta \sigma_x \sigma_y\} = e^{\beta N_l} \prod_{\langle xy \rangle} \left[(1 - p) + p \delta(\sigma_x, \sigma_y) \right],$$

where $N_l = 2N^2$ are the number of links of the lattice. We now introduce an auxiliary link field, which will help us later to define the spin clusters. We use the identity

$$a + b = \sum_{n=0,1} \left[a \delta(n, 0) + b \delta(n, 1) \right]$$

for every link $\langle xy \rangle$ of the lattice. We need N_l auxiliary fields for this, which we will call link fields. We are going to use the notation (x, μ) to specify the link $\langle xy \rangle$ with x the position and μ is the direction pointing to y. The link field $n_\mu(x) \in \{0, 1\}$ is hence associated with the link $\langle xy \rangle$ on the lattice. We thus obtain for the Boltzmann-Gibbs factor

$$\prod_{\langle xy \rangle} \exp\{\beta \sigma_x \sigma_y\} = e^{\beta N_l} \sum_{n_\mu(x)} \prod_{\langle xy \rangle} \left[(1-p) \delta(n_\mu(x), 0) + p \delta(n_\mu(x), 1) \delta(\sigma_x, \sigma_y) \right]$$

and our final result for the partition function (5.34) is:

$$Z = e^{\beta N_l} \sum_{\sigma_x, n_\mu(x)} \prod_{\langle xy \rangle} \left[(1-p) \delta(n_\mu(x), 0) + p \delta(n_\mu(x), 1) \delta(\sigma_x, \sigma_y) \right]. \quad (5.38)$$

The latter equation is an exact reformulation of the Ising model. We now have to sets of variables: $n_\mu(x)$ and σ_x. For the Monte Carlo simulation, we now devise a standard heat bath algorithm for both sets.

Let us assume that we want to update all the link variables $n_\mu(x)$ first. All the spins σ_x are considered as given for this step. For the n-update, we hence pick a link of the lattice, say $\langle xy \rangle$. If $\sigma_x \neq \sigma_y$, the Boltzmann-Gibbs factor is proportional to

$$(1 - p) \delta(n_\mu(x), 0) .$$

This means that we have to set $n_\mu(x) = 0$ to avoid a vanishing probability. On the other hand, f $\sigma_x = \sigma_y$, the Boltzmann-Gibbs factor is proportional to

$$(1 - p) \delta(n_\mu(x), 0) + p \delta(n_\mu(x), 1) .$$

Hence, we set the link to $n_\mu(x) = 1$ with probability p. We do this for all the links on the lattice. The n-update implies that all these spins σ on sites that are connected with $n_\mu(x) = 1$ links are equal. Hence, we can view this set of spins as one *cluster* .

Let us now consider the spin update in the 'heat bath' of all the given link fields $n_\mu(x)$. If we choose say σ_x for update, we consider all the spins σ_y that are connected to σ_x via links $n_\mu = 1$, i.e. we consider the whole cluster of which σ_x

is part. As seen above, the spins are necessarily equal. The Boltzmann-Gibbs factor is proportional to

$$p\,\delta(\sigma_x, \sigma_y)\,.$$

In order to avoid a vanishing all spins of the cluster must be equal. Note that the Boltzmann-Gibbs factor does infer a specific value, i.e. $+1$ or -1, for this cluster. Hence, we choose either of them with equal probability for the update. We then visit another site of the lattice that has not been yet part of a cluster that has been updated and repeat the update steps above. We continue until all clusters and, hence, all spins have been updated.

Note that this algorithm is exact since it can be viewed as standard heat bath Monte Carlo. The beauty of it is that a single MC step potentially updates many of the spins. This cluster update algorithm has been firstly proposed by Swendsen and Wang in 1987 [43].

Swendsen Wang Cluster algorithm:

1. Visit every link $(x, \mu) = \langle xy \rangle$ of the lattice and set $n_\mu(x) = 0$ if $\sigma_x \neq \sigma_y$. If $\sigma_x = \sigma_y$, choose $n_\mu(x) = 1$ with probability $p = 1 - \exp\{-2\beta\}$, i.e. choose a random number $u \in [0, 1]$ and set $n_\mu(x) = 1$ if $u \leq p$ and $n_\mu(x) = 0$ otherwise.

2. Choose a spin σ_x and construct the cluster to which σ_x belongs, i.e. find all other spins σ_y that are connected to σ_x by links $n_\mu(x) = 1$. Continue until all spins on the lattice belong to one cluster each (caution: there might be clusters that consist of a single spin).

3. Visit every cluster and assign $+1$ or -1 with equal probability to all spins of the cluster.

Since during the spin update, the spins of cluster are randomly assigned ± 1 with equal probability, on average 50% of the spins are *not* updated during one lattice sweep, which is slightly inefficient. We hence present a modified algorithms, which is also easier to implement, without proof of its exactness.

Wolff Cluster algorithm [44]:

1. Choose a randomly spin σ_x and construct a single cluster as follows: visit all sites connected to x and add the neighbouring spin to the cluster with probability $p = 1 - \exp\{-2\beta\}$. Continue to visit all neighbouring sites of spins in the cluster until no further spin is available to be added. Flipp the sign of all spins of the cluster.

2. Repeat with step 1 until a cluster flip exceeds the total of $N \times N$ spin updates. We call this one Wolff MC sweep.

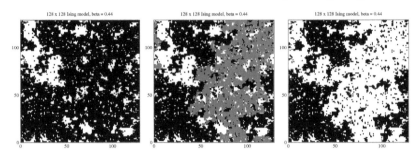

FIGURE 5.6: Stages of the Modified Swendsen Wang cluster update.

We will come back to the Wolff cluster algorithm in subsection 6.1.2 and convince ourselves that this algorithm is exact.

Note that we have defined the Swendsen Wang MC sweep in accordance with the "heat bath sweep": in either sweep, roughly the same amount of spins are updated. This makes it possible to compare the performance of both algorithms with respect to autocorrelation times.

The stages of step 1 are illustrated in Figure 5.6 for a 128×128 lattice at $\beta = 0.44$: the left panel shows the starting configuration while the picture in the middle shows one particular cluster in red. The right panel shows the configuration after all spins of the cluster have been flipped.

Finally, Figure 5.7 shows the integrated autocorrelation time for the cluster algorithm. We do not detect any autocorrelations with the current amount

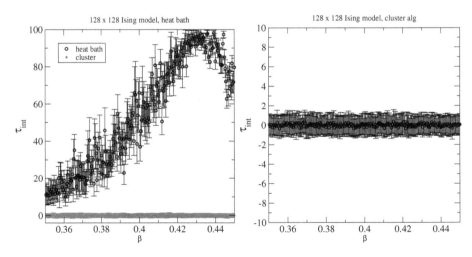

FIGURE 5.7: The integrated autocorrelation time for the 128×128 Ising model as a function of the inverse temperature β for the cluster algorithm (red symbols) in comparison with result from the Heat Bath algorithm.

of configurations, i.e. the autocorrelation time is compatible with zero within error bars. This is a striking contrast to the result from the heat bath algorithm. It is not expected that autocorrelations are completely absent but that it scales like

$$\tau_{\text{int}} \propto V^z, \qquad z \approx 0.3 - 0.6 \quad \text{(cluster algorithm)}.$$

5.2.3 Improved operators and correlation length

Before we introduce the concept of *improved operators* for the case of the Swendsen-Wang update algorithm (see subsection 5.2.2), we will study the algorithm in action.

We consider a 64×64 lattice and initialise the Markov Chain simulation by a a *hot start*, i.e., we set each spin σ_x of the lattice to ± 1 with equal probability. We then start the chain with $1,000$ Swendsen-Wang updates using the algorithm described in the box on page 203.

We then enter the stage where we take "measurements". We are interested in the average action per link:

$$A = \frac{1}{2N^2} \left\langle \sum_{\langle xy \rangle} \sigma_x \sigma_y \right\rangle.$$

We perform 20 Swendsen-Wang updates between measurements to further reduce auto-correlations. We take 400 independent measurements to estimate the above expectation value. For an illustration of the quality of the data, we quote here some selected results:

β	A	error
0.2	0.2133	0.000579
0.3	0.3525	0.000740
0.4	0.5527	0.000941
0.5	0.8732	0.000645

More findings are gathered in Figure 5.8, right panel. It shows the familiar linear rise proportional to β at small β and the usual non-linearity for larger β values induced by the phase transition.

Let us now switch to an important observable, which is notoriously difficult to estimate – the spin–spin correlation function:

$$C(r) = \langle \sigma_{x0} \sigma_{y0} \rangle, \qquad r = |x0 - y0|. \tag{5.39}$$

The reason for this is that the information about the status of the spin at point $x0$ rapidly decreases with distance r (at least for small $\beta < \beta_c$):

$$C(r) \propto \exp\left\{ -\frac{r}{\xi(\beta)} \right\}, \tag{5.40}$$

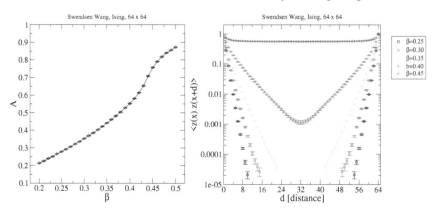

FIGURE 5.8: The average action A per link obtained from a Swendsen-Wang cluster simulation (left). The spin correlation function for several values of β as a function of the spin separation distance d using just 400 MCMC configurations and an improved operator (left).

where $\xi(\beta)$ is a measure for the distance over which spins still have a significant correlation and is hence called *correlation length*. Eq.(5.40) shows the dominant behaviour for large r, and one might expect power-law correction in r at small range.

The standard estimator for $C(r)$ in (5.40) is a sum over N_{MC} Markov Chain configurations:

$$C(r) \approx \frac{1}{N_{MC}} \sum_{i=1}^{N_{MC}} \sigma_{x0}^{(i)} \sigma_{y0}^{(i)} \,.$$

Note that each contribution to the sum above contributes ± 1. Hence, the exponentially small values of $C(r)$ arise from cancellations in the sum. If the result of the sum should be 10^{-5}, then at least $100,000$ MC configurations must be generated so that the addition of one more contribution ± 1 does not spoil the cancellation to this level.

Improved operator can be of help here. By definition, those operators have the same expectation value as the un-improved, original version:

$$D(x0, y0) = \sigma_{x0}\sigma_{y0} \,, \quad iD(x0, y0) = f(\{\sigma\}) \,, \quad \langle D \rangle = \langle iD \rangle \,. \tag{5.41}$$

There is a generic way to define an improved operator. If $W(\sigma \to \sigma')$ is the Markov chain transition probability for mapping the configuration $\{\sigma\}$ into $\{\sigma'\}$. These transition probabilities satisfy detailed balance (see exercise 5.3 in section 5.6):

$$P(\sigma)\, W(\sigma \to \sigma') = P(\sigma')\, W(\sigma' \to \sigma) \,,$$

where $P(\sigma)$ is Gibbs weight for the configuration $\{\sigma\}$. It implies that the Gibbs factor is an eigenvector with eigenvalue 1 to the transition matrix W:

$$\sum_\sigma P(\sigma)\, W(\sigma \to \sigma') \;=\; \sum_\sigma P(\sigma')\, W(\sigma' \to \sigma) \;=\; P(\sigma')\,, \quad (5.42)$$

where we have used ergodicity (see section 3.3), i.e.

$$\sum_{\sigma'} W(\sigma' \to \sigma) \;=\; 1\,.$$

We then define the improved version, i.e. iD, for the operator $D[\sigma]$ by

$$iD[\sigma] \;:=\; \sum_{\sigma'} W(\sigma \to \sigma')\, D[\sigma']\,. \quad (5.43)$$

Note that, for a given configuration σ, we sum over all possible target configurations σ' – not necessarily an easy feat. Using (5.42), we then observe:

$$\begin{aligned}
\langle iD[\sigma] \rangle &= \frac{1}{Z}\sum_\sigma iD[\sigma]\, P(\sigma) = \frac{1}{Z}\sum_{\sigma,\sigma'} P(\sigma)\, W(\sigma \to \sigma')\, D[\sigma'] \\
&= \frac{1}{Z}\sum_{\sigma'} D[\sigma']\, P(\sigma') = \langle D[\sigma] \rangle\,.
\end{aligned}$$

The latter equation only proves that the expectation values of the operators D and iD are the same, but that does not mean that iD is the better estimator. In some cases, however, exceptional gains can be made.

The calculation of the correlation function (5.39) is one of those cases. If we use the Swendsen-Wang algorithms, we first decompose the entire lattice into clusters C_i. We then independently flip the sign of all spins simultaneously in each of those clusters with 50% probability. For estimating the correlation function, we need to evaluate (see (5.38)):

$$\begin{aligned}
C(r) \;=\; & \frac{1}{Z}\, e^{\beta\, N_l} \sum_{\sigma_x,\, n_\mu(x)} \sigma_{x0}\, \sigma_{y0} \\
& \times \prod_{\langle xy \rangle} \Big[(1-p)\, \delta(n_\mu(x),0) \;+\; p\, \delta(n_\mu(x),1)\, \delta(\sigma_x,\sigma_y) \Big]\,.
\end{aligned}$$

To construct an improved operator, we start with a given configuration $\{\sigma\}$ and then try to average over the target configurations $\{\sigma'\}$. Assume that we found all the clusters C_i. Part of the average over the target configurations is an average over all random spin flips for each clusters. In case σ_{x0} and σ_{y0} are *not* in the same clusters, the later average would eliminate any contribution to the improved operator since the signs of σ_{x0} and σ_{y0} are ± 1 with equal probability. The only contribution to the improved operator arises from cases where σ_{x0} and σ_{y0} are in the *same* cluster. We hence define

$$iD(x0, y0) \;=\; \sum_{\text{clusters } i} \sigma_{x0}\, \sigma_{y0}\, \theta\,(C_i, \sigma_{x0})\, \theta\,(C_i, \sigma_{y0})\,,$$

where $\theta(C, \sigma_x) = 1$ if σ_x is part of the cluster C and $\theta(C, \sigma_x) = 0$ otherwise. Also note that if both spins are in the same cluster, they have the *same* orientation (because of the $\delta(\sigma_x, \sigma_y)$ term in the above formula), we find $\sigma_{x0}\,\sigma_{y0} = 1$ in any case. We therefore are led to our final definition:

$$iD(x0, y0) \;=\; \sum_{\text{clusters } i} \theta(C_i, \sigma_{x0})\,\theta(C_i, \sigma_{y0}) \;. \tag{5.44}$$

Note that this operator is always semi-positive. This implies a vast improvement over the operator D in (5.41). While in the former case, a large number of cancellations need to take place at large spin distance, the operator iD does not instigate any cancellations. If its expectation value is small, it is due to the fact that events rarely contribute at large distance.

Before we will discuss, let us briefly consider a possible implementation of the algorithm:

Step 1: Visit every link and set the link variables $n_\mu(x)$ according o their probability. This is a local procedure and can be performed link by link.

Step 2: Define the variable grid (x) which contains the cluster ID to which the site x belongs. To construct grid , visit every site x. if grid (x) >0, the site is already part of a cluster and move on. if grid (x) =0, we are defining a new cluster with this site as seed. Increase ID by one. Visit all sites y "connected" to x and assign grid (y) = ID.

Step 3: With equal probability, choose list$(ID) = \pm 1$, $ID = 1 \ldots ID_{\text{max}}$. Visit every site of the lattice and change the spin according to:

$$\sigma'(x) \;=\; list\Big(\text{grid}(x)\Big) \times \sigma_x \;.$$

We now have arrived at a new configuration. If we would like to estimate correlation function using iD, we continue with Step 4.

Step 4: Assume $C(d)$ is the variable that is the estimator for the spin-spin correlation function at distance d, $d = 1 \ldots N$. Visit every site x. Choose y, which is d units away from x in either the 1 or 2 direction. Increase $C(d)$ by 1 if

$$\text{grid}(x) \;==\; \text{grid}(y).$$

We have already seen one outcome of this simulation in Figure 5.8, left panel, namely the average action per link as a function of β, which painted a familiar picture. Figure 5.8, right panel, shows the correlation function as a function of the distance d for several β values. Note the logarithmical scale of the y-axis. For $\beta < \beta_c$, we indeed observe an exponential decrease of the correlation with distance. Note, however that the slope of this decrease, i.e. the correlation length $\xi(\beta)$, increases when we approach β_c from below. At some value β between 0.4 and $\beta_c = 0.4406867\ldots$, the correlation length ξ becomes comparable with the lattice size $N = 64$. In this case, all spins on the lattice are significantly correlated.

where $\theta\left(C, \sigma_x\right) = 1$ if σ_x is part of the cluster C and $\theta\left(C, \sigma_x\right) = 0$ otherwise. Also note that if both spins are in the same cluster, they have the *same* orientation (because of the $\delta(\sigma_x, \sigma_y)$ term in the above formula), we find $\sigma_{x0}\,\sigma_{y0} = 1$ in any case. We therefore are led to our final definition:

$$iD(x0, y0) \;=\; \sum_{\text{clusters } i} \theta\left(C_i, \sigma_{x0}\right)\,\theta\left(C_i, \sigma_{y0}\right) \;. \tag{5.44}$$

Note that this operator is always semi-positive. This implies a vast improvement over the operator D in (5.41). While in the former case, a large number of cancellations need to take place at large spin distance, the operator iD does not instigate any cancellations. If its expectation value is small, it is due to the fact that events rarely contribute at large distance.

Before we will discuss, let us briefly consider a possible implementation of the algorithm:

Step 1: Visit every link and set the link variables $n_\mu(x)$ according o their probability. This is a local procedure and can be performed link by link.

Step 2: Define the variable `grid (x)` which contains the cluster ID to which the site x belongs. To construct `grid` , visit every site x. if `grid (x) >0`, the site is already part of a cluster and move on. if `grid (x) =0`, we are defining a new cluster with this site as seed. Increase ID by one. Visit all sites y "connected" to x and assign `grid (y) = ID`.

Step 3: With equal probability, choose $\text{list}(ID) = \pm1$, $ID = 1\ldots ID_{\text{max}}$. Visit every site of the lattice and change the spin according to:

$$\sigma'(x) \;=\; list\!\left(\text{grid}(x)\right) \times \sigma_x \;.$$

We now have arrived at a new configuration. If we would like to estimate correlation function using iD, we continue with Step 4.

Step 4: Assume $C(d)$ is the variable that is the estimator for the spin-spin correlation function at distance d, $d = 1\ldots N$. Visit every site x. Choose y, which is d units away from x in either the 1 or 2 direction. Increase $C(d)$ by 1 if

$$\text{grid}(x) \;==\; \text{grid}(y).$$

We have already seen one outcome of this simulation in Figure 5.8, left panel, namely the average action per link as a function of β, which painted a familiar picture. Figure 5.8, right panel, shows the correlation function as a function of the distance d for several β values. Note the logarithmical scale of the y-axis. For $\beta < \beta_c$, we indeed observe an exponential decrease of the correlation with distance. Note, however that the slope of this decrease, i.e. the correlation length $\xi(\beta)$, increases when we approach β_c from below. At some value β between 0.4 and $\beta_c = 0.4406867\ldots$, the correlation length ξ becomes comparable with the lattice size $N = 64$. In this case, all spins on the lattice are significantly correlated.

where $P(\sigma)$ is Gibbs weight for the configuration $\{\sigma\}$. It implies that the Gibbs factor is an eigenvector with eigenvalue 1 to the transition matrix W:

$$\sum_{\sigma} P(\sigma)\, W(\sigma \to \sigma') \;=\; \sum_{\sigma} P(\sigma')\, W(\sigma' \to \sigma) \;=\; P(\sigma')\,, \quad (5.42)$$

where we have used ergodicity (see section 3.3), i.e.

$$\sum_{\sigma'} W(\sigma' \to \sigma) \;=\; 1\,.$$

We then define the improved version, i.e. iD, for the operator $D[\sigma]$ by

$$iD[\sigma] \;:=\; \sum_{\sigma'} W(\sigma \to \sigma')\, D[\sigma']\,. \quad (5.43)$$

Note that, for a given configuration σ, we sum over all possible target configurations σ' – not necessarily an easy feat. Using (5.42), we then observe:

$$
\begin{aligned}
\langle iD[\sigma] \rangle &= \frac{1}{Z} \sum_{\sigma} iD[\sigma]\, P(\sigma) = \frac{1}{Z} \sum_{\sigma,\sigma'} P(\sigma)\, W(\sigma \to \sigma')\, D[\sigma'] \\
&= \frac{1}{Z} \sum_{\sigma'} D[\sigma']\, P(\sigma') = \langle D[\sigma] \rangle\,.
\end{aligned}
$$

The latter equation only proves that the expectation values of the operators D and iD are the same, but that does not mean that iD is the better estimator. In some cases, however, exceptional gains can be made.

The calculation of the correlation function (5.39) is one of those cases. If we use the Swendsen-Wang algorithms, we first decompose the entire lattice into clusters C_i. We then independently flip the sign of all spins simultaneously in each of those clusters with 50% probability. For estimating the correlation function, we need to evaluate (see (5.38)):

$$
\begin{aligned}
C(r) \;=\; & \frac{1}{Z}\, e^{\beta N_l} \sum_{\sigma_x,\, n_\mu(x)} \sigma_{x0}\, \sigma_{y0} \\
& \times \prod_{\langle xy \rangle} \Big[(1-p)\, \delta(n_\mu(x),0) \;+\; p\, \delta(n_\mu(x),1)\, \delta(\sigma_x,\sigma_y) \Big]\,.
\end{aligned}
$$

To construct an improved operator, we start with a given configuration $\{\sigma\}$ and then try to average over the target configurations $\{\sigma'\}$. Assume that we found all the clusters C_i. Part of the average over the target configurations is an average over all random spin flips for each clusters. In case σ_{x0} and σ_{y0} are *not* in the same clusters, the later average would eliminate any contribution to the improved operator since the signs of σ_{x0} and σ_{y0} are ± 1 with equal probability. The only contribution to the improved operator arises from cases where σ_{x0} and σ_{y0} are in the *same* cluster. We hence define

$$iD(x0,y0) \;=\; \sum_{\text{clusters } i} \sigma_{x0}\, \sigma_{y0}\, \theta\,(C_i,\sigma_{x0})\, \theta\,(C_i,\sigma_{y0})\,,$$

5.2.4 Flux algorithms

The cluster algorithm of the previous subsection certainly provides an acceptable solution to the critical slowing down issue of the Ising model. Please note that is efficiency is by no means guaranteed: although our reformulation involving link fields is exact, there is a chance that the cluster that emerge have nothing in common with the "physical" clusters of the Ising system, which would swart the efficiency of the approach. Secondly, the definition of clusters in more sophisticated statistical models is far from obvious. An example a system with a gauge symmetry, for which the gauge freedom mask physical clusters. So far, attempts to devise a cluster update for gauge systems have failed. In this subsection, we will therefore another non-local update method, which is less effective in the case of the Ising model than cluster algorithms, but more widely applicable: the flux algorithms.

We will again study those in the context of the Ising model, but more sophisticated models have been successfully simulated. Selected recent achievements have been presented in [45, 46, 47, 48]. We again start with a reformulation of the Gibbs factor. Exploiting that the product $\sigma_x \sigma_y$ only takes values ± 1, we can write:

$$
\begin{aligned}
\exp\{\beta\,\sigma_x\sigma_y\} &= \frac{1}{2}\left(e^\beta + e^{-\beta}\right) + \frac{1}{2}\left(e^\beta - e^{-\beta}\right)\sigma_x\sigma_y \\
&= \cosh(\beta)\left[1 + \tanh(\beta)\,\sigma_x\sigma_y\right].
\end{aligned}
$$

The Ising partition function is then given by:

$$
Z = \sum_{\sigma_x} [\cosh(\beta)]^{N_l} \prod_{\langle xy\rangle}\left[1 + \tanh(\beta)\,\sigma_x\sigma_y\right]. \tag{5.45}
$$

We are now going to sum over all spin configurations. Firstly note that since $\sigma \in \{-1,1\}$, we find

$$
\sum_\sigma \sigma^n = \begin{cases} 2 & \text{for } n \text{ even including } n = 0 \\ 0 & \text{for } n \text{ odd} \end{cases}.
$$

We expand brackets in (5.45). There is one bracket for each link of the lattice. We avoid that any term that would vanish whence the sum over the spins σ_x is performed. This means that upon expanding we can disregard any terms that have an odd power of one of the spins. We could pick all ones from the square brackets, and the sum over all spins would yield 2^V, $V = N^2$. Once we have chosen a term $\tanh(\beta)\,\sigma_{x1}\sigma_{x2}$, we need to pick from the factors another terms that features σ_{x2}:

$$
\tanh^2(\beta)\,\sigma_{x1}\sigma_{x2}\,\sigma_{x2}\sigma_{x3}.
$$

Now, σ_{x3} is unpaired. We also need to avoid that the starting spin σ_{x1} stands alone. The only way to do this is

$$
\tanh^n(\beta)\,\sigma_{x1}\sigma_{x2}\,\sigma_{x2}\sigma_{x3}\ \cdots\ \sigma_{xn}\sigma_x.
$$

On the lattice, this can be depicted as a sequence of links forming a closed loop of length n stretching from $x1$ over $x2 \ldots x_n$ back to $x1$ again. Since every spin s now paired and since $\sigma^2 = 1$, we find

$$\tanh^n(\beta)\, \sigma_{x1}\, \sigma_{x2}\, \sigma_{x2}\, \sigma_{x3}\, \ldots\, \sigma_{xn}\, \sigma_x \;=\; \tanh^n(\beta).$$

We now could pick the 1 from the remaining square brackets and this would give a non-vanishing contribution to the partition function. If we instead pick $\tanh(\beta)\, \sigma_{y1}\, \sigma_{y2}$, we would start another closed loop. We therefore find the so-called *world line* representation of the Ising partition function:

$$Z \;=\; [\cosh(\beta)]^{N_l}\, 2^V \sum_{L} \Big[\tanh(\beta)\Big]^{n(L)}, \tag{5.46}$$

where we sum over all closed loops L and where $n(L)$ is the length of the loop. If we consider every link that is part of a loop as a *flux line*, we can say that *flux is conserved*: open flux strings would terminate at an un-paired spin, and the contribution would vanish upon the summation over spin configurations. Algorithms that directly update closed flux lines are called *flux algorithms*.

Two questions immediately arise:

- How do we update closed flux lines in accordance with the Gibbs factor?

- Even if we can generate statistically relevant flux line configurations, how do we calculate observables of the underlying Ising model?

We address the second item first. Generically, it is not straightforward to relate expectation values in either formulation. We illustrate the method for the action. Defining the action by

$$A \;=\; \sum_{\langle xy \rangle} \sigma_x \sigma_y \,,$$

its expectation value can be obtained from

$$\langle A \rangle \;=\; \frac{1}{Z}\, \frac{\partial Z}{\partial \beta}\,, \tag{5.47}$$

where we have used the partition function Z (5.34) in standard spin representation. We now use the flux representation (5.46) to obtain:

$$\langle A \rangle \;=\; \tanh(\beta)\, N_l \;+\; \frac{\langle n(L) \rangle_L}{\sinh(\beta)\, \cosh(\beta)}\,, \tag{5.48}$$

where we have introduced the average loop length by

$$\langle n(L) \rangle_L \;=\; \frac{\sum_L [\tanh(\beta)]^{n(L)}\, n(L)}{\sum_L [\tanh(\beta)]^{n(L)}}\,.$$

Let us now come back to the all important question: how do we estimate the theory (5.46) with closed strings as degrees of freedom? To set the scene, we define the link variable u_l for a link $\ell = (x, \mu)^7$ by:

$$u_\ell = \begin{cases} 1 & \text{if the link } \ell \text{ is part of a closed string,} \\ 0 & \text{else.} \end{cases}$$

A straightforward idea is to devise a Markov chain, transforming one configuration of closed strings into the next. As always, we would start with an initial "allowed" configuration such as a lattice with no loops at all. Crucially, our Markov step needs to transform one legitimate configuration into the next. This can be achieved as follows:

- Choose any closed loop L' consisting of links u'_ℓ.

- Carry out one Markov step: $\forall \ell$ with $u'_\ell = 1$, set $u_\ell = 1$ if $u_\ell = 0$, and set $u_\ell = 0$ if $u_\ell = 1$. All links with $u'_\ell = 0$ remain unchanged.

The latter step has a graphical interpretation: visualise all closed loops of a valid configuration in red. Superimpose the closed "Markov loop" in blue. If a blue link overlays a red one, remove both. If a blue link is "on its own", turn its colour in red. An illustration can be found in Figure 5.9. A "small" blue loop transforms a configuration of two red loops into a configuration with one large red loop.

Let's check that this udate step necessarily generates closed loops. In order to all loops being closed, we need to have that the number of loop links (those with $u_\ell = 1$) attached to any particular site is even. We leave it to an exercise to check that the above algorithm implies that the number of "red" links remains even.

How should we choose the "blue" transformation loops. We could choose the smallest possible loop around one plaquette. However, the corresponding Markow step is then a minor local change to the configuration, and we might run into ergodicity problems again. We thrive for a "large" non-local update as perhaps generated by a random walk, which is terminated when the walker returns to the start point. We then would invoke an accept–reject step in order to keep to old loop configuration or to install the new one.

This approach would be exact but, at the same time, could be very inefficient. Assume that we start with an empty lattice ($n(L) = 0$) and that the random walk suggests a loop with 100 links. In this case, the probability for accepting the change would be (for $\beta = 0.5$)

$$\tanh(\beta)^{100} \approx 2.9868 \times 10^{-34} .$$

The example shows that, while the algorithm is non-local, we would hardly

---

[7]The variable x specifies the lattice site and μ the direction.

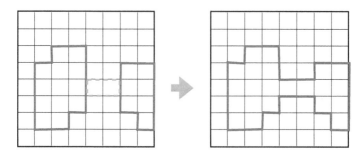

FIGURE 5.9: Transforming one configuration of close (red) loops into another one by superimposing the blue, dashed closed loop.

accept any change. Clearly, importance smapling is needed here. We proceed as follows:

Flux loop update step:

1. Randomly choose x_0 as start for the Markov update loop and store the existing loop configuration. Set $x = x_0$.

2. Choose randomly one of the $2D$ directions attached to x. Call the link ℓ. Set $p = \tanh \beta$ if $u_\ell = 0$ and $p = 1/\tanh \beta$ if $u_\ell = 1$. Choose a random number $u \in [0, 1]$.

3. Accept–reject step: If $u \leq p$, accept the change: set $u_\ell = 1$ if $u_\ell = 0$, and $u_\ell = 0$ if $u_\ell = 1$. Move along the link to the new site x. If x coincides with the start point x_0, return from the subroutine. You have successfully performed one Markov step. If not, continue with step 2.

   If $u > p$, reject the change and the whole attempt. Reinstate the stored configuration and return from the subroutine.

Once we can generate a Markov sequence of closed flux loops, we can estimate expectation values depending on loop variables such as $n(L)$ in 5.48. We illustrate the method for the 64×64 Ising model. With an "empty" lattice with $u_\ell = 0 \; \forall \ell$. We then perform $10,000$ elementary Markov steps as detailed in the box below for thermalisation. Subsequently, we perform 100 elementary updates to arrive at a new element of the Markov chain. We then consider chains with $10,000$ elements to estimate expectations values. For an error estimate, we repeat the whole process 40 times and retrieve statistical errors from a bootstrap analysis.

Our numerical findings are summarised in Figure 5.10 for several β. At small β, the presence of loops is exponentially suppressed. This is intuitive since the

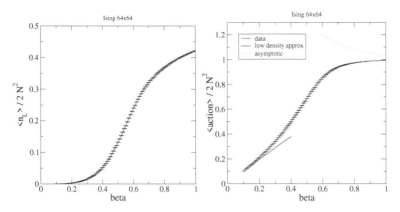

FIGURE 5.10: Left: average number $\langle n(L) \rangle$ of loop links with $u_\ell = 1$ in relation ot the number $2N^2$ of links for the $N \times N$ lattice. Right: reconstructed average action using eq.(5.48).

relative probability for the smallest loop ($N(L) = 4$) as compared to no loop at all ($N(L) = 0$) is given by

$$\tanh(\beta)^4 / \tanh(\beta)^0 \approx \beta^4 \, ,$$

which is a small number. We therefore expect

$$\langle A \rangle / 2N^2 \approx \tanh(\beta) \qquad \beta \approx 0 \, .$$

At large β values, $\tanh(\beta)$ approaches 1 and having loops is not restricted by then Gibbs factor anymore. Lattice loops appear in abundance, and we observe $n(L)/2N^2 \to 1/2$ for large β. We hence expect the following behaviour for the average action:

$$\langle A \rangle / 2N^2 \approx \tanh(\beta) + \frac{1}{2 \sinh \beta \cosh \beta} \qquad \beta \gg 0 \, .$$

Both approximations are shown in Figure 5.10, right panel in combination with the estimated average action using loop ensembles.

5.3   Micro-canonical simulations

The Monte Carlo simulations discussed so far are based upon a probability density function and simulate the so-called *canonical ensemble* in the context of Statistical physics. An equivalent description of the ensemble statistics is provided by the *micro-canonical* ensemble. We now harness this equivalence and derive an entirely different approach to Monte Carlo simulations.

5.3.1   What are micro-canonical ensembles?

Some significant gains can be made by specialising in the Monte Carlo simulation to the particular system under investigation. In statistical physics, some general underlying properties of the probability measure provide such a specialisation. Here, the probability measure, the so-called Gibbs factor,

$$\exp\Big\{\beta\, S(\phi)\Big\}\,, \qquad \beta = \frac{1}{kT}, \qquad T: \text{ temperature}$$

is expressed in terms of the action S (also called "energy" depending on the context), which is a function of the number V of degrees of freedom summarised in ϕ. The specification is that S is an *extensive* quantity, i.e. the action expectation value, for a given value β, scales with number V of degrees of freedom, i.e.

$$\langle S \rangle \;=\; s(\beta)\, V\,,$$

where the action density s is of order one, i.e. independent of V for large V. We here remind the reader that expectation values are defined by

$$\langle S \rangle \;=\; \frac{1}{Z} \int \mathcal{D}\phi\, S(\phi)\, \exp\Big\{\beta\, S(\phi)\Big\}\,, \tag{5.49}$$

where the partition function is given by

$$Z \;=\; \int \mathcal{D}\phi\, \exp\Big\{\beta\, S(\phi)\Big\}\,. \tag{5.50}$$

Central to micro-canonical simulations is the so-called *density-of-states*:

$$\rho(E) \;=\; \int \mathcal{D}\phi\, \delta\Big(E - S(\phi)\Big)\,, \tag{5.51}$$

where δ is the Dirac δ-function, which we have introduced in (1.8). The density-of-states has a geometrical interpretation: if $\int \mathcal{D}\phi$ measures the total volume of the configuration space (if the domain of each variable ϕ_k is compact and, thus, this volume exists), the density-of-states $\rho(E)$ quantifies the volume of the hyper-surface of constant action specified by:

$$E \;=\; S(\phi)\,.$$

Generically, the density-of-states is a smooth function of E, which admits the Taylor expansion (we will explore this property further in section 5.5):

$$\ln \rho(E) \;=\; \ln \rho(E_0) + \frac{d \ln \rho}{dE}\bigg|_{E=E_0} (E - E_0) + \dots\,. \tag{5.52}$$

The derivative again is of order one for many degrees of freedom:

$$\frac{d \ln \rho}{dE} = \mathcal{O}(1) \qquad \text{for} \qquad V \to \infty .$$

It is (usually) a decreasing function for increasing $E > 0$. Let us understand this for the Ising model: the maximum action of a $V = N \times N$ Ising model is

$$E_{\max} = 2N^2 = 2V .$$

There are only two spin configurations that realise $E = E_{\max}$: all spins are$+1$ or all spins equal -1. On the other hand, if we choose $E = 0$, we only to construct configurations for which the number of parallel spins equal the number of anti-parallel spins, and there are plenty of them.

The partition function Z (5.50) can be recovered from the density-of-states by performing a one-dimensional integration:

$$Z = \int dE \, \rho(E) \, \exp\{ \beta \, E\} = \int dE \, \exp\Big\{\ln \rho(E) + \beta \, E\Big\} . \tag{5.53}$$

For degrees of freedom ϕ with compact domain, the above integral truncates since there are not any states with $E > E_{\max}$. $\rho(E)$ is decreasing, and the Gibbs factor $\exp\{\beta E\}$ is monotonically increasing. The product of both has a maximum, let us say, at $E = E_0$. Since E and E_0 are extensive quantities, this maximum is very sharp for even medium-sized systems such as the 128×128 Ising model. Expanding the density-of-states for energies around the maximum energy E_0 using (5.52), we find a connection between temperature and E_0:

$$\ln \rho(E_0) + \frac{d \ln \rho}{dE}\bigg|_{E=E_0} (E - E_0) + \beta \, E \to \text{extremal}$$

$$\Leftrightarrow \qquad \frac{d \ln \rho}{dE}\bigg|_{E=E_0} + \beta = 0 . \tag{5.54}$$

How can we estimate expectation values in a micro-canonical simulation?

Assume that $M(\phi)$ is the observable. Think of the ϕ as Ising spins and M the magnetisation. Inserting

$$1 = \int dE \, \delta\Big(E - S(\phi)\Big)$$

into the definition of the expectation value, we find:

$$\begin{aligned}
\langle M \rangle &= \frac{1}{Z} \int \mathcal{D}\phi \, M(\phi) \, e^{\beta S(\phi)} \\
&= \frac{1}{Z} \int dE \int \mathcal{D}\phi \, M(\phi) \, \delta\Big(E - S(\phi)\Big) \, e^{\beta S(\phi)} \\
&= \frac{1}{Z} \int dE \, e^{\beta E} \int \mathcal{D}\phi \, M(\phi) \, \delta\Big(E - S(\phi)\Big) , \tag{5.55}
\end{aligned}$$

where we have used that the δ-function vanishes for $E \neq S(\phi)$. Defining the expectation value over the micro-canonical ensemble by

$$\langle M \rangle_{\text{micro}}(E) := \frac{1}{\rho(E)} \int \mathcal{D}\phi \, M(\phi) \, \delta\left(E - S(\phi)\right), \qquad (5.56)$$

we derive the identity from (5.55):

$$\langle M \rangle(\beta) = \frac{1}{Z} \int dE \, \rho(E) \, e^{\beta E} \, \langle M \rangle_{\text{micro}}(E), \qquad (5.57)$$

$$Z = \int dE \, \rho(E) \, e^{\beta E}.$$

For large number V of degrees of freedom, the combination $\rho(E) \exp\{\beta E\}$ sharply peaks at $E = E_0$. This might justify a classical approximation of the above integrals replacing the E integrations by the respective integrands evaluated at $E = E_0$ (times dE) to finally obtain:

$$\langle M \rangle(\beta) \approx \langle M \rangle_{\text{micro}}(E_0), \qquad \frac{d \ln \rho}{dE}\bigg|_{E=E_0} + \beta = 0. \qquad (5.58)$$

In the following, we will discuss a potential implementation of a micro-canonical simulation. Configruations ϕ, which are generated under the con-strained $E = S(\phi)$ are said to belong to the *mirco- canonical ensemble*.

5.3.2   The demon method

An efficient algorithm for a micro-canonical simulation has been given 1983 by Mike Creutz [49]: an additional degree fo freedom, the so-called *demon*, is added to the statistical system. Since the demon is only one out of V such degrees of freedom, the change to ensemble averages is suppressed by the volume and irrelevant for large systems. The only interaction of the demon is to store and relive energy (or action depending on the context). The algorithm proceeds as follows:

The demon algorithm:

1. Perform a random change of the configuration, and calculate the change in total energy (action) ΔE.

2. If ΔE is negative, i.e. the energy after the change is smaller than energy before the change, accept the change and increase the energy (action) of the demon by $|\Delta E|$.

   If $\Delta E \geq 0$ is smaller or equal to the demon energy, accept the change and *decrease* the energy (action) of the demon by ΔE.

   If $\Delta E \geq 0$ is bigger than the demon energy, reject the change and continue with item 1 above.

Apparently, the demon energy is (semi-) positive, and the total energy (action) is constant by construction. Let us explore this approach with the now well-explored Ising model. As a start, we prepare a configuration with total energy S_{total}. To this aim, we proceed as follows: set the demon energy E_D to zero, and set all Ising spins on the lattice to 1. This generates the a configuration with maximal energy. Then, randomly pick a site x, and flip the corresponding spin $\sigma_x \to -\sigma_x$. If the energy of the spin system falls below S_{total} (for the first time), stop and keep the actual energy of the system as S_{total}. Note that the action of a spin system is given

$$S \;=\; \text{constant} \;+\; \sum_{y\in\langle xy\rangle} \sigma_x\sigma_y \;,$$

where we have single out the spin at site x. Flipping this spin changes the energy of the system by

$$\Delta E \;=\; b\left(-\sigma_x\right) \;-\; b\sigma_x \;=\; -2b\,\sigma_x \;, \qquad b = \sum_{y\in\langle xy\rangle} \sigma_y \;.$$

The sum over the neighbouring spins is at most ± 4 implying that

$$|b| \leq 4 \qquad \Rightarrow \qquad |\Delta E| \leq 8 \;,$$

and that the achieved S is at most 8 units below the original design value. Take $S_{\text{total}} - S$ as the demon energy. This creates a system with the desired target value. Note also that randomly flipping spins creates disorder and can be viewed as a heat-bath simulation at infinite temperature ($\beta = 0$). Ultimately, it is possible to generate configurations with vanishing total energy, but not with significant *negative* ones. Is this is what is desired, one could start with a configuration for which any pair of spins has negative signs. This is called a *ferromagnetic* configuration in statistical physics. This configuration has minimal energy, and with randomly flipping spins, one should be able to generate a configuration with any negative energy and zero energy.

The spin system and the demon will reach a thermal equilibrium and the demon energy will be distributed according to

$$P(E_D) \;\propto\; \exp\left\{-\beta\,E_D\right\} \;, \qquad \beta \;=\; \frac{1}{k\,T} \;, \tag{5.59}$$

where T is the temperature of the system. We stress that the design energy S_{total} is the parameter that characterises the simulation, and it is temperature that emerges as an observable in a micro-canonical simulation. The result for the $V = 128 \times 128$ Ising model for two design energies is shown in Figure 5.11. An exponential fit to (5.59) yields $\beta \approx 0.312$ for $S_{\text{total}} = 3/8V$ and $\beta \approx 0.417$ for $S_{\text{total}} = 5/8V$. Note that the demon energy changes in units of 4. Rather than relying on a histogram and a fit procedure, we could measure the average

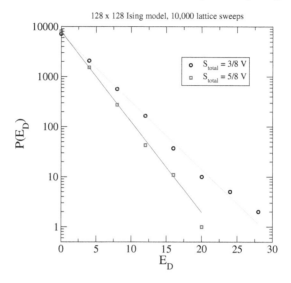

FIGURE 5.11: Probability distribution of the demon energy for values for the design energy S_{total}.

demon energy $\langle E_D \rangle$ of the micro-canonical simulation. Assuming (5.59), we expect the following relation to β:

$$
\begin{aligned}
\langle E_D \rangle &= \frac{1}{z} \sum_{n=0}^{\infty} 4n \, \exp\{- \beta \, 4\,n\} \,, \qquad z = \sum_{n=0}^{\infty} \exp\{- \beta \, 4\,n\} \\
&= -\frac{d}{d\beta} \ln z = \frac{d}{d\beta} \ln\left(1 - \exp\{-4\beta\}\right) = \frac{4}{\exp\{4\beta\} - 1} \,.
\end{aligned}
$$

Solving for β, we finally obtain:

$$
\beta = \frac{1}{4} \ln\left(1 + \frac{4}{\langle E_D \rangle}\right) . \tag{5.60}
$$

The later equation offers the possibility to use a Monte Carlo estimate for $\langle E_D \rangle$ and therefore to obtain an estimate of β with error bars. For large demon energies, i.e. for many degrees of freedom, we find the expression

$$
\beta \approx \frac{1}{\langle E_D \rangle} \, .
$$

We stress that (5.60) also holds for small system sizes and only assumes thermalisation. In this case, however, deviations between a standard heat-bath simulation and a micro-canonical simulation can be sizeable.

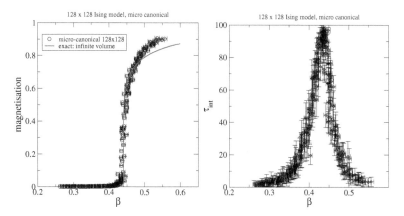

FIGURE 5.12: Magnetisation as a function of the inverse temperature for a 128×128 Ising model (micro-canonical simulation) compared to the exact infinite volume result (left). Integrated autocorrelation time for the magnetisation and the micro-canonical update (right).

5.3.3 Performance and auto-correlations

By specifying the design energy S_{total}, we can get estimates for β via (5.60) and for the magnetisation $\langle M \rangle_{\text{micro}}$ (5.56) in the usual way as an average over the spin configurations generated. This enables us to generate a graph of the implicit function $\langle M \rangle_{\text{micro}}$ depending on the inverse temperature β. Note, however, that β inherits a statistical error from the estimate of $\langle E_D \rangle$.

We studied the Ising model on a square lattice with $N \times N = 128 \times 128$ sites. For each design energy S_{total}, we performed $10\,000$ lattice sweeps. One sweep is defined as follows: choose randomly a site for the spin flip and, in case, update the demon energy if applicable. Repeat this $N \times N$ for one complete lattice sweep. Our findings are shown in Figure 5.12, left panel. The simulation correctly identifies the phase transition. There are deviations to the exact result close to the critical value β_c and at large values of β. Since we are calculating at finite volume, those deviations are expected when compared to the infinite volume result. Figure 5.12, right panel, shows the integrated autocorrelation time. Again, we observe critical slowing down near the criticality. Since the update is "local" as in the case of the heat-bath algorithm and not informed by the spin clusters of the model, this is expected.

The micro-canonical algorithm is easy to implement and a very fast algorithm. Since, however, the total action needs to be known for every demon step, an efficient parallel version is difficult to construct. The micro-canonical simulation is a popular choice in physical chemistry, where it is, e.g. used to study the melting of metals [50].

5.4 Flat histogram methods

In subsection 5.2.1, we have discussed critical slowing down and introduced cluster algorithms as a way to mitigate the long autocorrelation times that characterise local updates. While cluster algorithms are an efficient answer to critical slowing down, they have been devised only for a restricted number of systems. In the absence of a generally applicable cluster algorithm, which depicts the current situation in the field, we can ask whether alternative approaches may work to mitigate long autocorrelation times, and, equally importantly, if these methods exist, what are their limitations.

At first-order phase transitions, long autocorrelation times are generated by free energy barriers. If we find a way to eliminate those barriers, we will end up with a flat energy landscape, which will result in a constant probability density function for generating a configuration. In such a scenario, all configurations will be equally probable. The central question here is how we would map such a random walk in configuration space into the target distribution. In this section, we will discuss flat histogram methods, which achieve this objective numerically.

5.4.1 Reweighting

Let us step back and let us recall the logic of studying a system with a Markov Chain Monte Carlo method. The typical setting consists in computing integrals of the form

$$\langle O(\beta)\rangle = \frac{1}{Z}\int (\mathcal{D}\phi)\, O(\phi)e^{-\beta S(\phi)} , \qquad Z = \int (\mathcal{D}\phi)\, e^{-\beta S(\phi)} . \qquad (5.61)$$

In a standard Monte Carlo approach, to compute O we would generate N configurations $\{\phi\}_i$, $i = 1, \ldots, N$, distributed according the probability density

$$p(\phi) = \frac{1}{Z}e^{-\beta S(\phi)} \qquad (5.62)$$

and then compute the integral as

$$\langle O(\beta)\rangle = \frac{1}{N}\sum_i O(\phi_i) . \qquad (5.63)$$

Note that Eq. (5.62) enables us to rewrite (5.61) as

$$\langle O\rangle = \int (\mathcal{D}\phi)\, O(\phi)p(\phi) , \qquad (5.64)$$

where we have dropped the dependence on β, as now the probability density function $p(\phi)$ can be considered in a wider context. If we have a second probability distribution $p'(\phi)$, we can formally recast the previous equation as

$$\langle O(\beta)\rangle = \int (\mathcal{D}\phi)\, O(\phi)\frac{p(\phi)}{p'(\phi)}p'(\phi)\ . \tag{5.65}$$

Using the normalisation

$$\int (\mathcal{D}\phi)\, p(\phi) = 1\ , \tag{5.66}$$

we can rewrite Eq. (5.65) as

$$\langle O\rangle = \frac{\int (\mathcal{D}\phi)\, O(\phi)\frac{p(\phi)}{p'(\phi)}p'(\phi)}{\int (\mathcal{D}\phi)\,\frac{p(\phi)}{p'(\phi)}p'(\phi)}\ . \tag{5.67}$$

Using the definition of averaging over the distribution $p'(\phi)$, the previous expression becomes

$$\langle O\rangle = \frac{\langle O\frac{p(\phi)}{p'(\phi)}\rangle'}{\langle \frac{p(\phi)}{p'(\phi)}\rangle'}\ , \tag{5.68}$$

where $\langle\rangle'$ indicates the expectation over the distribution $p'(\phi)$. This equation is a central result: it tells us that the expectation value of an observable over a probability distribution $p(\phi)$ is equal to the ratio of two expectation values over a second distribution $p'(\phi)$, the numerator being the expectation value of the product $Op(\phi)/p'(\phi)$ and the denominator the expectation of $p(\phi)/p'(\phi)$. This observation, originally due to Ferrenberg and Swendsen [51], leads to an alternative simulation strategy: instead of generating configurations with the original probability distribution $p(\phi)$, we can generate them with an arbitrary probability distribution $p'(\phi)$ and then use Eq. (5.68) for reconstructing the wanted averages from averages with respect to the simulated distribution. For the strategy to be effective, three conditions must be verified:

1. $p'(\phi)$ (and $p(\phi)$) must be known up to a normalisation factor, for the numerator and the denominator used in the reconstruction to be computable;

2. there must be a numerical advantage in using $p'(\phi)$ instead of $p(\phi)$;

3. dominant configuration for the distribution $p(\phi)$ are also dominant configurations for $p'(\phi)$.

We comment briefly on the last point. Eq. (5.68) is exact, but Monte Carlo methods are stochastic and enable us to generate only finite samples. The practical implication is that Monte Carlo methods will never provide robust tails of the sampling probability distribution if the latter has support in the whole real axis. Therefore, if the target distribution $p(\phi)$ has an important part of its support in the tail of $p'(\phi)$, then relevant configurations for the former will not appear correctly in the sampling of the latter, giving rise to a large systematic error in the measurement. This issue is known as the overlap problem, and will be illustrated more in detail in section 7.2.

5.4.2   Multicanonical sampling

Let us now go back to our original problem: long autocorrelation times at first order phase transitions. In that case, on a finite lattice, in the critical region a doubly peaked probability distribution characterises the system. The relative height of the two peaks can be tuned by varying β. We can define a pseudocrical value $\beta_c(L)$ as the value of β at which the two peaks have the same height. In the limit $L \to \infty$, $\beta_c(L)$ will go to the critical value β_c.

To simplify the scenario of a first order phase transition, let's take a bimodal distribution, such as the one sketched in Figure 5.13. Robust sampling from that distribution will reproduce the two peaks. In practical applications, though, the variable to be sampled, S (whose value has been indicated with E, as usual), scales with the volume. Therefore, we have that the distance between the two peak positions, $E_2 - E_1$, is proportional to V. A local update touches one degree of freedom at a time, which means that a local update produces a change in S that is of order one. To cover an energy span of order V with stochastic updates of order one, one would need at the very least a time $\tau \propto V^2$. This can be understood intuitively by thinking of the Markov chain producing the configurations as a random walk, which is the case when we have a flat histogram. In the presence of a bimodal histogram, an update with changes in the action of order one would need to move the action value through the region with suppressed probability. Therefore, the correlation time (which in this case can be taken as the average number of sweeps that enables us to go from one peak to the other and back, and for this reason is also known as the *tunneling time*) will scale as

$$\tau \propto \frac{P_M}{P_m} f(V) \;, \tag{5.69}$$

where $f(V)$ is a polynomial in V that diverges at least as V^2, P_M is the maximum value of the probability distribution $P(E)$ and P_m is the minimum value of $P(E)$ between the two degenerate maxima.

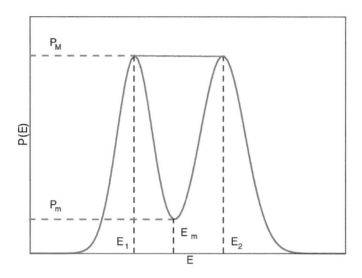

FIGURE 5.13: A bimodal probability distribution, characterised by two degenerate maxima at E_1 and E_2, where the function takes the value P_M, and a minimum P_m between the two peaks, at value E_m of the independent variable. The multicanonical algorithm aims to remove the valley between the two peaks, replacing it with the constant value P_M. The original probability distribution can be obtained through reweighting.

In addition to the dependency on V in $f(V)$, for a physical system at a first order phase transition, there is a dependency on V also in the ratio P_M/P_m, which is exponential:

$$\frac{P_M}{P_m} \propto e^{-cL^{d-1}} , \qquad (5.70)$$

where d is the dimensionality of the system and L its linear size. This dependency is generated by the physics driving the phase transition, and in particular by the presence of bubbles of one phase within the other. More in detail, the height of the valley is determined by the surface free energy density of those bubbles, which, when the bubbles have a size that is comparable with the system size, scale as the surface of the system itself. In fact, the exponent in Eq. (5.70) is nothing but this surface free energy.

An exponentially rising autocorrelation time is in general a huge impediment to the investigation of systems that are of sufficient size for the asymptotic scaling form at criticality to be applicable. Therefore, the availability of algorithms who can mitigate and possibly completely remove this exponential scaling of τ is crucial. The multicanonical method aims to achieve this objective by flattening the probability distribution between the peaks. To this end, we define the multicanonical weight [52, 53] as

$$W_{MC}(E) = \begin{cases} e^{-\beta E}, & E < E_1 \text{ and } E > E_2 \\ \frac{P_M}{P(E)} e^{-\beta E}, & E_1 \leq E \leq E_2 \end{cases} \quad . \tag{5.71}$$

where $P(E) = \rho(E)e^{-\beta E}$, with $\rho(E)$ the density of states for action E. The resulting multicanonical probability distribution coincides with the canonical probability distribution for $E < E_1$ and $E > E_2$, and is constant for $E_1 < E < E_2$. Continuity at $E = E_1$ and $E = E_2$ sets the value of the constant to P_M. The multi-canonical ensemble, as defined by the multicanonical weight, Eq. (5.71), has all the energies between E_1 and E_2 appearing with the same probability in a simulation, which results in the lowest expected tunnelling time between the two degenerate equilibrium state, which, we recall, is proportional to V^2. Distributions of observables in the canonical system can then be obtained by using reweighting, with $p'(E) = W_{MC}(E)$.

At this stage, Eq. (5.71) is a formal definition, since we do not know $P(E)$ before doing the simulation (and in fact if we did we would not need a simulation). In practical applications, one divides the interval between E_1 and E_2 in B bins and performs a first rough simulation, which might employ a smaller lattice, to determine the multicanonical weight by measuring the frequency of occupation of the bins. This process enables us to obtain a first implementation of sampling with the multicanonical weights, which will be more efficient than the original one. We can then iterate the process until we obtain a satisfactorily flat histogram. While this is certainly not the most efficient nor the most practical technique for obtaining the desired flattening of the diagram between the maxima, due to its simplicity, it is of relative easy implementation and understanding.

5.4.3 Wang-Landau sampling

Let us give another look at the multicanonical weight between the peaks:

$$P_{MC} \propto \frac{e^{-\beta E}}{P(E)} = \frac{e^{-\beta E}}{\rho(E)e^{-\beta E}} = \frac{1}{\rho(E)}, \tag{5.72}$$

i.e. a flat histogram between the maxima can be generated by sampling with weight $\rho(E)^{-1}$. This fact is more general. In fact, if we consider the partition function with weight $W(E)$,

$$Z = \int dE \rho(E) W(E), \tag{5.73}$$

we note that a flat histogram is obtained when $W(E) = 1/\rho(E)$. We can reverse the argument: if we are able to pick a weight $W(E)$ such that the resulting histogram is flat, then $W(E)^{-1} = \rho(E)$, i.e. we have determined the density of states, which is a crucial quantity describing the physics of the system. Moreover, if we know $\rho(E)$, we can in principle compute the partition function and observable expectations at any β semianalytically, performing a numerical integration. For instance, we could compute explicitly

$$Z = \int dE \rho(E) e^{-\beta E} . \tag{5.74}$$

We could think about determining ρ with a Monte Carlo method, but this would result in a computational problem of exponential complexity. In fact, in terms of the entropy S, we can write

$$\rho = e^{S} . \tag{5.75}$$

Since S, being an extensive quantity, is subject to fluctuation of order $V^{1/2}$, rho would fluctuate exponentially with the square root of the volume. The resulting statistical noise would overwhelm the signal.

An efficient algorithm for the determination of ρ is due to Wang and Landau [54]. Let us consider the case of a system with discrete energy levels E_n. The partition function is given by

$$Z = \sum_n \rho_n e^{-\beta E_n} . \tag{5.76}$$

In order to determine ρ, we sample with probability distribution $p(E_n) = g_n$, i.e. we consider the ensemble given by

$$\mathcal{Z} = \sum_n \rho_n g_n . \tag{5.77}$$

The purpose of the algorithm is to determine recursively g_n such that $g_n^{-1} = \rho_n$, so that the resulting sampling has a flat histogram. In order to measure the flatness, during a simulation we record the number of entries at E_n, which we call h_n, and define

$$\overline{h} = \frac{1}{M} \sum_n h_n , \tag{5.78}$$

where M is the number of possible energy levels. We define the histogram to be flat if

$$\inf_n h_n \geq k\overline{h} , \tag{5.79}$$

with k a factor order one. In [54], $k = 0.8$ was chosen. Note that a measure of flatness had to be introduced in order to account for fluctuations and for the finiteness of the simulation, and Eq. (5.79) is just a possible choice. Another a priori reasonable choice would be ensuring that the difference between the maximum and the minimum value of the histogram normalised to the former is less than some predetermined value. Numerical experiments will determine the best measure of flatness for a given dynamics.

An important ingredient of the algorithm is how to update the configuration. For a proposed change that drives the system from energy E_l to energy E_m, we accept the configuration with the metropolis probability

$$\pi(E_l \rightarrow E_m) = \min(g_m/g_l, 1) . \tag{5.80}$$

This equation tends to drive the system towards larger g_n, i.e. lower ρ_n, therefore compensating the entropic term which is instead exactly ρ_n.

There is one crucial difference between the Wang-Landau algorithm and more conventional Monte Carlo sampling: unlike in the latter, in the Wang-Landau sampling g_n is not constant, but it is also updated during the simulation: each time an energy level n is visited, the probability g_n is updated according to

$$g_n \rightarrow g_n/f , \tag{5.81}$$

with $f > 1$. States which are visited more often end up with a smaller g, i.e. with a larger ρ, as it is expected.

We have now identified the three main components of the Wang-Landau algorithm: (a) the histogram of entries at each energy level, which is used as a measure of flatness; (b) an update probability that drives the system towards the lowest densities of states, to ensure that the latter are sampled correctly; (c) an update rule for the (inverse) density of states itself, to adjust the latter towards the exact value. With these components, we can now build the Wang-Landau algorithm for determining ρ_n. The algorithms start with an initialisation step:

Initialisation of the Wang-Landay algorithm

1. Choose a sequence f_j of updating factors for g such that

$$f_0 > 1 ; \qquad f_{j+1} > f_j ; \qquad \lim_{j \to \infty} f_j = 1 .$$

2. Initialise g_n with a trial set of values, e.g. the flat distribution

$$g_n = 1 \;\; \forall n .$$

At fixed recursion j, the Wang-Landau algorithm updates g_n according to the following steps:

Updates of g at iteration j:

Set
$$h_n = 0 \quad \forall n \ ,$$

then repeat

1. Propose an update of the elementary variables, and call E_l the current value of the energy and E_m the proposed one.

2. Accept the proposed configuration with probability
$$\pi(E_l \to E_m) = \min(g_m/g_l, 1) \ .$$

3. If the configurations have been accepted, set
$$g_m \leftarrow f g_m \ , \qquad h_m \leftarrow h_m + 1 \ ,$$

   otherwise set
$$g_l \leftarrow f g_l \ , \qquad h_l \leftarrow h_l + 1 \ ,$$

until
$$\inf_n h_n \geq k \overline{h} \ .$$

The above algorithm is iterated for j iterations, depending on the required precision on the density of states that one aims to obtain. For instance, one can stop when $f_{\text{final}} \simeq 10^{-8}$ (i.e. the value at which $f = 1$ within single precision). The final value of g_n gives inverse density of states, not normalised.[8]. The latter can then be used in computing averages of observables depending on E as a function of β without running new simulations, according to the formula:

$$\langle O(E) \rangle = \frac{\sum_n O(E_n) g_n^{-1} e^{-\beta E_n}}{\sum_n g_n^{-1} e^{-\beta E_n}} \tag{5.82}$$

or used as weights in a multicanonical simulation. The latter approach has the advantage of not resulting in systematic errors due to the approximation of ρ_n with g_n^{-1}.

In order to obtain the correct result, the choice of the factors f_j with which g_n is updated is crucial. In the original work [54], the authors started with $f_1 = e$ and decreased f following the recursion relation $f_{j+1} = \sqrt{f_j}$. It was later realised in [55] that this leads to systematic errors, due to the high

---

[8]This is an important remark, as, while the normalisation of the density of states does not play any role in the computation of observables that can be expressed as ensemble averages, normalisation does enter in quantities such as the free energy and the entropy. The density of states can be correctly normalised if one knows the exact density of states in a particular limit.

velocity of convergence of the sequence. The authors of [55] also showed that the correct convergence is provided by a sequence that converges more slowly, choosing for instance $f_j \propto 1/j$.

5.4.4 Do flat histogram methods work?

To test the very important claim that flat histogram methods enable simulations in polynomial times in situations in which one would expect exponential complexity with more standard approaches, algorithms and prescription for simulating with a flat histogram have been inspected very closely from a broad angle. Using the exact density of states of the Ising model, in [56] the authors showed that a perfect flat histogram method is still bound to have a residual exponential scaling. In a similar spirit, the authors of [57] showed that the multicanonical algorithm is bound to have a residual exponential slowing down at a first-order phase transition. This is related to the complexity of the dynamics at the transition, which results in free energy barriers that are not overcome by the multicanonical approach.

These studies suggest that physical phenomena that determine critical slowing down and exponential transition times can be of such a high level of complexity that no current technique eliminates completely the corresponding algorithmic difficulties related to the exponentially growing computational cost. Yet, in practical applications, flat histogram methods have proved to be able to provide a significant improvement in terms of efficiency and precision in cases where importance sampling methods fail spectacularly.

Several improvements have been proposed in the literature that increase the efficiency of flat histogram methods, enabling either faster convergence or better accuracy of the result. Besides, additional prescriptions based on the flat histogram idea have been provided, resulting in new approaches that are full of physical insights. Given the fast pace at which the literature in the field is moving, it would be impossible to provide an accurate and current picture in this book. Therefore, we have restricted our treatment to the foundational concepts that enable the reader to engage with the current scientific literature.

5.5 The Linear Logarithmic Relaxation (LLR) method

While the Wang-Landau techniques are based on histograms, controlling the induced uncertainties is not always straightforward: when is the histogram "flat enough"? How does the bin size affect the precision? The LLR method is a modified Wang-Landau approach that offers improvements with respect to those questions.

5.5.1 The Linear Logarithmic Relaxation formulation

The density-of-states approach falls into the class of multi-canonical or Wang-Landau techniques. It offers a high-performing alternative for cases in which Monte Carlo importance sampling methods are inefficient. We will be discussing the field theoretical setting of a potentially large set of degrees of freedom denoted by ϕ, which are randomly distributed according to the Gibbs factor and with the partition function Z as normalisation.

$$\exp\left\{\beta \, S[\phi]\right\} / Z(\beta) \,, \qquad Z(\beta) = \int \mathcal{D}\phi \, e^{\beta S[\phi]} \,.$$

The density-of-states is the marginal distribution of composite observable, say $M(\phi)$:

$$\rho(m) = \int \mathcal{D}\phi \, \delta\Big(M(\phi) - m\Big) \, e^{\beta S[\phi]} \,.$$

Once we have an estimate of ρ at our finger tips, the partition function and the expectation value of any function of M, $O(M)$, can be easily calculated by simple integration:

$$Z(\beta) = \int dm \, \rho(m) \,, \qquad \langle O \rangle = \frac{1}{Z(\beta)} \int dm \, \rho(m) \, O(m) \,.$$

In the following, we will study the case when the marginal is the action $S[\phi]$ itself. In this case, we can define the density-of-states in a slightly different manner:

$$\rho(E) = \int \mathcal{D}\phi \, \delta\Big(S[\phi] - E\Big) \,, \qquad Z(\beta) = \int dE \, \rho(E) \, e^{\beta \, E} \,. \tag{5.83}$$

In any case, the density-of-states is the probability distribution of a marginal,

$$\rho(m) \quad \text{for} \quad M, \qquad \rho(E) \, e^{\beta E} \quad \text{for} \quad E = S[\phi] \,,$$

and hence a postive quantity. We will also assume that it is a sufficiently smooth function so that we can trade $\rho(E)$ for the so-called LLR-coefficient $a(E)$ by the mapping

$$\rho(E) = \exp\left\{ - \int_0^E dE' \, a\,(E') \right\} \,, \tag{5.84}$$

where we arbitrarily have chosen the normalisation such that $\rho(0) = 1$. Under the smoothness assumption, the above equation is invertible:

$$a(E) = - \frac{d}{dE} \ln \rho(E) \,.$$

The so-called Linear Logarithmic Relaxation (LLR) method [58, 59] iteratively estimates $a(E)$ for a discrete set of values E_k using a sequence of Monte Carlo simulations.

Let us assume we would like to estimate $a(E_k)$. The tools for this programme are the "double bracket" expectation values:

$$\langle\langle f \rangle\rangle (a) \;=\; \frac{1}{\mathcal{N}} \int \mathcal{D}\phi \; f(\phi) \; W\left(S[\phi] - E_k, \delta\right) \; \exp\{a\,(S[\phi] - E_k)\} \quad (5.85)$$

$$\mathcal{N} \;=\; \int \mathcal{D}\phi \; W\left(S[\phi] - E_k, \delta\right) \; \exp\{a\,(S[\phi] - E_k)\} \;.$$

Thereby, $W(x, \delta)$ restricts the integration range to a slice of thickness δ around the hyper-surface defined by $S[\phi] - E_k = 0$. Popular choices are

$$W(x, \delta) \;=\; \begin{cases} 1 & \text{for } x \in [-\delta, \delta] \,, \\ 0 & \text{else} \,. \end{cases} \quad (5.86)$$

$$W(x, \delta) \;=\; \exp\left\{-\frac{x^2}{\delta^2}\right\}\,. \quad (5.87)$$

Note that, despite the double-bracket notation, $\langle\langle f \rangle\rangle (a)$ in (5.84) are standard expectation values, which can be estimated with MCMC methods.

Let us consider a particular type:

$$g(a) \;=\; \langle\langle S[\phi] - E_k \rangle\rangle (a) = \frac{1}{\mathcal{N}} \int \mathcal{D}\phi \left(S[\phi] - E_k\right)$$
$$W\left(S[\phi] - E_k, \delta\right) e^{a(S[\phi] - E_k)}. \quad (5.88)$$

Using the shorthand notation $\Delta S := S[\phi] - E_k$ and $\bar{S} := \langle\langle \Delta S \rangle\rangle$, we can show that $g(a)$ is a monotonically increasing function of a:

$$\frac{d}{da} g(a) \;=\; \langle\langle \Delta S^2 \rangle\rangle - \langle\langle \Delta S \rangle\rangle^2 = \langle\langle (\Delta S - \bar{S})^2 \rangle\rangle \;>\; 0.$$

Inserting the unit element

$$1 \;=\; \int dE \; \delta\left(S[\phi] - E\right)$$

into the expression (5.84), we can replace $S[\phi]$ by E_k by virtue of the Dirac δ-function:

$$\langle\langle S[\phi] - E_k \rangle\rangle (a) \;=\; \frac{1}{\mathcal{N}} \int dE \int \mathcal{D}\phi \left(E - E_k\right) W\left(E - E_k, \delta\right) e^{a(S[\phi] - E_k)}$$
$$=\; \frac{1}{\mathcal{N}} \int dE \, \rho(E) \left(E - E_k\right) W\left(E - E_k, \delta\right) e^{a(E - E_k)}.$$

For illustration purposes, we use the window function (5.86). Changing then integration variable introducing $e = E - E_k$, we find

$$\langle\langle S[\phi] - E_k \rangle\rangle (a) \;=\; \frac{1}{\mathcal{N}} \int_{-\delta}^{\delta} de \, \rho(E_k + e) \, e \, e^{a\,e} = \frac{\rho(E_k)}{\mathcal{N}} \int_{-\delta}^{\delta} de \, \frac{\rho(E_k + e)}{\rho(E_k)} \, e \, e^{a\,e}$$
$$=\; \frac{\rho(E_k)}{\mathcal{N}} \int_{-\delta}^{\delta} de \, e \, \exp\left\{a\,e - \int_{E_k}^{E_k + e} dE' \, a(E')\right\}. \quad (5.89)$$

Along the same lines, we find

$$\mathcal{N} = \rho(E_k) \int_{-\delta}^{\delta} de \, \exp\left\{ a\,e \, - \int_{E_k}^{E_k+e} dE' \, a(E') \right\} .$$

We are now looking for a solution of the equation:

$$\langle\langle S[\phi] - E_k \rangle\rangle(a) = 0 . \tag{5.90}$$

We first prove that the solution exists and is unique. To this aim, note that, for large $a \gg 0$, the integrals will be dominated by the upper boundary. For large negative a, i.e. $a \ll 0$, the dominant contributions rise from the lower boundary. Thus, we obtain

$$\lim_{a\to\infty} \langle\langle S[\phi] - E_k \rangle\rangle(a) = \delta , \qquad \lim_{a\to-\infty} \langle\langle S[\phi] - E_k \rangle\rangle(a) = -\delta .$$

We have already shown that $\langle\langle S[\phi] - E_k \rangle\rangle(a)$ is smooth and monotonically increasing with a. Therefore, a solution exists and is unique.

For small enough $\delta > 0$, we can calculate the solution. Expanding $a(E)$ around $E = E_k$,

$$a(E) = a(E_k) + a'(E_k)\,(E - E_k) + \dots ,$$

we find after a systematic expansion in e:

$$\langle\langle S[\phi] - E_k \rangle\rangle(a) = \frac{\delta^2}{3}\left(a - a(E_k) \right) + \mathcal{O}(\delta^4) .$$

The desirable solution is given by

$$a = a(E_k) . \tag{5.91}$$

If we set $a = a(E_k)$, we find

$$\langle\langle S[\phi] - E_k \rangle\rangle\left(a = a(E_k) \right) = -\frac{d^2 a(E)}{dE^2}\Big|_{E=E_k} \frac{\delta^4}{30} .$$

and, thus, the criterion for a good approximation with respect to the expansion in δ:

$$\delta \ll \left| \frac{1}{30} \frac{d^2 a(E)}{dE^2}\Big|_{E=E_k} \right|^{-1/4} .$$

We finally arrive at the key ingredient to the LRR approach: Solve the stochastic non-linear equation (5.90) and you obtain the estimate (5.91) of the LLR coefficient at the $E = E_k$.

5.5.2 The Robbins-Monro solution

Our task is now to solve the stochastic non-linear equation (5.88) to estimate the LLR coefficient at a given value $E = E_k$. We already established that $g(a)$ is a smooth and bounded function of a with exactly one solution. An iterative approach to finding the solution to non-linear equations is the standard Newton-Raphson method. Starting with an initial guess a_1 for the solution, the method iteratively refines the solution:

$$a_{n+1} = a_n - \frac{g(a_n)}{g'(a_n)} \,. \tag{5.92}$$

Using (5.89), we can write $g(a)$ as the expectation value with the distribution $W(a)$:

$$g(a) = \langle e \rangle_w = \frac{\int de\, e\, W(e)}{\int de\, W(e)} \,, \quad W(e) = \exp\left\{ ae - \int_{E_k}^{E_k+e} dE'\, a(E') \right\} \,,$$

which yields

$$g'(a) = \langle e^2 \rangle_w - \langle e \rangle_w^2$$

We can estimate this derivative at the solution a_∞ and using the expansion with respect to δ:

$$\langle e \rangle_w(a_\infty) = 0 \,, \quad \langle e^2 \rangle_w = \frac{1}{3}\delta^2 - \frac{2}{45}\frac{da(E)}{dE}\Big|_{E=E_k} \delta^4 + \mathcal{O}(\delta^6) \,.$$

Instead of the Newton-Raphson iteration (5.92), the LLR method uses a *fixed point* iteration instead by replacing $g'(a_n)$ by $g'(a_\infty)$ (in the small δ approximation):

$$a_{n+1} = a_n - \frac{3}{\delta^2} \langle\!\langle S[\phi] - E_k \rangle\!\rangle(a_n) \,. \tag{5.93}$$

Please note that coefficient in front of the double expectation value generically is proportional to $1/\delta^2$. The additional numerical constant, however, does depend on the Window function. For example, for the Gaussian Window function (5.87), the pre-factor would be $2/\delta^2$.

The of the stochastic non-linear equation by the iteration of (5.93) is not straightforward. The issue is that we only can obtain *estimates* of the expectation value involved, i.e.

$$\langle\!\langle S[\phi] - E_k \rangle\!\rangle(a_n) \,,$$

which is afflicted by stochastic noise. In practice, we invoke a MCMC process for any a and generate a sequence of random variables

$$(S[\phi] - E_k)_1 \to (S[\phi] - E_k)_2 \to (S[\phi] - E_k)_3 \to \dots \to (S[\phi] - E_k)_N \,.$$

and estimate

$$\langle\!\langle S[\phi] - E_k \rangle\!\rangle(a_n) \approx \frac{1}{N} \sum_{i=1}^{N} (S[\phi] - E_k)_i.$$

We need to truncate the sequence. The stochastic error decreases like $1/\sqrt{N}$ for sufficiently large N. Using large enough N would certainly vive a good estimate for the "drift term" $\langle\!\langle S[\phi] - E_k \rangle\!\rangle(a_n)$, but that only leads us to a better approximation a_{n+1}. For small n during the iteration, we probably do not need to have a high accuracy estimate since we are anyhow only interested in a_∞, and large N are computationally expensive. Clearly, we need to strike a balance, but which?

Fortunately, the question was answered by Robbins and Monroe in a seminal paper in the fifties [60]. The topic "solution of non-linear stochastic equation" has been further investigated (see [61, 62, 63] for further reading). We sum- marise their findings. Robbins and Monro propose an under-relaxed iterative approach:

$$a_{n+1} = a_n - \alpha_n \langle\!\langle S[\phi] - E_k \rangle\!\rangle(a_n).$$

with a sequence of postive weights α_n. These coefficients satisfy

$$\sum_{n=1}^{\infty} \alpha_n \to \infty, \qquad \sum_{n=1}^{\infty} \alpha_n^2 \to \text{finite}.$$

A popolular choice is $\alpha_n \propto 1/n$. For a fixed finite length N of the iteration, we repeat from scratch (i.e. starting with a_1 and using a different set of random numbers for the MCMC simulation) the Robbins Monroe iteration a number of times, which generates the independent random variables

$$a_N^{(1)}, a_N^{(2)}, a_N^{(3)}, a_N^{(4)}, \dots.$$

Robbins and Monro proved that the set of variables above is normally dis- tributed with the correct mean a_∞.

We are now in the position to piece together the LLR approach:

1. Choose a start value a_1, n_{ita}, i.e. the number of random variables used to estimate the double-expectation values, and n_{rm} – the number of Robbins Monroe iterations. Choose the set of values E_k, $k = 1 \dots n_x$ for which we are going to estimate the LLR coefficient.

2. Carry out the Robbins Monroe iteration

$$a_{n+1} = a_n - \frac{3}{\delta^2 n} \langle\!\langle S[\phi] - E_k \rangle\!\rangle(a_n), \qquad n = 1 \dots n_{\text{rm}} \qquad (5.94)$$

for each value E_k.

3. Repeat steps 1–2 N_{rept} times and store the LLR coefficients

$$a_{n_{\text{rm}}}^{(1)}, \ a_{n_{\text{rm}}}^{(2)}, \ a_{n_{\text{rm}}}^{(3)}, \ \ldots \ a_{n_{\text{rm}}}^{(N_{\text{rept}})}.$$

4. Use the data to reconstruct the density of states and observables including error making use of the copies of the LLR coefficient, e.g. via the bootstrap approach.

Subsequently, other methods have been developed to solve the non-linear stochastic equation (5.90). We here only refer to the *Functional Fit Approach* [64, 65, 66]. The method starts with the assumption that the LLR coefficient in the integral (5.89) can be approximated by a constant

$$a(E) \ = \ a_1 \ .$$

We then find with a short calculation:

$$\langle\!\langle S[\phi] - E_k \rangle\!\rangle(a) \ = \ \delta \ \frac{1}{\tanh((a - a_1)\delta)} \ .$$

By numerical calculating the left side for a set of a values, the method fits the functional form on the right to obtain the desired value a_1. In practice, the method performs very well. However, it lacks the information on the stochastic error propagation when compared with the approach using the Robbins-Monro iteration (5.94).

5.5.3   LLR and observables

From the definition of the density-of-states (5.83), it is clear that the expectation value of any function of the action $S[\phi]$ can be reduced to a one-dimensional integral if the density $\rho(E)$ is known:

$$\langle f(S[\phi]) \rangle \ = \ \frac{\int \mathcal{D}\phi \ f(S[\phi]) \ e^{\beta S[\phi]}}{\int \mathcal{D}\phi \ e^{\beta S[\phi]}} \ = \ \frac{\int dE \ f(E) \ \rho(E) \ e^{\beta E}}{\int dE \ \rho(E) \ e^{\beta E}} \ . \quad (5.95)$$

What, however, if we are interested in more generic observables $h(\phi)$ that are not necessarily function of the action only?

Making again use of the unit element

$$1 \ = \ \int dE \ \delta\Big(S[\phi] - E\Big) \ , \quad (5.96)$$

we find

$$\int \mathcal{D}\phi \ h(\phi) \ e^{\beta S[\phi]} \ = \ \int dE \ \mathcal{D}\phi \ h(\phi)\delta\Big(S[\phi] - E\Big) \ e^{\beta E}$$

$$= \ \int dE \ e^{\beta E} \int \mathcal{D}\phi \ \delta\Big(S[\phi] - E\Big) \ \frac{\int \mathcal{D}\phi \ h(\phi) \ \delta\Big(S[\phi] - E\Big)}{\int \mathcal{D}\phi \ \delta\Big(S[\phi] - E\Big) \ .}$$

We define the average of $h(\phi)$ over the hyper-surface of $S[\phi] = E$ by

$$\left(h(\phi)\right)_E = \frac{\int \mathcal{D}\phi \, h(\phi) \, \delta\left(S[\phi] - E\right)}{\int \mathcal{D}\phi \, \delta\left(S[\phi] - E\right)}. \tag{5.97}$$

We then use the definition of the density-of-states and obtain the important formula:

$$\langle h(\phi) \rangle = \frac{\int \mathcal{D}\phi \, h(\phi) \, e^{\beta S[\phi]}}{\int \mathcal{D}\phi \, e^{\beta S[\phi]}} = \frac{\int dE \, \left(h(\phi)\right)_E \, \rho(E) \, e^{\beta E}}{\int dE \, \rho(E) \, e^{\beta E}}. \tag{5.98}$$

The LLR method provides an excellent approach to the function $\rho(E)$, but how do we calculate $(h)_E$ in (5.97)?

To this aim, we study the double-bracket expectation value of $h(\phi)$ for a particular value $E_k = S[\phi]$:

$$\langle\langle h(\phi) \rangle\rangle(a; E_k) = \frac{1}{\mathcal{N}} \int \mathcal{D}\phi \, h(\phi) \, W\left(S[\phi] - E_k, \delta\right) \, e^{a(S[\phi] - E_k)}.$$

Inserting (5.96), we find after a short calculation:

$$\begin{aligned}
\langle\langle h(\phi) \rangle\rangle &= \frac{1}{\mathcal{N}} \int dE \, W(E - E_k, \delta) \, e^{a(E - E_k)} \int \mathcal{D}\phi \, h(\phi) \, \delta\left(S[\phi] - E\right) \\
&= \frac{1}{\mathcal{N}} \int dE \, W(E - E_k, \delta) \, e^{a(E - E_k)} \, \rho(E) \, \frac{\int \mathcal{D}\phi \, h(\phi) \, \delta\left(S[\phi] - E\right)}{\int \mathcal{D}\phi \, \delta\left(S[\phi] - E\right)}. \\
&= \frac{1}{\mathcal{N}} \int dE \, W(E - E_k, \delta) \, e^{a(E - E_k)} \, \rho(E) \, \left(h(\phi)\right)_E. \tag{5.99}
\end{aligned}$$

It is time to look at the normalisation $\mathcal{N}$:

$$\begin{aligned}
\mathcal{N} &= \int \mathcal{D}\phi \, W\left(S[\phi] - E_k, \delta\right) \, e^{a(S[\phi] - E_k)} \\
&= \int dE \, W(E - E_k, \delta) \, e^{a(E - E_k)} \int \mathcal{D}\phi \, \delta\left(S[\phi] - E\right) \\
&= \rho(E_k) \int dE \, W(E - E_k, \delta) \, e^{a(E - E_k)} \, \frac{\rho(E)}{\rho(E_k)}.
\end{aligned}$$

Using the step function as Window function, see (5.86), we find the familiar result upon expanding with respect to δ and setting $a = a(E_k)$:

$$\begin{aligned}
\mathcal{N} &= \rho(E_k) \int_{-\delta}^{\delta} de \, \exp\left\{a \, e - \int_{E_k}^{E_k + e} dE' \, a(E')\right\} \\
&= 2\delta \, \rho(E_k) \left[1 - \frac{1}{4} \frac{da(E)}{dE}\Big|_{E = E_k} \delta^2 + \dots\right]. \tag{5.100}
\end{aligned}$$

We now assuming that $(h)_E$ is a smooth function of E and admits a Taylor expansion around $E = E_k$. Along the same lines, we expand the numerator of (5.99):

$$\int dE \, W(E - E_k, \delta) e^{a(E - E_k)} \, \rho(E) \left(h(\phi) \right)_E$$

$$= 2 \, \delta \, \rho(E_k) \left\{ (h)_{E_k} + \frac{1}{6} \left[\frac{d^2 h}{dE^2} - \frac{da}{dE} \right] \delta^2 + \ldots \right\} . \quad (5.101)$$

Inserting (5.100) and (5.101) into (5.99), we obtain the central result:

$$\left(h(\phi) \right)_{E_k} = \langle\!\langle h(\phi) \rangle\!\rangle \left(a = a(E_k); E_k \right) + \mathcal{O}(\delta^2) . \quad (5.102)$$

The latter equation implies that we can estimate the desirable function (5.97) using standard MCMC simulations and the double-bracket expectation value (5.102).

The calculation of the expectation value of an observable such as $h(\phi)$ using the LLR method now proceeds as follows:

1. Choose a value $E_k = S[\phi]$ and calculate the LLR coefficient $a(E_k)$ using the Robbins-Monro iteration truncated at n_{rm}.

2. Keep $a_{n_{rm}}$ fixed and estimate the double-bracket expectation value $\langle\!\langle h \rangle\!\rangle$ using n_{obs} MCMC samples.

3. Repeat steps 1–2 N_{rept} times, and store the tuples

$$\left(a_{n_{rm}}, \langle\!\langle h \rangle\!\rangle \right)^{(1)}, \quad \left(a_{n_{rm}}, \langle\!\langle h \rangle\!\rangle \right)^{(2)}, \quad \ldots \quad \left(a_{n_{rm}}, \langle\!\langle h \rangle\!\rangle \right)^{(N_{rept})} .$$

4. Repeat 1–3 for all values E_k of the set.

5. Using the data set to reconstruct the density-of-states and the expectation value $\langle h \rangle$ (5.98).

Note that the only β dependence is that explicitly shown in (5.98). This means that we can obtain the expectation value for a continuous spectrum of β values by performing two one-dimensional integrals.

5.5.4   LLR showcase: the Ising model

Let us illustrate the LLR techniques in action for the Ising model in two dimensions. We will large work with $L = 32$ lattice size. The degrees of freedom are now Ising spins, i.e. $\phi = \sigma = \{\pm 1\}$. The action is given by

$$S[\sigma] = \sum_{\langle xy \rangle} \sigma_x \sigma_y , \quad (5.103)$$

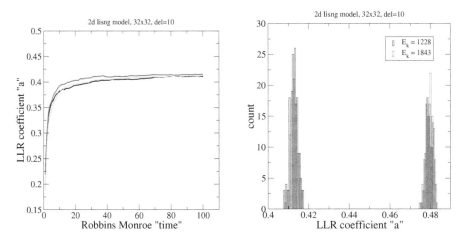

FIGURE 5.14: Left: Three Robbins Monro trajectories for the 2d Ising model.

where x and y are neighbouring sires of the lattice. Maximal action is achieved when all spins are the same, i.e. $S_{max} = 2 L^2$.

For the LLR approach, we first pick an action hyper-surface by choosing ten value E in $E = S[\sigma]$. For the illustration here, we have chosen $E = 1228 \approx 0.6 S_{max}$. For the simulation, we have chosen $N_{ita} = 100$ elements in the Monte–Carlo chain for the estimate of the double-bracket expectation values (5.94). The width of the Window function is set to $\delta = 10$. We start the Robbins Monroe iteration with the start value $a_1 = 0$. Our result for three random starts of the iterations is shown in 5.14. We only show the first 100 Robbins Monroe steps. In actual simulations, a total number of iterations of $n_{RM} = 4,000$ or $n_{RM} = 10,000$ is a good choice for the current setting. We observe that the three trajectories start converging for large MC times. For two action slices defined by E, we have studied the distribution of $a_{n_{RM}}$ by creating a histogram from 200 Robbins Monroe trajectories. The result is also shown in Figure 5.14, right panel. We can see that the distributions are sharply localised in line with the expectations form the Robbins Monroe paper that those are Gaussian distributions. Needless to say that the width of those decrease with increasing n_{RM}.

We observed in the last subsection that we recover expectation values with the LLR method with a systematic error of order $\mathcal{O}(\delta^2)$, see (5.102). In practice, this means that we need to check that our results are already sufficiently independent of δ by exploring a small set of them. We are not only interested in the density-of-states ρ, derived from the LLR coefficient a, but also in observable, which is not a sole function of the action S. We have chosen here

$$h(\phi) \; \equiv \; h(\sigma) \; = \; \left(\frac{1}{L}\sum_x \sigma_x\right)^2 . \tag{5.104}$$

Note that due to the square in (5.104), the Z2 symmetry *does not* imply that the corresponding expectation value vanishes. In fact, we expect that for $\beta < \beta_c$, the Ising spins on the lattice are disordered implying

$$\frac{1}{L}\sum_x \sigma_x \approx 0 ,$$

while, in the broken phase for $\beta > \beta_c$, we expect

$$\left(\frac{1}{L}\sum_x \sigma_x\right)^2 \; = \; \mathcal{O}(1) .$$

For a systematic study of the δ dependence, we have chosen

$$n_{\mathrm{RM}} = 4,000, \qquad E \; = \; 1638 \approx 0.8\, S_{\mathrm{max}}.$$

For δ as large as 50, we observe a mild dependence on δ, which can be modelled by a quadratics. The choice of δ is uncritical and we have chosen $\delta = 10$ or $\delta = 20$ for most simulations shown here.

Once we have identified a suitable parameter range for δ and n_{RM}, the task is to estimate the LLR coefficient a and the observable $(h)_E$ for a set of action values E_i. In oder to obtain error bars for our mean values, we repeat each Robbins Monro trajectory N_{rept} times. With $N_{\mathrm{rept}} = 1,000$, our final result for 380 values of E is shown in Figure 5.15, right panel. We observe that a is an odd function of E, which makes $\rho(E)$ (5.84). This is expected due to ferromagnetic and anti-ferromagnetic symmetry for "un-frustrated" lattices (L even). Also shown is $(h)_E$, defined in (5.97) and our choice (5.104).

In order to reconstruct the density-of-states $\rho(E)$ via (5.84), we expand the LLR coefficient into a Taylor polynomial in E of order. To this aim, we increase the order of the Taylor polynom until we achieve a satisfactory representation of the numerical data within error bars. For the present case, we use the fact that $a(E)$ is an odd function and only use odd powers of E in the Taylor expansion. Once the taylor polynomial is obtained, we can calculate ρ from (5.84) by analytically integrating the polynomial. We then have an analytical expression of $\rho(E)$ at our finger tips, where the information of the underlying theory of interest is encoded in the taylor coefficients. Further details of the approach are presented in subsection 7.2.2.

For data at hand, a polynomial of grade 17 yields good results. The probability density function ofr he marginal E is then simply given by

$$\rho(E) \; \exp\{\beta E\} .$$

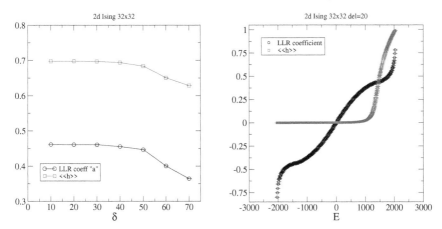

FIGURE 5.15: Left: δ-dependence of the LLR coefficient and $\langle\langle h \rangle\rangle$ (5.104). Right: the final result for the LLR coefficient a and the observable $(h)_E$ as a function of the action E.

Figure 5.16 shows this marginal distribution as a function of β for the 2d Ising model of size $L = 32$. The normalisation is chosen such that the distribution is 1 at maximum. We observe that the marginal distribution is sharply peaked at some value $E_{\max}$. It is expected that the width in relation to the volume $V = L^2$ decrease like $\sqrt{V}/V$ so that domain for which the distributions take significant values even shrinks further for increasing V. Having $\rho(E)$ and $(h)_E$ now at our disposal, it would be up to the 2 one-dimension integrals in (5.98) to calculate any expectation value $\langle h \rangle(\beta)$. We would like to pursue a different

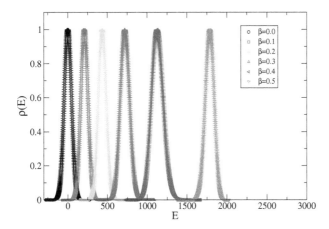

FIGURE 5.16: Marginal distribution $\rho(E) \exp\{\beta E\}$ for the 32×32 Ising model for six β values.

route and introduce an excellent approximation, the quality of which increases further with increasing volume.

Using that the marginal distribution is sharply peaked, we use the so-called *classical approximation*:

$$\langle h(\phi) \rangle \;=\; \frac{\int dE \left(h(\phi) \right)_E \rho(E)\, e^{\beta E}}{\int dE\, \rho(E)\, e^{\beta E}} \;\approx\; \left(h(\phi) \right)_{E_{\max}} . \qquad (5.105)$$

It remains to find $E_{\max}$ and the corresponding value for $(h(\phi))_E$. To this aim, we note with the defintion of ρ (5.84) that

$$d(E) := \rho(E)\, \exp\{\beta E\} \;=\; \exp\left\{ \int_0^E dE' \left(\beta - a(E') \right) \right\} .$$

We assume that there is only *one* maximum for the marginal distribution. This assumption is not always justified. In fact, for theories with a first order phase transition, the marginal has a double peak structure for parameters close to the phase transition. This is, however, not the case here.

Under the above assumption, we find

$$\frac{d}{dE} d(E) \;=\; \rho(E)\, \exp\{\beta E\} \left[\beta - a(E) \right] \;=\; 0$$

We come to the simple solution

$$a\,(E_{\max}) \;=\; \beta\,. \qquad (5.106)$$

These findings now allow to plot any observable as a function of β using the LLR method in classical approximation. There are the steps:

1. Choose a value E for the constraint $E = S[\phi]$ and take it as $E_{\max} = E$.

2. Calculate the corresponding LLR coefficient $a(E_{\max})$ using the Robbins Monro iteration (5.94) and obtain $(h)_{E_{\max}}$ using (5.102).

3. Repeat step 2 to generate a sequence of tuple for $(a, (h))$. Use this sequence to calculate the average of a and (h) including their errors.

4. Identify the average of a as β (This is the classical approximation (5.106)) and plot a point (h) versus a including error bars for both.

5. Repeat steps 1–4 and thus generate an implicit plot of $\langle h \rangle$ as a function of β.

Choosing the average spin density squared as observable (see (5.104)), the outcome is shown in Figure 5.17 for the 2d Ising model of size 32×32. The LLR classical approximation clearly reproduces the critical value β_c. Also shown is the Onsager solution for comparison. Note that the Onsager result is valid for $L \to \infty$. Deviations of our numerical result at $\beta < \beta_c$ and for larger β values are due to the finite volume effects.

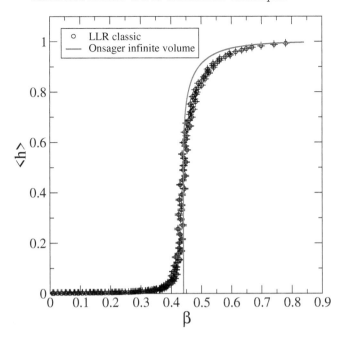

FIGURE 5.17: Expectation value of $\langle m^2 \rangle$ as a function of β obtained from the LLR approach in classical approximation. Also shown is the Onsager solution for the infinite volume case.

5.6 Exercises

Exercise 5.1 *Reconsider the example of subsection 5.1.4 for a Hybrid-Monte Carlo simulation of a simple system of two variables (x, y), say coordinates.*

a. *Choose $\delta t = 0.01$ and vary $n_h = 1 \ldots 200$. For a starting point of $(2, 0)$, plot several HMC trajectories in the xy-plane. Note that those trajectories depend on the initial (random choice) of momentum $(p_x, p_y)_0$.*

b. *Estimate and visualise the autocorrelation function for the position vector $\vec{x} = (x, y)$:*

$$C_x(\tau) = \left\langle \vec{x}(0)\vec{x}(\tau) \right\rangle / \left\langle \vec{x}^2 \right\rangle.$$

Why is $\langle \vec{x} \rangle = 0$?

Exercise 5.2 *Reconsider again the example of subsection 5.1.4. The purpose is now to make a fair comparison with the Metropolis algorithm, in terms of best computational effort.*

a. *Give an (at least rough) estimate of the number of floating point oper-ations that are needed for each step of Metropolis, and for each step of HMC (in terms of the chosen number of integration steps for the latter case).*

b. *Repeat the numerical exercise, and in particular Figure 5.3 (right), rescaling the horizontal axis by the number of floating point operations needed for each step, so that the comparison is made in terms of com-putational effort.*

c. *Now try to find the best possible parameters, which optimize the inte-grated autocorrelation time, both for Metropolis and for HMC, so as to reach a definite assessment about the suggested algorithm for this model.*

Exercise 5.3 *Consider the problem of numerically integrating the equations of motion for a particle of mass m moving in presence of an attractive Coulomb potential, i.e. with Hamiltonian*

$$H = \frac{1}{2m}\left(p_r^2 + \frac{p_\theta^2}{r^2} + \frac{p_\phi^2}{(\sin\theta)^2 r^2} \right) - \frac{\alpha}{r}$$

where p_r, p_θ and p_ϕ are the momenta conjugate respectively to the radial co-ordinate and to the polar and azimuthal angles, while α is a positive constant.

Write down the elementary leapfrom step for this system, and try to make it explicit, i.e. ready to implement in your integration algorithm.

Exercise 5.4 *Show that the Swendsen-Wang Update algorithm from subsec-tion 5.2.2 is ergodic.*

To this aim, we need to convince ourselves that any spin configuration $\{\sigma_x\}$ can be generated with a non-zero probability. We will do this by showing that we can flip any single spin with non-zero probability while all the others remain the same.

Consider an arbitrary spin σ_x on the lattice. Consider the cases that this spin has $0, 1, 2, 3, 4$ neighbouring spins that are the same and calculate for each of the cases the probability p_k, $k = 0 \ldots 4$ that σ_x forms a single spin cluster.

Exercise 5.5 *Show that the Swendsen-Wang Update algorithm from subsection 5.2.2 satisfies detailed balance.*

If $W(\sigma \to \sigma')$ is the Markov chain transition probability for two given lattice configurations σ and σ', show

$$\exp\left\{\beta \sum_{\langle xy \rangle} \sigma_x \sigma_y\right\} W(\sigma \to \sigma') = \exp\left\{\beta \sum_{\langle xy \rangle} \sigma'_x \sigma'_y\right\} W(\sigma' \to \sigma).$$

Chapter 6

From Statistical Systems to Quantum Field Theory*

One of the main targets for the stochastic methods treated in this Volume is the study of interacting statistical systems, like the Ising model introduced in Chapter 4, and other similar models for which Monte–Carlo methods represent one of the most effective investigation tools. There is another major application, which has fostered the development of high-performance computing over the last few decades, and consists in the numerical simulation of quantum field theories (QFTs), which represent our present tool to describe fundamental interactions, i.e. the behaviour of Nature at the most fundamental level. In this context, Monte–Carlo methods are mostly used, within the so-called Path-Integral formulation of QFT, to investigate situations of physical interests where standard analytical tools, like perturbation theory, fail.

DOI: 10.1201/9781315156156-6

The fact that similar methods apply to both cases does not come by chance, but is related to a more profound correspondence between QFT and statistical field theory, that will be introduced and discussed in this chapter.

We will start, as an informal invitation to the issue, by a discussion and illustration of algorithms and numerical results for the $O(2)$ model, looking at its properties both from the point of view of Statistical Mechanics and from the point of view of QFT. The correspondence will then be treated and derived in more details in the following Sections, with a special focus on the discretisation and numerical simulation of quantum gauge theories, which are the particular QFTs used to describe fundamental interactions. We will illustrate the major algorithms used in the context of the numerical simulation of lattice gauge theories, and of quantum chromo dynamics (QCD) in particular, which has been one of the major stimulus for the development of new algorithms (like HMC) and of new computational hardware over the last few decades.

6.1 Invitation: The O(2) model

The $O(2)$ model has degrees of freedom, say spins, associated with the sites of a lattice with a homogeneous spacing a. As we will see below, the correlation length ξ of the spins diverges for certain parameter choices, $\xi/a \to \infty$. As we will discuss later on, this is a necessary condition for a statistical model to have a QFT correspondence. The $O(2)$ model in three dimensions is one of the simplest models that shows this behaviour.

6.1.1 Model and Symmetry

The $O(2)$ model here serves a dual purpose: it is of great importance in statistical physics where it provides a theoretical basis for superfluidity of pure ^4He [67], where it describes some electromagnetic properties of certain high-T_c superconducting materials [68] and where it sheds light onto the Bose-Einstein condensation of atomic vapour (hard sphere bosons) [69]. Secondly, it will illustrate for us the transition from statistical model to an interacting quantum field theory.

We will study the $O(2)$ model in three dimensions. Underlying the model, is a regular cubic grid of size $V = N_t N_x N_y$. The degrees of freedom are unit vectors $\vec{n}_x$ that are located at the sites of the cubic lattice specified by the vector x. These unit vectors are vectors of an internal two-dimensional space and parameterised by the angles ϕ_x:

$$\vec{n}_x = \Big(\cos(\phi_x), \sin(\phi_x)\Big)^T .\tag{6.1}$$

As in the case of the Ising model, the dynamics of the model is dictated by its energy, which we prefer to call *action*, denoting it by S, in view of the correspondence with QFT. Action and partition function are given by

$$S = - \sum_{\ell=\langle x,\nu\rangle} \vec{n}_x^T \vec{n}_{x+\nu} = - \sum_{\ell=\langle x,\nu\rangle} \cos(\phi_x - \phi_{x+\nu}), \tag{6.2}$$

$$Z = \int \mathcal{D}\phi_x \, \exp\left\{-\beta \, S[\phi]\right\}, \tag{6.3}$$

In statistical physics, the parameter is inversely proportional to the temperature. In a quantum field theory setting, β is called the *coupling constant*. We have introduced the short-hand notation where $x + \nu$ is the neighbouring site of x in ν direction. The sum in (6.2) extends over all links $\ell = (x, \mu)$ of the lattice. The symmetries of the model largely influence its properties. Note that, in our case, the interaction between neighbouring vectors, i.e. $\vec{n}_x^T \vec{n}_{x+\nu}$ is a scalar with respect to the internal space. Hence, the model possesses a global $O(2)$ symmetry,

$$\vec{n}_x \to O(\varphi)\, \vec{n}_x \;\Rightarrow\; \phi_x \to \phi_x + \varphi, \tag{6.4}$$

where $O(\varphi) \in O(2)$. Emmy Noether first proved in 1915 that every differentiable symmetry of the action of a physical system has a corresponding conserved current (see, e.g. [70]). In the present case and following Noether's derivation of this current [70], we would conclude that

$$j_\mu(x) = \sin(\phi_x - \phi_{x+\mu}), \tag{6.5}$$

In a classical field theory, Noether's theorem would imply that for classical fields, i.e. those satisfying the "equation of motion", the divergence of the current, i.e.

$$\Delta_\mu j_\mu(x) = \sum_\mu \left[j_\mu(x) - j_\mu(x - \mu) \right], \tag{6.6}$$

vanishes. In a statistical physics context, fields are randomly fluctuating rather than bound to satisfy a PDE. In this case, the Noether theorem still makes an impact: any number of insertions of the divergence (6.6) into the partition function causes the integral to vanish, e.g.

$$\int \mathcal{D}\phi_x \, \Delta_\mu j_\mu(x_0) \exp\left\{-\beta \, S[\phi]\right\} = 0, \qquad \forall x_0.$$

In order to prove the latter equation, we start with the identity (for any x_0)

$$\int \mathcal{D}\phi_x \, \frac{\partial}{\partial \phi_{x_0}} \exp\left\{-\beta \, S[\phi]\right\} = 0,$$

which holds since $\phi_x \in [0, 2\pi[$ and since the action is a smooth function of the fields ϕ_x. Using the action (6.2), we find that any derivative $\partial/\partial\phi_x$ produces an insertion $\Delta_\mu j_\mu(x_0)$. We therefore prove any number of insertions of the divergence (6.6) into the partition function causes the integral to vanish, e.g.

$$
\begin{aligned}
0 &= \int \mathcal{D}\phi_x \, \frac{\partial}{\partial\phi_{x_0}} \cdots \frac{\partial}{\partial\phi_{x_n}} \exp\left\{-\beta\, S[\phi]\right\} \\
&= \int \mathcal{D}\phi_x \, \Delta_\mu j_\mu(x_0) \cdots \Delta_\mu j_\mu(x_n) \exp\left\{-\beta\, S[\phi]\right\} = 0 \,.
\end{aligned}
$$

If the vectors $\vec{n}_x$ are considered as small magnets, which tend to align[1], we can also introduce an external magnetic field $\vec{h}$ by modifying the action to

$$
S[\phi] \;\to\; S[\phi] - \sum_x \vec{h}\,\vec{n}_x \,.
$$

We choose a homogeneous external field and, without any loss of generality, we have pointing it in the internal one direction: $\vec{h} = (j, 0)^T$. The generalised partition function (with external field) is then given by:

$$
Z[j] \;=\; \int \mathcal{D}\phi_x \exp\left\{-\beta\, S[\phi] + \beta j \sum_x \cos(\phi_x)\right\}. \tag{6.7}
$$

Note that the enhanced theory does not possess the $O(2)$ symmetry any more since the external magnetic field $\vec{h}$ introduces a preferred direction. However, for small external fields j we can study the symmetry properties of the original theory by exploring its response to the explicit symmetry breaking by $\vec{h}$.

6.1.2 Wolff cluster algorithm

The Wolff cluster algorithm [71] updates many spins (those of a cluster) at a time and is therefore highly efficient in regions of the parameter space where local update algorithms are severely hampered by critical slowing down. Note that fields $\vec{n}_x$ are functions of the three-dimensional space (coordinate x) and vectors of 2-dimensional internal vector space. The algorithm is based upon reflections: randomly choose a unit vector $\hat{r}$, $\hat{r}^2 = 1$ of the two-dimensional vector space and "reflect" the vector $\vec{n}$ as illustrated in Figure 6.1. This defines the mapping:

$$
\vec{n} \to \vec{n}' : \qquad \vec{n}' = \mathrm{R}(\hat{r})\,\vec{n} \;=\; \vec{n} - 2\,(\vec{n}\hat{r})\,\hat{r} \,. \tag{6.8}
$$

---

[1] Note that parallel oriented neighbouring magnets have $\vec{n}_x \vec{n}_{x+\mu} = 1$ and thus a high probabilistic measure.

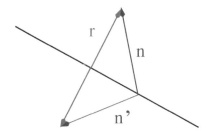

FIGURE 6.1: Reflection of the vector $\vec{\phi}$ along the line perpendicular to $\hat{r}$.

The mapping has some important properties: if we reflect twice, we recover the vector $\vec{n}$. This is easy to show

$$\mathrm{R}(\hat{r})\left[\mathrm{R}(\hat{r})\vec{n}\right] \;=\; \vec{n} - 2\,(\vec{n}\hat{r})\,\hat{r} - 2\left[(\vec{n} - 2\,(\vec{n}\hat{r})\,\hat{r})\cdot\hat{r}\right]\hat{r} \;=\; \vec{n}$$

leading to

$$\mathrm{R}^2(\hat{r}) \;=\; 1\,. \tag{6.9}$$

We also have the useful relation

$$\left[\mathrm{R}(\hat{r})\,\vec{n}\right]\vec{b} \;=\; \vec{n}\left[\mathrm{R}(\hat{r})\,\vec{b}\right], \tag{6.10}$$

which is easily seen by noting that

$$\left[\mathrm{R}(\hat{r})\vec{n}\right]\vec{b} \;=\; \vec{n}\vec{b} - 2(\hat{r}\vec{n})\,(\hat{r}\vec{b})$$

is symmetric under the interchange $\vec{n} \leftrightarrow \vec{b}$. With the help of the latter equations (6.9) and (6.10), we can show that if two neighbouring vectors are reflected, their contribution to the action remains the same:

$$\left[\mathrm{R}(\hat{r})\,\vec{n}_x\right]\left[\mathrm{R}(\hat{r})\,\vec{n}_{x+\mu}\right] \;=\; \vec{n}_x\left[\mathrm{R}^2(\hat{r})\,\vec{n}_{x+\mu}\right] \;=\; \vec{n}_x\,\vec{n}_{x+\mu}\,. \tag{6.11}$$

This is intuitive from the geometric point of view: the contribution to the action only depends on the scalar product and hence on the angle between the two vectors. This angle, however, stays the same if both vectors are flipped simultaneously.

The Wolff update algorithms work as follows [71]:

1. Setup a reflection by randomly choosing $\hat{r}$ and choose randomly a lattice site x, which will become the first point of the cluster c.

2. Flip $\vec{n}_x \rightarrow \vec{n}'_x = R(\hat{r})\,\vec{n}_x$, add x to the cluster c and mark x as a site already visited.

3. Visit all links connecting x to its nearest neighbours y that are not yet marked as visited. Activate the bond with probability

$$P(\vec{n}'_x, \vec{n}_y) = 1 - \exp\Big\{ \min[0,\, \beta\,\vec{n}'_x\,(1 - R(\hat{r}))\,\vec{n}_y] \Big\} \quad (6.12)$$
$$= 1 - \exp\Big\{ \min[0,\, 2\beta\,(\hat{r}\vec{n}'_x)\,(\hat{r}\vec{n}_y)] \Big\}.$$

4. If this happens, flip $\vec{n}_y \rightarrow \vec{n}'_y = R(\hat{r})\,\vec{n}_y$, add y to the cluster c and mark y as a site already visited.

5. Continue recursively in the same way for all links leading to unmarked neighbours until the process stops. Continue with step 1 as next cluster update step.

We point out that the transfers from one configuration to the next are only performed by discrete flips despite the fact the underlying degrees of freedom are continuous. How can this algorithm be exact?

To establish the exactness of the approach, we are now going to investigate two main properties for exact algorithms: (i) ergodicity (actually, regularity) and (ii) detailed balance.

Ergodicity: Here, we have to show that any configuration $\{\vec{n}_x\}$ can be generated with a non-vanishing probability. Our strategy will be as follows: firstly, we will show that we can create a single spin cluster c with a finite probability, let us say at site x. We then show that we can find a $\hat{r}$ such that we can map any spin $\vec{n}$ into any other spin $\vec{n}'_x$. If we repeat this for all sites of the lattice, we can map any configuration $\{\vec{n}_x\}$ into any other $\{\vec{n}'_x\}$ with a non-vanishing probability. What is the probability that we create a single spin cluster at site x? If we choose a neighbouring site y, the probability that this site is *not* updated is:

$$1 - P(\vec{n}'_x, \vec{n}_y) = \exp\Big\{ \min[0,\, 2\beta\,(\hat{r}\vec{n}'_x)\,(\hat{r}\vec{n}_y)] \Big\}.$$

If $(\hat{r}\vec{n}'_x)\,(\hat{r}\vec{n}_y) \geq 0$, we find

$$1 - P(\vec{n}'_x, \vec{n}_y) = 1 - \exp\{0\} = 0,$$

and the spin at y is not flipped and not added to the cluster. If $(\hat{r}\vec{n}'_x)(\hat{r}\vec{n}_y) < 0$, the non-update probability is non-vanishing:

$$0 < 1 - P(\vec{n}'_x, \vec{n}_y) = \exp\left\{ 2\beta\,(\hat{r}\vec{n}'_x)(\hat{r}\vec{n}_y) \right\} < 1\,.$$

In any case, the probability that we *do not update* y, and hence any of the neighbours of x, is non-vanishing. We are now dealing with a cluster consisting of one single spin at position x. Now choose any $\vec{n}'_x$ different from $\vec{n}_x$. Can we find a $\hat{r}$ that maps $\vec{n}_x$ to $\vec{n}'_x$? An inspection of Figure 6.1 suggests that this is always possible. Indeed, first note that $\hat{r}\vec{n}_x \neq 0$, since, for $\hat{r}\vec{n}_x = 0$, we would have $\vec{n}_x = \vec{n}'_x$, which we have excluded. From the reflection definition (6.8), we find

$$2\,(\vec{n}\hat{r})\,\hat{r} = \vec{n} - \vec{n}'\,.$$

Hence, we choose $\hat{r} \propto \vec{n} - \vec{n}'$ and normalise to unit length. This completes the proof of ergodicity.

In a similar way, we can argue that $\hat{r}$ can be chosen so that the change of $\vec{n}$ vanishes, hence the algorithm is also aperiodic, since we have a non-zero probability of keeping the state of the system unchanged. Being ergodic and aperiodic, the Markov chain is regular: the last step is proving it satisfies Detailed Balance.

Detailed Balance: After one cluster update step, we have created the configuration $\{\vec{n}'_x\}$ for which $\vec{n}'_x = \mathrm{R}(\hat{r})\,\vec{n}_x$ for all $x \in c$ and $\vec{n}'_x = \vec{n}_x$ for all other sites. The probability for this transition is denoted by $W(\vec{n}_x \to \vec{n}'_x)$. We then consider the transition probability $W(\vec{n}'_x \to \vec{n}_x)$ that we reverse this step by a second cluster update step. Detailed balance requires that

$$\frac{W(\vec{n}_x \to \vec{n}'_x)}{W(\vec{n}'_x \to \vec{n}_x)} = \frac{\mathcal{P}(\{\vec{n}'_x\})}{\mathcal{P}(\{\vec{n}_x\})}\,, \tag{6.13}$$

where $\mathcal{P}$ is the probability for each of the configurations provided by the Gibbs factor. Let us calculate the probability that we have created $\{\vec{n}'_x\}$ out or $\{\vec{n}_x\}$. All spins at sites $x \in c$, we have flipped the spins and the probability for that to happen is given by the probability $Q_c(\vec{n} \to \vec{n}')$ that a cluster c has formed. It is very tedious to calculate this probability for any given sizeable c since it is a growth process during which we dynamically add sites to the cluster according to the "link" probability $P(\vec{n}'_x, \vec{n}_y)$ (here we have the y is already part of the c and we question whether we should add x to the cluster). There are many possibilities with which a (sizeable) cluster can form and $Q_c(\vec{n} \to \vec{n}')$ is a sum of probabilities of possible paths the cluster forms. Luckily, we do not need an explicit expression for Q_c for proving the Wolff algorithm is exact.

We now need also consider the probability that spins at the boundary of the cluster is *not* flipped. For x and y being neighbouring sites, we introduce the notation

$$\begin{aligned}
\langle xy \rangle &\in & c & \qquad \text{if } x \text{ and } y \text{ are in } c, \\
\langle xy \rangle &\in & \partial c & \qquad \text{if } x \in c \text{ and } y \notin c, \\
\langle xy \rangle &\notin & c & \qquad \text{if } x \notin c \text{ and } y \notin c.
\end{aligned}$$

For $\langle xy \rangle \in \partial c$, we must have that the spin at y is *not* flipped. The probability for *not flipping* is

$$1 - P(\vec{n}'_x, \vec{n}_y) \, .$$

The transition probability is therefore

$$W(\vec{n}_x \rightarrow \vec{n}'_x) = Q_c\left(\vec{n} \rightarrow \vec{n}'\right) \left(\prod_{\langle xy \rangle \in \partial c} [1 - P(\vec{n}'_x, \vec{n}_y)] \right) . \qquad (6.14)$$

Let us now consider the reverse process: all spins in c are now $\vec{n}'$ spins, and we need to find the probability that all spins are $\vec{n}$ spins after the flip. Let's again look at all spins of the cluster c. The probability that a cluster forms that flips back the spins is given by definition by

$$Q_c\left(\vec{n}' \rightarrow \vec{n}\right) .$$

Consider the formation of a cluster as a *dynamic process* where we probe at each time step whether we should add a site. Once we have created the cluster after some time, we can reverse the process. Because of the property

$$P(\vec{n}'_x, \vec{n}_y) = P(\vec{n}_y, \vec{n}'_x) \, ,$$

we find that for any process that has created the cluster c, we find a time reversed process with the *same* probability of the process happening. This leads to the *time reversal* symmetry:

$$Q_c\left(\vec{n} \rightarrow \vec{n}'\right) = Q_c\left(\vec{n}' \rightarrow \vec{n}\right) . \qquad (6.15)$$

For $\langle xy \rangle \in \partial c$, we have at x already $\vec{n}_x$ and do not want to flip the spin at y, which thus remains $\vec{n}_y$. The probability for this is

$$1 - P(\vec{n}_x, \vec{n}_y) \, .$$

Hence, the total transition probability becomes

$$W(\vec{n}'_x \rightarrow \vec{n}_x) = Q_c\left(\vec{n} \rightarrow \vec{n}'\right) \left(\prod_{\langle xy \rangle \in \partial c} [1 - P(\vec{n}_x, \vec{n}_y)] \right) . \qquad (6.16)$$

We find from (6.12) using (6.10) that

$$\begin{aligned}
P(\vec{n}_x, \vec{n}'_y) &= 1 - \exp\Big\{ \min[0, \, \beta \, (\vec{n}_x \vec{n}'_y - \vec{n}_x \, \vec{n}_y)] \Big\} & (6.17) \\
&= 1 - \exp\Big\{ \min[0, \, \beta \, (\vec{n}'_x \vec{n}_y - \vec{n}'_x \, \vec{n}'_y)] \Big\} = P(\vec{n}'_x, \vec{n}_y).
\end{aligned}$$

Secondly, we find

$$P(\vec{n}_x, \vec{n}_y) = 1 - \exp\left\{ \min[0, \beta\,(\vec{n}_x \vec{n}_y - \vec{n}_x\,\vec{n}'_y)] \right\} \tag{6.18}$$
$$= 1 - \exp\left\{ \min[0, -\beta\,\vec{n}'_x\,(1 - \mathrm{R}(\hat{r}))\,\vec{n}_y)] \right\}.$$

We are now well prepared to reconsider the ratio of transition probabilities (6.13):

$$\frac{W(\vec{n}_x \to \vec{n}'_x)}{W(\vec{n}'_x \to \vec{n}_x)} = \frac{Q_c\!\left(\vec{n} \to \vec{n}'\right)\left(\prod_{\langle xy\rangle \in \partial c}[1 - P(\vec{n}'_x, \vec{n}_y)]\right)}{Q_c\!\left(\vec{n}' \to \vec{n}\right)\left(\prod_{\langle xy\rangle \in \partial c}[1 - P(\vec{n}_x, \vec{n}_y)]\right)}$$

The first factors in denominator and numerator cancel because of (6.15). We are left with

$$\frac{W(\vec{n}_x \to \vec{n}'_x)}{W(\vec{n}'_x \to \vec{n}_x)} = \frac{\prod_{\langle xy\rangle \in \partial c} \exp\left\{ \min[0, \beta\,\vec{n}'_x\,(1 - \mathrm{R}(\hat{r}))\,\vec{n}_y)] \right\}}{\prod_{\langle xy\rangle \in \partial c} \exp\left\{ \min[0, -\beta\,\vec{n}'_x\,(1 - \mathrm{R}(\hat{r}))\,\vec{n}_y)] \right\}}. \tag{6.19}$$

We can simplify the latter expression by noting the following:

$$\text{For } a \geq 0: \quad \min[0, a] = 0, \quad \min[0, -a] = -a,$$
$$\text{for } a < 0: \quad \min[0, a] = a, \quad \min[0, -a] = 0.$$

These equations imply that for all a

$$\frac{\exp\{\min[0, a]\}}{\exp\{\min[0, -a]\}} = \exp\{a\}.$$

The latter identification helps to simplify (6.19)

$$\frac{W(\vec{n}_x \to \vec{n}'_x)}{W(\vec{n}'_x \to \vec{n}_x)} = \prod_{\langle xy\rangle \in c} \exp\left\{ \beta\,\vec{n}'_x\,(1 - \mathrm{R}(\hat{r}))\,\vec{n}_y \right\}$$
$$= \frac{\prod_{\langle xy\rangle \in c} \exp\left\{ \beta\,\vec{n}'_x\,\vec{n}_y \right\}}{\prod_{\langle xy\rangle \in c} \exp\left\{ \beta\,\vec{n}_x\,\vec{n}_y \right\}} = \frac{P(\vec{n}')}{P(\vec{n})}.$$

This completes the study of detailed balance using the Wolff algorithm. We will cast a different glance on the Wolff update in the next section, which will reveal a close connection to the Ising spin update.

6.1.3   Phase transition and symmetry breaking

Let su come back to the O(2) model with partition function (6.3). For vanishing coupling strength β (also called inverse temperature in a statistical physics context), the spins $\vec{n}_x$ are unconstrained by the probability density and purely randomly oriented. On the other hand, for large values of the coupling β, the spins are compelled to align: the probability density function in (6.3) penalises any deviation from the $\vec{n}_x$ parallel to its neighbour $\vec{n}_{x+\nu}$ with low values. At a certain intermediate values of β, a changeover occurs from a phase with disordered spins to a phase where spins are more or less aligned. Here, we will further explore how and at which value β this transition takes place.

For a study of this phase transition, we need suitable observables (i.e. expectation values) that are sensitive to the transition. The expectation value of the spin itself (or of one of its components) would not work since for a finite system size V, because

$$\Big\langle \cos(\phi_x) \Big\rangle = 0 . \tag{6.20}$$

This can be easily seen by performing the variable transformation

$$\phi_x \rightarrow \phi_x + \varphi \tag{6.21}$$

in the integral (6.3). In order to trace out any indications for a phase transition at finite volumes, we use the extended partition function (6.7) that adds a source term to the action which breaks the $O(2)$ symmetry explicitly. We then study the response of the expectation value in (6.20) to variations of the external source. We define the response function:

$$R_j(\beta) = \Big\langle \cos(\phi_x) \Big\rangle = \frac{1}{V} \frac{\partial}{\partial j} \ln Z[j] . \tag{6.22}$$

As long as $j \neq 0$, the above argument does not apply, and we find $R_j(\beta) \neq 0$. For any *finite* system, we already said $R_0(\beta = 0)$. However, if we consider the limit of infinitely many degrees of freedom $V \rightarrow \infty$, we cannot draw a definite conclusion. In fact, we face a choice there

$$\lim_{j \to 0} \lim_{V \to \infty} R_j(\beta) \begin{cases} = 0 & \text{Wigner-Weyl realisation} \\ \neq 0 & \text{spontaneous sym. breaking} \end{cases} ,$$

where the order of the limits is crucial. In the latter case, we say that the symmetry is *spontaneously broken*. Note that all quantum field theories operate with infinitely many degrees of freedom; hence, the concept of *Spontaneous Symmetry Breaking* (SSB), i.e. a symmetry is not realised despite the underlying action has that symmetry, has profound phenomenological consequences. Notice that, from the point of view of the probability distribution, this is still symmetric even in the infinite volume limit; however, it becomes non-zero only around spin configurations which break the symmetry, i.e. which have non-zero magnetisation.

We only can simulate systems with a finite number of degrees of freedom. Can we still infer whether or not this theory shows SSB in the limit of infinitely many degrees of freedom? The answer is yes! It is sufficient to study small perturbations j only, which leads us in leading order (and finite volumes) to

$$R_j = \chi(\beta)\,j + \mathcal{O}(j^2),\tag{6.23}$$

$$\chi(\beta) = \frac{1}{V}\frac{\partial^2}{\partial j^2}\ln Z[j]|_{j=0} = \left\langle\cos(\phi_{x_0})\sum_x\cos(\phi_x)\right\rangle,$$

where χ is (half of) the so-called *magnetic susceptibility*, x_0 is any reference lattice site, and we used the lattice translational invariance of the system and the fact that the magnetisation vanishes for finite volume and zero external field. If χ remains finite in the infinite volume limit, the response function vanishes in the limit of vanishing source j, and the system is in the Wigner-Weyl phase. Hence, a diverging magnetic susceptibility in the infinite volume limit is a necessary condition for the spontaneous symmetry breakdown. Observe that χ can be viewed as the expectation value for $\sum\cos(\phi)$ where the factor $\cos(\phi_{x_0})$ acts as a reference against which the spin orientation is counted. If all spins coherently align, it means that $\cos(\phi_{x_0})\approx(\cos\phi_x)$ leading us to:

$$\chi(\beta) \approx V\left\langle\cos^2(\phi_{x_0})\right\rangle,$$

producing the desired singularity for $V\to\infty$ and spontaneous symmetry breakdown. The magnetic susceptibility can be written in a manifestly $O(2)$ invariant form. To this end, we rewrite the expectation value on the right-hand side of (6.23) as

$$\frac{1}{2}\sum_x\left\{\left\langle\cos(\phi_x - \phi_{x_0})\right\rangle + \left\langle\cos(\phi_x + \phi_{x_0})\right\rangle\right\}.$$

Using the variable transform (6.21), we conclude that the latter expectation value vanishes for $j = 0$ (at any finite volume). Thus, we can write the magnetic susceptibility as

$$\chi(\beta) = \frac{1}{2}\sum_x\left\langle\cos(\phi_x - \phi_{x_0})\right\rangle,\tag{6.24}$$

which is manifestly $O(2)$ invariant. Introducing the spin correlation function, by

$$G(x - y) = \left\langle\vec{n}_x\cdot\vec{n}_y\right\rangle,\tag{6.25}$$

the magnetic susceptibility can be easily related to the integrated spin correlation function:

$$\chi(\beta) = \frac{1}{2}\sum_x G(x - x_0).\tag{6.26}$$

The latter identity explains the role of $\chi(\beta)$ as an order parameter for the spontaneous breaking of the $O(2)$ symmetry: in the disordered phase, the

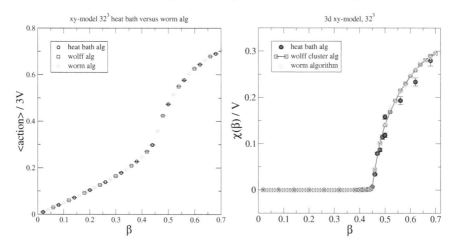

FIGURE 6.2: Left panel: The average action as a function of β for the 3d $O(2)$ model using a 32^3 lattice. Right panel: the magnetic susceptibility in units of the system volume as a function of β. Figures from [46].

spin correlation function exponentially decreases over the distance of the correlation length ξ, and $\chi(\beta)$ is independent of the system size. If the correlation length near criticality exceeds the system size, we find by means of the sum at the right-hand side of (6.26) that $\chi(\beta)$ diverges with the volume.

Let us study the model numerically. Without presenting the details here, we have carried simulations for a $V = 32^3$ lattice using a heat-bath algorithm, the Wolff cluster algorithm (see previous subsection 6.1.2) and a so-called Worm algorithm, which we will introduce in section 7.3.1. The action is shown as a function of β in Figure 6.2 (left panel). While there is a smooth changeover from an approximate linear behaviour at small β to a more saturated behaviour at large values, no drastic change in the graph points towards a phase transition. This picture changes when we look at the magnetic susceptibility χ as shown in Figure 6.2 (right panel). While we observe small values for χ/V for β, we observe a drastic rise at $\beta \approx 0.43$. For values higher than the critical values χ indeed becomes comparable in size with the volume V indicating a spontaneous breaking of the rotational symmetry at higher β.

6.1.4   Quantum Field Theory rising

There is a clear correspondence between a quantum field theory in $(d-1)$ spatial dimensions plus 1 time dimension and a statistical field theory in d dimension. Let us look at this correspondence for the case of the $O(2)$ model in three dimensions.

For a statistical field theory, the $O(2)$ model describes spins that are located at the sites of a 3d lattice with spacing a and total extent of $V = L^3$, $L = Na$, if n counts the sites in one direction. The only parameter of the model is β, which has the interpretation of the inverse temperature, i.e., $\beta = 1/T$. Computer simulation then reveals the thermodynamical properties of the spin system as a function of the temperature T. Here, the lattice spacing a is a measurable and fixed parameter, say a measure for the crystal dimension. The calculation, e.g. makes a prediction of the correlation length ξ in units of the lattice spacing for various temperatures,

$$F(\beta) = \xi/a, \qquad \xi(T) = F(1/T)\, a \ .$$

which immediately has a physical meaning: it provides insights into over what physical distance (say in units of the physical lattice spacing a) spins are correlated. We point out that $F(\beta)$ is a function that can be estimated using Monte Carlo simulations and is considered being available.

From a Quantum Field Theory perspective, the *same* numerical setup acquires a different interpretation: the theory has only two spatial dimensions and time. QFT calculations make statements about dynamical properties of the model. If we confine our interest to thermodynamical properties only, we can disregard time and introduce another (artificial) time direction often called *Euclidean time*. β has no longer a connection to temperature but is interpreted as a coupling strength between the (quantum) spins. Temperature re-enters the considerations via a connection to the extent L of the lattice in Euclidean time direction: $T = 1/L = 1/(N_t\, a)$. The major difference between a QFT and a lattice system is that the QFT operates in a space (time) continuum. The lattice is only facilitating calculations and a sensible limit $a \to 0$ must be implied. The lattice spacing is therefore considered a regulator (rather than a parameter), and the above limit is called the *continuum limit*. How do we achieve this limit in a sensible way?

Let us come back to the fundamental function $F(\beta)$:

$$F(\beta) = \xi/a, \qquad a(\beta) = \xi/F(\beta) \ .$$

We have merely shifted the interpretation of F from the statistical physics interpretation: ξ or its inverse $m = 1/\xi$ (mass gap) is the physical object and fixed, while we now have the lattice spacing a being dependent on the coupling strength. We now are in the position to realise why the correlation length (in units of the lattice spacing) needs to diverge to have achieve a QFT equivalence. This behaviour guarantees the continuum limit. Assume that $F(\beta)$ divergence for $\beta_{\rm crit}$. We then find

$$a\left(\beta \to \beta_{\rm crit}\right) = \lim_{\beta \to \beta_{\rm crit}} \frac{\xi}{F(\beta)} = 0 \ .$$

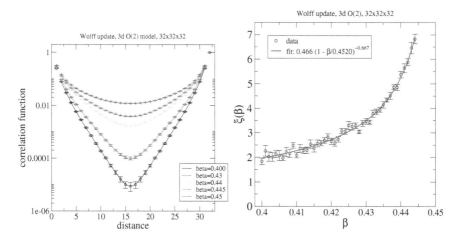

FIGURE 6.3: Left panel: The spin-spin correlation function as a function of the spin distance for several β; 3d $O(2)$ model using a 32^3 lattice. Right panel: The extracted correlation length ξ/a in units of the lattice spacing a as a function of β.

Clearly, the spin correlation function and the distilled fundamental function F is key for the QFT limit. It emerges from

$$\left\langle \vec{n}_x \vec{n}_y \right\rangle \propto \exp\left\{ -\frac{|x-y|}{\xi} \right\}. \tag{6.27}$$

The left panel of Figure 6.3 shows the spin correlation function as a function of $|x-y|$. Note that we are working with periodic boundary conditions, which explains the symmetry around the $L/2$ axis. We observe an exponential decrease for not to small distances, which allows to extract the correlation length. The result for the correlation length in units of the lattice spacing is also shown in Figure 6.3 (right panel). The data are well represented by the fit (red line):

$$\xi(\beta)/a = a_0 \left(1 - \frac{\beta}{\beta_{\mathrm{crit}}}\right)^{-\nu}. \tag{6.28}$$

A high precision study in [72] reports for the fit parameters:

$$\beta_{\mathrm{crit}} = 0.454157(14), \qquad \nu = 0.6711(16), \qquad a_0 = 0.4866(26).$$

Let us now study how QFT results can be obtained from the lattice-regularised statistical model. We will focus on the temperature dependence of the magnetic susceptibility and target the dimensionless function $\chi(T)/\chi_0$ (with $\chi_0 = \chi(T = 0)$). In the QFT setting, there are two ways to change the

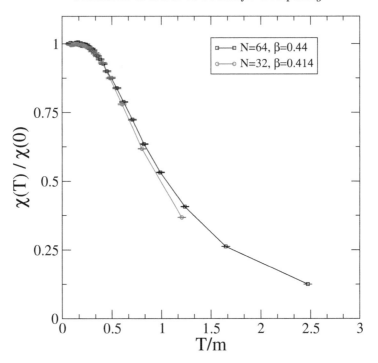

FIGURE 6.4: The magnetic susceptibility in the QFT arising from the O(2) model for two values of the lattice regulator.

temperature: (i) to change the lattice spacing $a(\beta)$ or (ii) to change the number N_t of lattice sites in time direction. We are going to measure any energy dimension in units of the mass gap m

$$T/m = \frac{1}{N_t\,a(\beta)}\,\frac{a(\beta)}{F(\beta)} = \frac{F(\beta)}{N_t}.$$

If the lattice spacing is "small enough", we would expect that physical observable do not depend anymore on a. To put this assumption to the test, we carry out calculations at $\beta = 0.44$ using a $64^2 \times N_t$ lattice. For a study of the lattice spacing (in-)dependence, we also do simulations utilising a $32^2 \times N_t$ lattice and $\beta = 0.414$. Note that both simulations implement roughly the same spatial volume since

$$a(\beta = 0.414) \approx 2\,a(\beta = 0.44).$$

The result for $\chi(T)/\chi_0$ is shown in Figure 6.4 as a function of the temperature T in units of the physical mass gap m. We find that the collapse of the data to one curve is quite satisfactorily.

6.2   The Bridge to QFT: the Feynman path-integral

There is a bridge connecting the realm of Classical Statistical Physics to the realm and Quantum Field Theory (QFT), i.e. to the language we adopt today to describe Nature at its most fundamental level. This bridge has been a fertile crossing of ideas in the last few decades, bringing to fluorishing progress in both realms, and is based on the Feynman path-integral formulation of Quantum Mechanics, which we are going to briefly review in this section. We will discuss in particular equilibrium physics, i.e. the formulation of thermal QFT, which leads directly to the Euclidean path-integral, for which the application of numerical Monte Carlo methods is particularly well suited. We will start from a simple Quantum Mechanical example (a particle in a one-dimensional potential) then moving to the more complex case of a Quantum Field Theory, which will find application in the remaining part of this chapter.

Let us consider a non-relativistic quantum particle described by the Hamiltonian $H = H(q, p)$, where q and p are respectively the coordinate and the momentum operators. Physical states constitute a linear space, called Hilbert space, on which physical observables act as Hermitian linear operators. One can choose different equivalent complete sets of states, for instance, eigenstates of q, of p or of the Hamiltonian H:

$$p|k\rangle = k|k\rangle ; \quad q|x\rangle = x|x\rangle ; \quad H|n\rangle = E_n|n\rangle$$

and completeness of each set is expressed as follows:

$$\int \frac{dk}{2\pi\hbar}|k\rangle\langle k| = \int dx|x\rangle\langle x| = \sum_n |n\rangle\langle n| = \text{Id}$$

where Id is the identity operator, $|\cdot\rangle$ represents a state (ket) and $\langle\cdot|$ its dual or hermitian conjugate (bra). We are assuming the energy spectrum to be discrete, but it could be continuous as well. The wave function of a given physical state is obtained by scalar product with the corresponding position engeinstate, $\psi(x) = \langle x|\psi\rangle$, the wave function of a momentum eigenstate is a plane wave: $\langle x|k\rangle = \exp(ikx/\hbar)$. The temporal evolution of a given physical state is fixed by the Hamiltonian operator,

$$|\psi(t)\rangle = e^{-iHt/\hbar}|\psi(0)\rangle \tag{6.29}$$

if H does not explicitly depend on time, otherwise a time order integral must be considered in the exponential.

We have already discussed, in Chapter 4, how to write down thermodynamics for such a system in terms of the density matrix operator, which for the Gibbs canonical ensemble is given by $\rho = \exp(-\beta H)$, where $\beta = 1/(k_B T)$. Thermal averages are expressed in terms of the density operator as follows

$$\langle O \rangle = \frac{\text{Tr}\,(O\rho)}{\text{Tr}\,(\rho)} = \frac{\text{Tr}\,\left(Oe^{-\beta H}\right)}{Z} \tag{6.30}$$

where $Z = \text{Tr}\,\left(e^{-\beta H}\right)$ is the partition function and the trace can be taken over any complete set of states (it is invariant). Choosing the Hamiltonian eigenstates, one obtains the usual Gibbs distribution over energy eigenstates:

$$\langle O \rangle = \sum_n P_n O_n \;\; ; \;\; O_n \equiv \langle n|O|n \rangle \;\; ; \;\; P_n = \frac{e^{-\beta E_n}}{Z} \;\; ; \;\; Z = \sum_n e^{-\beta E_n} \;\; ; \tag{6.31}$$

in the zero temperature limit, one recovers the expectation value over the ground energy (or vacuum, in field theory) state.

Equation (6.31) hints by itself at a possible numerical evaluation of thermal averages by Monte Carlo sampling of the Gibbs distribution over energy eigenstates, since P_n is a well defined positive probability distribution. There is however a little, non-trivial problem hindering that: the energy eigenstates and eigenvalues are generally unknown, so sampling over them is not well defined in practice, i.e. in a way which can be easily implemented on a computer. As a possible way out, we can try changing the complete set of states over which the trace is computed, going to some standard, well-known basis, which is usually known as the *computational basis*. A working example is given by position eigenstates. For instance, we can write

$$\langle O \rangle = \frac{\int dx \langle x|Oe^{-\beta H}|x \rangle}{\int dx \langle x|e^{-\beta H}|x \rangle} \;, \tag{6.32}$$

the non-trivial problem is now the computation of matrix elements over position eigenstates, which however can be systematically treated within the path-integral approach.

The path-integral formulation is usually introduced by considering the probability amplitude (i.e. the complex quantity whose squared modulus returns the actual probability) that the particle is observed at position x_b at time t, if it has been observed in x_a at time zero: $K(x_b, x_a, t, 0) \equiv \langle x_b|e^{-iHt/\hbar}|x_a \rangle$. This is usually known as the *propagator* and permits to reconstruct how the wave function evolves at any time. Moreover, it is strictly connected to the relevant matrix element of the thermal density operator by substitution $t \to -i\beta\hbar$ and by setting $|x_a\rangle = |x_b\rangle = |x\rangle$. The rotation in the complex time plane is usually known as the *Wick rotation* from Minkowski to Euclidean time: we will do that at the end; however, all steps shown below could be done in Euclidean time right from the beginning.

We have in mind a non-relativistic Hamiltonian $H = p^2/(2m) + V(q)$, where $V(q)$ is the potential. The propagator can be easily computed just for some specific cases, like the free particle, or the harmonic oscillator, $V(q) = m\omega^2 q^2/2$. For the free particle, one has

$$K(x_b, x_a, t_b, t_a) = \langle x_b | e^{-ip^2 t/(2m\hbar)} | x_a \rangle = \int \frac{dk}{2\pi\hbar} \langle x_b | e^{-ip^2 t/(2m\hbar)} | k \rangle \langle k | x_a \rangle$$

$$= \int \frac{dk}{2\pi\hbar} e^{-ik^2 t/(2m\hbar)} e^{ik(x_b - x_a)/\hbar} = \sqrt{\frac{m}{2\pi i\hbar t}} \exp\left(\frac{im(x_b - x_a)^2}{2\hbar t}\right) \quad (6.33)$$

where we have inserted a complete set of momentum eigenstates and performed the Gaussian integral in the complex plane.

In the general case, using the expression $e^{-iHt/\hbar} = (e^{-iH\Delta t/\hbar})^N$, where $\Delta t \equiv t/N$, one can formally rewrite the propagator as a sum over the amplitudes of processes in which the particle passes through a sequence of positions at $N - 1$ intermediate times, as follows:

$$K(x_b, x_a, t, 0) = \int dx_1 dx_2 \ldots dx_{N-1} \langle x_b | e^{-iH\Delta t/\hbar} | x_{N-1} \rangle$$

$$\langle x_{N-1} | e^{-iH\Delta t/\hbar} | x_{N-2} \rangle \ldots \langle x_1 | e^{-iH\Delta t/\hbar} | x_a \rangle, \quad (6.34)$$

such sequences of positions become, in the $N \to \infty$ limit, the different paths over which the functional integral is performed.

Equation (6.34) is still not particularly useful, since the matrix element $\langle x_{j+1} | e^{-iH\delta t/\hbar} | x_j \rangle$ is still not exactly computably, like the original one. However, in the $N \to \infty$ limit one can exploit the Lie-Trotter formula $\exp(A + B) = \lim_{N \to \infty} (\exp(A/N) \exp(B/N))^N$, where A and B are any pair of non-commuting operators, so as to substitute

$$\langle x_{j+1} | e^{-i\Delta t(p^2/(2m)+V(q))/\hbar} | x_j \rangle \simeq \langle x_{j+1} | e^{-i\Delta t p^2/(2m\hbar)} e^{-i\Delta t V(q)/\hbar} | x_j \rangle$$

$$= \left(\frac{m}{2\pi i\hbar\Delta t}\right)^{1/2} \exp\left(\frac{i}{\hbar}\left(\frac{m}{2}\frac{(x_{j+1} - x_j)^2}{\Delta t^2} - V(x_j)\right)\Delta t\right) \quad (6.35)$$

where we have used the expression for the free particle propagator in equation (6.33). It is easy to recognize, as $\Delta t \to 0$, the appearance in the exponential of the kinetic energy $m\dot{x}^2/2$ computed along the path.

Putting all together and taking the $N \to \infty$ limit, we arrive at the following expression

$$K(x_b, x_a, t, 0) = \mathcal{N} \int \mathcal{D}x \exp\left(\frac{iS[x]}{\hbar}\right) \quad (6.36)$$

where $\mathcal{D}x$ stands for the functional integration over all possible paths such that $x(0) = x_a$ e $x(t) = x_b$, while $S[x]$ is the so-called *action functional*, which is nothing but the time integral of the lagrangian:

$$S[x] = \int_0^t dt'\left(m\frac{\dot{x}^2(t')}{2} - V(x(t'))\right) = \int_0^t dt' \mathcal{L}(x(t'), \dot{x}(t')). \quad (6.37)$$

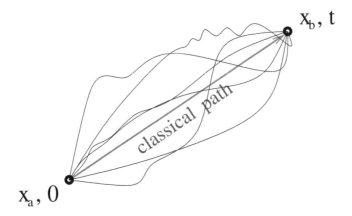

FIGURE 6.5: A pictorial representation of the path-integral representation of the quantum probability amplitude for a particle moving from point x_a to point x_b in a time t. In the classical limit, $\hbar \to 0$, only paths close to the classical path satisfying Eulero-Lagrange equations gives a significant contribution to the amplitude.

Finally, $\mathcal{N}$ is a divergent normalisation factor, $\mathcal{N} = \lim_{N \to \infty} \left(\frac{m}{2\pi i\hbar\Delta t} \right)^{N/2}$. Equation (6.36) is the famous path-integral expression for the particle propagator, which also gives an easy bridge to Classical Mechanics. The amplitude is written as a sum over all possible paths leading from x_a to x_b, each path being weighted by the phase factor $e^{iS/\hbar}$. In the $\hbar \to 0$ limit or, viceversa, for macroscopic objects having a large action compared to $\hbar$, the phase factor is rapidly oscillating, leading to large cancellations between paths which are close to each other. The only exception is represented by paths which are close to a stationary path, on which the action functional S reaches an extremum, since those paths will sum their phases coherently and will give the main contribution to the probability amplitude; on the other hand, the stationary path is the classical trajectory, satisfying Eulero-Lagrange equations. So, if some obstruction hinders the classical path, eliminating nearby paths from equation (6.36), the amplitude will rapidly approach zero and the particle will barely reach x_b: this is the classical limit. A schematic view of the path-integral is reported in Figure 6.5.

The integral in equation (6.36) is the sum of rapidly oscillating objects, hence does not seem suitable for Monte Carlo sampling. However, we were originally interested in the matrix element $\langle x | e^{-\beta H} | x \rangle$. It is easily checked, either by repeating all steps again or simply by changing $t \to -i\beta\hbar$ in the final result, that

$$\langle x | e^{-\beta H} | x \rangle = \mathcal{N} \int_{x'(0) = x'(\beta\hbar) = x} \mathcal{D}x' \exp\left(\frac{-S_E[x']}{\hbar} \right) \qquad (6.38)$$

where now $\mathcal{N} = \lim_{N \to \infty} \left(\frac{mN}{2\pi\beta\hbar^2} \right)^{N/2}$ and

$$S_E[x(\tau)] \equiv \int_0^{\beta\hbar} d\tau' \left(\frac{m}{2} \left(\frac{dx}{d\tau'} \right)^2 + V(x(\tau')) \right) \equiv \int_0^{\beta\hbar} d\tau' \mathcal{L}_E . \qquad (6.39)$$

In previous equations, we have introduced the *Euclidean action* S_E and the corresponding Euclidean Lagrangian $\mathcal{L}_E$. Turning back to equation (6.32), we can now give the following path-integral representation of the partition function and of thermodynamical expectation values:

$$Z = \int dx \langle x | e^{-\beta H} | x \rangle = \mathcal{N} \int_{x(0)=x(\beta\hbar)} \mathcal{D}x \exp \left(\frac{-S_E[x]}{\hbar} \right) \qquad (6.40)$$

$$\langle O \rangle = \frac{\mathrm{Tr} \left(e^{-\beta H} O \right)}{\mathrm{Tr} \left(e^{-\beta H} \right)} = \int \mathcal{D}x \, P[x] \, O[x] ; \quad P[x]\mathcal{D}x = \frac{e^{-S_E[x]/\hbar} \mathcal{D}x}{\int \mathcal{D}x e^{-S_E[x]/\hbar}} \qquad (6.41)$$

where the functional integrations is now over paths running over a compactified Euclidean time with periodic boundary conditions, $x(\tau + \beta\hbar) = x(\tau)$, and $O[x]$ is a proper functional representation of the original observable.

Equation (6.41) is our final result. If $S_E[x]$ is a real-valued and lower bounded functional, as it is the case in this simple example if the potential is lower bounded itself, such a result gives a representation of the original quantum thermodynamical average in terms of an average over a probability distribution over paths. After a proper discretisation, the numerical problem is well defined: paths are well-defined objects that we can treat on our computer, in order to perform a Monte Carlo sampling of the Euclidean path-integral: this topic and the extension to QFT are briefly discussed in the following.

Note however that, in some cases, life is harder. While the original Gibbs distribution over energy eigenstates, equation (6.31), is always positive definite, what we are doing here, i.e. rewriting the partition function in a different way, as a sum of amplitudes over a well defined computational basis, is not guaranteed to lead to positive path-integral, i.e. it may happen that the integrand has its sign or even its phase not well defined. This is generally known as a *sign problem* and affects many fields, going from condensed matter physics to fundamental interactions. In these cases, one may hope in a quantum, rather than classical, computer, encoding the quantum Hamiltonian and thus being able to perform the sampling over quantum energy eigenstates by itself: this possibility has been dreamt long ago by Richard Feynman and could be realized by future progress in Quantum Computation. Alternatively, one may try to approach the problem within a classical computational scheme, as we will discuss in more detail in Chapter 7.

6.2.1 A working example: the harmonic oscillator and Brownian motion

The harmonic oscillator represents a simple exercise, where one can compare numerical computations with exact results, like it happens for the Ising model in two dimensions. However, one should be aware that the extension to more complex systems, which are not solvable analytically and not easily approachable by other approximation schemes, like perturbation theory, is typically straightforward in the path-integral approach. The reader is encouraged to consider further examples like, e.g. a quartic potential.

For the harmonic oscillator, $V(q) = m\omega^2 q^2/2$, we have

$$-\frac{1}{\hbar}S_E[x] = -\frac{1}{\hbar}\int_0^{\beta\hbar} d\tau \left[\frac{m}{2}\left(\frac{dx}{d\tau}\right)^2 + \frac{m\omega^2 x(\tau)^2}{2}\right]. \tag{6.42}$$

We need to discretize the Euclidean time so as to work in terms of a computable finite number of stochastic variables. Let us take, in place of the continuous interval $[0, \beta\hbar]$, a discrete lattice with N sites, equally spaced by $a = \beta\hbar/N$: the path is now represented by a vector of N coordinates, moreover it is convenient to rescale them by the typical length of the system, $\sqrt{\hbar/(m\omega)}$, introducing the dimensionless stochastic variables

$$y_j \equiv \left(\sqrt{\frac{\hbar}{m\omega}}\right)^{-1} x(ja); \qquad j = 0, 1, \dots N-1.$$

The remaining step is to write down a lattice version $S_L[y]$ of $S_E/\hbar$, so that the probability distribution for the y variables is proportional to $e^{-S_L[y]}\prod_j dy_j$: in doing so one needs to provide discretized versions for integration and derivatives. There are many (actually infinite) possibilities; however, one should be aware that the discretized version should always find correspondence with the product of finite matrix elements like $\langle x_{j+1}|T_N|x_j\rangle$, where T_N is the exponential of a suitable hermitian operator such that $\lim_{N\to\infty}(T_N)^N = e^{-\beta H}$, otherwise fundamental properties of the quantum theory, like *reflection positivity*, could be invalid. Further details on this can be found in standard textbooks specialized in lattice field theories [73, 74, 75, 76].

We will adopt the lowest order approximation for integration and the forward discretisation for the derivative:

$$\int_0^{\beta\hbar} d\tau \ \to \ \sum_{n=0}^{N-1} a; \qquad \frac{dx}{d\tau} \ \to \ \left(\sqrt{\frac{\hbar}{m\omega}}\right)^{-1}\frac{y_{j+1}-y_j}{a}$$

so that, after a few substitutions, one finds:

$$e^{-S_E[x]/\hbar} \to e^{-S_L[y]} = \exp\left(-\sum_{j=0}^{N-1}\left[\frac{(y_{j+1}-y_j)^2}{2\eta} + \frac{\eta}{2}y_j^2\right]\right) \tag{6.43}$$

where we have introduced the dimensionless parameter $\eta \equiv a\omega$, which tells us how fine the discretisation is, compared to the typical time scale of the system, i.e. the oscillation period $2\pi/\omega$.

The distribution in equation (6.43) reveals a mapping of the original quantum system onto a classical one-dimensional statistical system of spin-like variables y_j, with the Euclidean action playing the role of the energy and characterized by nearest neighbours or self-interactions. We are not going into more details regarding the possible numerical sampling algorithms: given the locality of the interaction, one can easily devise a local (e.g. Metropolis) algorithm, but this is left as an exercise for the reader. We are going, however to make some further relevant comments about the distribution and the method itself.

First of all, in this case one does not deal just with statistical Monte Carlo errors. For finite values of a, one is simulating an approximation of the original quantum system, hence systematic errors should be expected, vanishing only in the continuum limit. Such limit is realized sending $\eta \to 0$ while keeping the product $N\eta = \beta\hbar$ fixed, so that one needs systems of increasing size N approaching the limit. In practice, what one should do is to repeat the numerical experiment for a few values of η and then extrapolate results to $\eta = 0$ by fitting systematic errors by some power law behavior in η (in most cases, corrections are $O(\eta^2)$ if η is sufficiently small).

Second, it is clear that as $\eta \to 0$ the distribution in equation (6.43) is dominated by the kinetic term: this is not completely unexpected, since also in classical mechanics on very small scales, and for regular potentials, every motion goes almost straight. In this case, however, that means that $y_{j+1} - y_j$ is distributed almost like a gaussian variable with variance η: that makes Euclidean paths similar to those that would be realized in the Brownian motion discussed in Chapter 2, i.e. their distribution is in fact dictated by the Wiener measure. More specifically, one has that $|y_{j+1} - y_j| \sim O(\sqrt{\eta})$, implying that as $\eta \to 0$ only continuous paths contribute to the path-integral. However, that also means that $|\dot{x}| \propto |y_{j+1} - y_j|/\eta \sim O(\eta^{-1/2})$, i.e. it is divergent. Continuous paths are therefore continuous but non-differentiable. This is a general phenomenon of the path-integral formulation, common both to simple quantum mechanical systems like this and to Quantum Field Theory, which leads to the appearance of divergent contributions which need proper subtractions: we are not going to further discuss this aspect.

6.2.2   Extension to Quantum Field Theory

QFTs represent a particular case of quantum many body models, which are common in condensed matter physics. Instead of a single degree of freedom, one considers a collection of objects, e.g. quantum spins, oscillators, or atoms, typically sitting on the sites of a lattice and interacting among them: that already makes the system challenging, unless it can be treated in the small oscillations (linear) regime. The particular feature of QFT is that we deal with a continuous medium, i.e. the lattice spacing goes to zero and we have an infinite amount of degrees of freedom per unit volume, and no upper bound for the system frequencies. We consider here one of the simplest possible cases: a scalar field theory in 3+1 dimensions (three spatial plus e temporal), switching to natural units, $\hbar = c = 1$, as well as to the covariant relativistic notation, as usual in QFT[2].

The system is made up of hermitian quantum operators $\phi(\vec{x})$ sitting on each spatial point, and of their conjugate momenta $\pi(\vec{x})$, satisfying the canonical commutation relations $\left[\hat{\phi}(\vec{x}), \hat{\pi}(\vec{y})\right] = i\delta^3(\vec{x} - \vec{y})$. The Hamiltonian is written as the spatial integral of a Hamiltonian density

$$H = \int d^3x \mathcal{H} = \int d^3x \left(\frac{1}{2}\pi^2 + \frac{1}{2}|\vec{\nabla}\phi|^2 + \frac{1}{2}m^2\phi^2\right), \qquad (6.44)$$

where m is the mass of scalar field, which corresponds to the actual mass of the identical and non-interacting relativistic particles described by this QFT. As usual one can consider different complete sets of physical states, like Hamiltonian eigenstates, or eigenstates $|\phi\rangle$ of the field operators, which are the analogous of position eigenstates for the single particle. The path-integral formulation can be approached also in this case starting from the transition amplitude between different field eigenstates, which after a treatment similar to single particle case is rewritten as

$$\langle \phi_b | \exp(-iHt)|\phi_a\rangle = \int \mathcal{D}\phi \, e^{iS[\phi]} \qquad (6.45)$$

where the action integral is now defined as

$$\begin{aligned}
S &= \int dt \int d^3x \, \mathcal{L} = \int dt \int d^3x \left(\frac{1}{2}(\partial_t\phi)^2 - \frac{1}{2}|\vec{\nabla}\phi|^2 - \frac{1}{2}m^2\phi^2\right) \\
&= \int d^4x \left(\frac{1}{2}\partial_\mu\phi\partial^\mu\phi - \frac{1}{2}m^2\phi^2\right) \qquad (6.46)
\end{aligned}$$

where $\mathcal{L}$ is the Lagrangian density and the functional integration is over all field configurations with fixed boundary conditions, $\phi(\vec{x}, 0) = \phi_a(\vec{x})$ and $\phi(\vec{x}, t) = \phi_b(\vec{x})$.

The physics of the system at thermodynamical equilibrium is again described in terms of traces involving the thermal density operator $e^{-\beta H}$, and the traces can be taken either over the energy eigenstates or over field operator eigenstates. In the second case, we arrive at the path-integral formulation:

$$Z(T) = \int \mathcal{D}\phi_{(s)} \, \langle \phi_{(s)}|e^{-\beta H}|\phi_{(s)}\rangle = \mathcal{N}\int \mathcal{D}\phi \, e^{-S_E[\phi]} \qquad (6.47)$$

---

[2] In high-energy physics it is usual to consider natural units where $\hbar = h/(2\pi) = 1$, with h the Planck constant, and the speed of light $c = 1$ as well, so that space and time are measured with the same units, which are the inverse of energy and mass units.

where the index "s" in the first integral indicates that functional integration is over spatial configurations of the field, while in the second case it is over all Euclidean space-time configurations, with a compactified Euclidean time $\tau \in [0, \beta]$ and periodic boundary conditions (p.b.c.) in τ, $\phi(\vec{x}, \beta) = \phi(\vec{x}, 0)$. S_E is instead the Euclidean action, obtained after analytic continuation $t \to -i\tau$:

$$S_E = \frac{1}{2} \int_0^\beta d\tau \int d^3x \left(\partial_\mu \phi \, \partial_\mu \phi + m^2 \right)$$

where from now we consider just lower indexes, since the metric is the standard Euclidean one. $\mathcal{N}$ is the usual divergent normalisation factor, disappearing in the definition of thermal averages:

$$\langle \hat{O} \rangle_T = \int \mathcal{D}\phi \, P[\phi] O[\phi] \; ; \qquad P[\phi] \, \mathcal{D}\phi \equiv \frac{\mathcal{D}\phi \, e^{-S_E[\phi]}}{\int \mathcal{D}\phi \, e^{-S_E[\phi]}} . \qquad (6.48)$$

In this case, in order to make the system suitable for numerical simulations, one needs to discretize not only the Euclidean time, but also the spatial directions. A simple and standard discretisation is to consider a hypercubic four-dimensional lattice. In natural units ($\hbar = c = 1$), space and time can be measured with the same units, which are the inverse of energy and mass unites; therefore, one has the option to take a different lattice spacings in the temporal and spatial directions, a_t and a_s, or to take them equal, $a_t = a_s = a$; we consider the latter case, i.e. a isotropic lattice. The finite spacing in the spatial directions sets a lower (upper) limit on the possible wavelengths (wavenumbers) propagating in the medium, which is usually called *ultraviolet (UV) cutoff*. That also removes the UV divergencies typical of a QFT, so that the lattice acts also as a *regulator* making the theory well defined.

Of course, in order to make the number of degrees of freedom finite, one also needs to limit to take a finite number of lattice spacings also in the spatial directions: the corresponding boundary conditions are not constrained as in the temporal direction, however periodic boundary conditions (pbc) are usually taken also in space, in order to minimize finite size effects. In order to work with dimensionless quantities, in this case one rescales everything in lattice spacing units: taking into account that in $3+1$ dimension a scalar field has mass dimension 1, we can set

$$\hat{\phi} \equiv a\phi \; ; \quad \hat{m} = am \qquad (6.49)$$

where the "hat" is used for dimensionless quantities. Every lattice site will be associated with a vector of $3 + 1$ integer numbers, n, and we will adopt the notation $n + \hat{\mu}$ to indicate the lattice site displaced by one spacing in forward μ direction with respect to n. The discretisation can now proceed similarly to what was done for the single-particle case. Rewriting the four-dimensional integration as $\int d^4x \; \to \sum_n a^4$ and adopting the forward discretisation for

$\partial_\mu \hat{\phi}$, we find

$$S_{E,L} = \frac{1}{2}\sum_n \left[\hat{m}^2\,\hat{\phi}(n)^2 + \sum_\mu (\hat{\phi}(n+\hat{\mu}) - \hat{\phi}(n))^2\right]. \qquad (6.50)$$

Therefore, we end up with a statistical model of $N_x \times N_y \times N_z \times N_t$ real stochastic variables, where N_μ is the number of lattice sites in the μ direction, with the following probability distribution

$$\mathcal{P}[\hat{\phi}] \prod_n d\hat{\phi}(n) = \exp(-S_{E,L}[\hat{\phi}]) \prod_n d\hat{\phi}(n). \qquad (6.51)$$

The only free parameter left is $\hat{m} = ma$; therefore, we expect to approach the continuum limit as $\hat{m} \to 0$, keeping the quantity $N_t\hat{m} = m/T$ fixed, where T is the physical temperature.

What we have found in this rather simple example is quite general. Every QFT in D spatial dimensions can be put in correspondence, through the path-integral formalism, with a classical statistical system in $D + 1$ dimensions. Also in this case the correspondence leads to a system with local interactions, i.e. involving up to nearest neighbours sites, meaning that one can devise simple and efficient local updating algorithms, whose implementation is left as an exercise for the reader.

6.2.3 Energy gaps, continuum limit and critical points

We are not going to discuss the various interesting examples of physical observables that are usually studied in Monte Carlo simulations of QFTs. There is however one significant case, which is relevant to the overall discussion, since it puts a strict link between the energy spectrum of the D-dimensional QFT and the correlation functions of the corresponding $D + 1$ dimensional statistical system, and because of that between the continuum limit of the former and the critical points of the latter.

Consider the thermal two-point function in Euclidan time of a generic operator O, function of the coordinate operator q or of the field operators in QFT, $\langle O(\tau)O^\dagger(0)\rangle_T$, where $O(\tau) \equiv e^{H\tau}Oe^{-H\tau}$. We can give a path integral representation of it, which takes the form of a correlation function in the corresponding statistical system:

$$\langle O(\tau)O^\dagger(0)\rangle_T = \frac{\mathrm{Tr}\left(e^{-\beta H}O(\tau)O^\dagger(0)\right)}{\mathrm{Tr}\left(e^{-\beta H}\right)} = \frac{\mathrm{Tr}\left(e^{-(\beta-\tau)H}Oe^{-H\tau}O^\dagger\right)}{\mathrm{Tr}\left(e^{-\beta H}\right)}$$

$$= \frac{\int \mathcal{D}\phi\, e^{-S_E[\phi]}O(\tau)O^*(0)}{\int \mathcal{D}\phi\, e^{-S_E[\phi]}}$$

On the other hand, in the $\beta \to \infty$ limit, we recover the vacuum expectation value, which after inserting a complete set of energy eigenstates can be expressed as follows

$$\langle 0|O(\tau)O^\dagger|0\rangle = \sum_n \langle 0|e^{H\tau}Oe^{-H\tau}|n\rangle\langle n|O^\dagger|0\rangle = \sum_n e^{-(E_n-E_0)\tau}|\langle n|O^\dagger|0\rangle|^2 .$$

Now, if we subtract the $n = 0$ contribution (assuming the vacuum state is unique), which in terms of the path integral means taking the connected correlation function, we obtain the following large time behavior for the two-point function

$$\lim_{\tau\to\infty} \left(\langle 0|O(\tau)O^\dagger|0\rangle - \langle 0|O|0\rangle^2\right) = e^{-(E_{\bar n}-E_0)\tau}|\langle \bar n|O|0\rangle|^2 \qquad (6.52)$$

where $|\bar n\rangle$ is the physical state of the lowest energy which is coupled to the vacuum by the creation operator $O^\dagger$. Therefore, the large distance behaviour of the connected correlation function of the statistical system gives information on the energy spectrum of the quantum system. In particular, the energy gap $\Delta E = E_{\bar n} - E_0$ fixes the long range decay of the correlation function, i.e. its correlation length, $\xi = 1/\Delta E$.

In a relativistic QFT, the lowest lying excitations over the vacuum correspond to the masses of the physical states (particles), because the energy of a relativistic particle at rest equals its mass. The set of different correlation lengths ξ_O, corresponding to different operators O, will be in correspondence with the masses $m_O = 1/\xi_O$ of the lightest particles having the same quantum numbers of the corresponding operators.

That leads us to a further important consideration: in a lattice simulation we will always measure correlations lengths in dimensionless lattice units, $\hat\xi_O = \xi_O/a$, where a is the lattice spacing. If the discretized QFT admits a well defined continuum limit, in which the UV cutoff is removed ($a \to 0$) keeping physical quantities, like the particle masses, fixed, then the dimensionless correlation lengths must diverge. This has various implications: the QFT admits a continuum limit, or in other words can be *renormalized*, only in correspondence of a critical point of the statistical system, where its correlation lengths diverge; if the statistical system does not present any critical behaviour, it will never correspond to a well defined continuum QFT. On the other hand, that has also practical implications from the numerical point of view: around a critical point, autocorrelation times grow according to critical slowing down exponents, therefore we expect the need for a growing computational effort as we approach the continuum limit, which adds to that related to the increasing density of stochastic variables.

6.3   Gauge Theories

The numerical simulation of quantum gauge theories is a major target for scientific computing. The standard model of particle physics, which is our present theoretical tool to describe the world around us at the most fundamental level, is based on quantum gauge theories. However in some cases, like the study of the theory of quark and gluons, QCD, in the low energy regime, the proper tool to obtain reliable predictions based on it relies on Monte Carlo simulations of the theory discretized on a space-time lattice. As a matter of fact, Lattice QCD simulations have represented one of the main motivations for hardware and algorithmic development in high performance computing since a few decades.

Just to make a few examples, we rely on numerical simulations to predict the origin of most of the visible mass of the Universe, which is contained into nucleons (protons and neutrons), because these particles are composed of almost massless quarks, so that most of their mass comes, through the famous Einstein relation $E = mc^2$, from the QCD interaction energy involving quarks and gluons. In the same way, it is just by numerical simulations that we have known that, up to one microsecond after the Big Bang, the Universe contained a hot plasma of deconfined quarks and gluons, i.e. something completely different from the confined world of nucleons we know today.

This is by itself an excellent motivation to dedicate part of our discussion to this issue. In addition to that, lattice gauge theories provide us with a wider and more general perspective on statistical models, which finds application in many other fields. Indeed, until now we have considered models built in terms of spin-like variables living on lattice sites, and typically interacting through the lattice links connecting nearest neighbours sites; the interaction is in some cases characterized by some global symmetry, i.e. it is invariant if we perform some kind of rotation on the internal degrees of freedom of all spin variables at the same time.

In lattice gauge theories, one requires symmetry under local trasformations (called *gauge transformations*) made independently on each lattice site. That leads to the introduction of a new kind of stochastic variables, the so-called *gauge links variables*, which live on the lattice links and take the memory of how differently the local gauge transformation acts on two nearest neighbour sites. The interaction among spin-like variables must necessarily involve the *gauge links*, in order to be gauge invariant; however, the new kind of variables can also interact among themselves: the interaction lives on higher dimensional geometrical entities, namely, on two-dimensional elementary closed loops made up in terms of the lattice links.

It is interesting to notice that these new concepts, even if usually imported from QFT, naturally arise in the context of Statistical Physics itself, when one considers generalized duality transformations. The duality transformation was originally introduced by Kramers and Wannier to show that a two-dimensional Ising model at a given temperature can be exactly mapped onto another Ising model at a different temperature, with the high-T and low-T limit exchanged. This has been extensively reviewed in Chapter 4: the mapping goes through the corresponding elements of the dual lattice, so that the interaction living on the lattice links of the original model translates into an interaction living on the dual links of the corresponding dual model. When one generalizes the duality transformation to higher dimensional lattices, one easily ends up with gauge theories, or even beyond. For instance, considering the Ising model in three dimensions, one has that the dual element of a lattice link is an elementary square (plaquette) of the dual lattice: as we will shortly see, an interaction living on plaquettes is exactly what we expect for a gauge theory, in this case it turns out to be a Z_2 gauge theory in three dimensions. For a review on the fascinating subject of duality transformations, we refer the reader to Ref. [77].

6.3.1 Continuum Yang-Mills theories and their lattice counterpart

Gauge symmetry arises as a useful option in the formulation of Classical Electrodynamics, then becomes a necessity when going to the Quantum Theory. It is well known that the homogeneous Maxwell equations can be solved automatically by re-expressing the electric and magnetic fields, $\vec{E}$ and $\vec{B}$, in terms of the vector and scalar potentials, $\vec{A}$ and Φ: $\vec{B} = \vec{\nabla} \wedge A$, $\vec{E} = -\vec{\nabla}\Phi - \partial A/\partial t$, however this formulation has an inherent redundancy, since the electric and magnetic fields are invariant under a *gauge transformation* of the potentials, $\vec{A} \to \vec{A} + \vec{\nabla}\Lambda$, $\Phi \to \Phi - \partial\Lambda/\partial t$, where Λ is any differentiable function of space and time. Adopting the Lorentz covariant formalism, one would write the field strength tensor as $F_{\mu\nu} = \partial_\mu A_\nu - \partial_\nu A_\mu$, where $A_\mu = (\Phi, -\vec{A})$, and the gauge transformation as $A_\mu \to A_\mu - \partial_\mu\Lambda$.

Now, it turns out that the introduction of the potential is the only possible way to obtain a Hamiltonian or Lagrangian formulation of the theory: since quantisation necessarily goes through Hamilton or Lagrange, the gauge redundancy, or symmetry, is inherited as an essential foundational aspect by Quantum Electrodynamics (QED). Moreover, in the quantum theory, the meaning of the gauge transformation as a local symmetry transformation emerges more clearly, since it happens that the wave function of a particle of charge q transforms, in correspondence with the gauge trasformation, as follows: $\psi(\vec{x}, t) \to e^{iq\Lambda(\vec{x},t)}\psi(\vec{x}, t)$. A phase factor is an element of the $U(1)$ Abelian group[3], this is why QED is said to be an Abelian gauge theory based on the $U(1)$ gauge group.

The concept of gauge symmetry and of quantum gauge theories was generalized to the case of non-Abelian gauge groups by Yang and Mills in 1954 [78]. This generalisation laid the foundation for the formulation of the standard model of particle physics a few decades later. In the following we will briefly summarize continuum Yang-Mills theories, then moving to their lattice formulation and numerical simulation.

---

[3]In group theory, the term Abelian or non-Abelian distinguishes among groups whose composition rule is, respectively, commutative or non-commutative.

Let us consider a theory containing some multicomponent complex field that we indicate by $\psi(x)$, where x stands for a space-time point. Space-time will be considered as Euclidean in the following, since this is the standard case in which path-integral Monte Carlo simulations are feasible. The field $\psi(x)$ transforms under some internal symmetry group acting locally and linearly on its components, $\psi(x) \to G(x)\psi(x)$, where G is some matrix representation of the symmetry group element depending on x. The conjugate of the field, $\bar{\psi}$, is assumed to transform as $\bar{\psi}(x) \to \bar{\psi}(x)G^{-1}(x)$, which is natural if the gauge group is unitary, so that the local bilinear $\bar{\psi}(x)\psi(x)$ is gauge invariant.

It is easy to check that the standard derivatives of $\psi(x)$ transform in a different way, because derivation takes into account also how the local internal reference frame changes

$$\partial_\mu \psi \to G(\partial_\mu \psi + G^{-1}(\partial_\mu G)\psi) \tag{6.53}$$

however, that can be cured by introducing the *covariant derivative*, defined in terms of the gauge connection, or gauge field A_μ:

$$D_\mu \equiv \partial_\mu + igA_\mu \tag{6.54}$$

which transforms as the field $\psi(x)$ itself, $D_\mu\psi \to GD_\mu\psi$, provided that the gauge field A_μ transforms as follows:

$$A_\mu \to GA_\mu G^{-1} - \frac{i}{g}G(\partial_\mu G^{-1}). \tag{6.55}$$

Here g is a coupling constant, which could be reabsorbed in the definition of A_μ, but is left out here in agreement with the standard definition given in Electrodynamics. Starting from the gauge field, one can construct the field strength tensor and its transformation law

$$F_{\mu\nu} \equiv -\frac{i}{g}[D_\mu, D_\nu] \quad ; \quad F_{\mu\nu} \to GF_{\mu\nu}G^{-1} \tag{6.56}$$

where $[,]$ stands for the commutator of the two operators (considering both their matrix and differential actions). For a generic group, both A_μ and $F_{\mu\nu}$ are elements of the group algebra, i.e. they can be written as linear combinations of the group generators T^a:

$$A_\mu = A_\mu^a T^a \quad ; \quad F_{\mu\nu} = F_{\mu\nu}^a T^a \tag{6.57}$$

where

$$F_{\mu\nu}^a = \partial_\mu A_\nu^a - \partial_\nu A_\mu^a - gf^{bca} A_\mu^b A_\mu^c \quad ; \quad [T^a, T^b] \equiv if^{abc}T^c \tag{6.58}$$

and f^{abc} are the structure constants of the group. For the Abelian gauge group $U(1)$, for which $G(x) = \exp(ig\Lambda(x))$, A_μ e $F_{\mu\nu}$ simplify to standard real fields transforming as follows:

$$A_\mu \to A_\mu + \partial_\mu\Lambda \qquad F_{\mu\nu} \to F_{\mu\nu}. \tag{6.59}$$

One can easily construct local gauge invariant interaction terms by combining $\bar{\psi}$, ψ and their covariant derivatives. However, one can also consider gauge invariant pure gauge terms, built in terms of traces of local products of the field strength tensor. In order to respect standard requirements of renormalisability and invariance under Lorentz transformations and time reversal, one usually considers just the term $\mathrm{Tr}\,(F_{\mu\nu}F_{\mu\nu})$, where a sum over repeated indexes is understood. The Euclidean path-integral representation of the pure gauge theory then takes the form of a statistical distribution $P[A]$ over gauge configurations weighted by the pure gauge action term:

$$P[A]\,\mathcal{D}A \propto \mathcal{D}A\, e^{-S_G[A]}\; ; \quad S_G[A] \equiv \frac{1}{2}\int d^4x\, \mathrm{Tr}\,(F_{\mu\nu}F_{\mu\nu})\,. \qquad (6.60)$$

The integration measure is a standard functional integration over gauge fields living in the group algebra, which is affected by the gauge redundancy: in perturbative analytic approaches, this must be cured by a proper gauge fixing procedure. This is however usually not required in the lattice formulation, since, as we are going to show, the integration moves from the algebra to the group itself, and it is finite for compact groups.

When bringing the concept of gauge theory to a lattice formulation, it is natural to associate a different gauge transformation to each lattice site, by defining a transformation matrix $G(n)$ acting on fields living at lattice site n as usual, i.e. $\psi(n) \to G(n)\psi(n)$, $\bar{\psi}(n) \to \bar{\psi}(n)G^{-1}(n)$. In the following, we will usually consider, as the simplest possible discretisation, a isotropic hypercubic lattice, so that n is a D-dimensional vector of integer numbers, where D is the space-time dimension, and the correspondence with continuum coordinates is $x_\mu = a\,n_\mu$, where a is the lattice spacing.

In this context, one has the problem of building a gauge invariant theory and a lattice generalisation of the gauge connection. The discretisation of derivative terms in the action will unavoidably lead to lattice interactions involving some degree of non-locality, for instance involving fields living on nearest neighbours sites, hence transforming in different ways. The problem was elegantly solved in terms of parallel transports by Kenneth Wilson in 1974 [79], in a seminal paper that laid the foundations of Lattice QCD.

The concept of *parallel transport* arises, as an alternative way of defining the gauge connection, already in the continuum theory. The idea is that, in order to compare two fields defined in different points, $\psi(x)$ and $\psi(y)$, we first need to "transport" them onto each other, in order to factorize differences due to different local gauge frames. The transport unavoidably depends on the path along which it is done, i.e. one defines the parallel transport $U_C(x,y)$ connecting point x to point y along path C, which is an element of the gauge group and is defined according to its transformation and composition properties,

$$U_C(x,y) \to G(x)U_C(x,y)G^{-1}(y)\,; \quad U_{C_1+C_2}(x,z) = U_{C_1}(x,z)U_{C_2}(z,y) \quad (6.61)$$

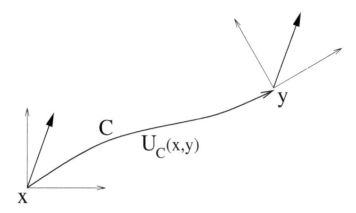

FIGURE 6.6: A pictorial representation of the parallel transport of a vector field along a given path.

so that $\bar{\psi}(x) \, U_C(x, y) \, \psi(y)$ is gauge invariant. A practical realisation for infinitesimal parallel transports in terms of the gauge field A_μ is easily verified (see equation (6.55)) to be given by

$$U(x, x + dx) \simeq \mathrm{Id} + ig A_\mu dx_\mu \qquad (6.62)$$

and generalizes to finite transports as follows:

$$U_C(x, y) \equiv \mathcal{P} \exp \left(ig \int_{x,C}^{y} dx'_\mu \, A_\mu(x') \right) \qquad (6.63)$$

where $\mathcal{P}$ indicates a *path-ordering* prescription in the definition of the exponential, which is needed for non-Abelian gauge fields, and the line integration is performed along path C. A pictorial representation of a parallel transport is sketched in Figure 6.6.

The main idea of lattice gauge theories is that the elementary gauge degrees of freedom are the parallel transports connecting nearest-neighbours sites, in terms of which every other parallel transport can be obtained by the composition rule. There is no need to specify the path for the elementary transport, since they are naturally associated with lattice links and are therefore named as *gauge link variables*, or simply *gauge links*.

In the following we will take the convention of indicating by $n + \hat{\mu}$ the lattice site displaced by one unit in direction μ with respect to site n. The elementary gauge link and its transformation properties are then defined as follows:

$$U_\mu(n) \equiv U_{C=\mathrm{link}}(n, n + \hat{\mu}); \quad U_\mu(n) \to G(n) U_\mu(n) G^{-1}(n + \hat{\mu}) \qquad (6.64)$$

and the reverse transport is of course $U_{link}(n + \hat{\mu}, n) = U_\mu^{-1}(n)$.

One can state a formal correspondence between gauge links and the vector potential, which becomes exact in the $a \to 0$ limit and reads as follows:

$$U_\mu(n) = \exp\left(ig \int_n^{n+\hat{\mu}} dx_\nu \, A_\nu\right) \simeq e^{iga A_\mu(n)}. \qquad (6.65)$$

Moreover, gauge links allow for a straightforward discretisation of the covariant derivative, for instance, one can consider its *forward* version

$$D_\mu \psi(x) \quad \to \quad \frac{1}{a}\left(U_\mu(n)\psi(n+\hat{\mu}) - \psi(n)\right). \qquad (6.66)$$

However, since we will first deal with a lattice version of the pure gauge theory, we are interested now in how to build gauge invariant interaction terms just in terms of gauge link variables. The most general, gauge invariant pure gauge operator is the group trace of some parallel transport over a closed loop, $U_{loop}(n, n)$, indeed one has

$$\mathrm{Tr}(U_{loop}(n, n)) \quad \to \quad \mathrm{Tr}(G(n)U_{loop}(n, n)G^{-1}(n)) = \mathrm{Tr}(U_{loop}(n, n)). \qquad (6.67)$$

The smallest possible non-trivial loop is that around an elementary square, or *plaquette*, of the lattice

$$\Pi_{\mu\nu}(n) \equiv U_\mu(n)\, U_\nu(n+\hat{\mu})\, U_\mu^{-1}(n+\hat{\nu})\, U_\nu^{-1}(n) \qquad (6.68)$$

which is illustrated in Figure 6.7 and was considered by K. Wilson in order to write down the first discretized version of a pure gauge Yang-Mills action, also known as plaquette (or Wilson) action. In the following, we are going to check that indeed the trace of the plaquette is a good starting point to discretize the continuum pure gauge action in equation (6.60). We will focus on $SU(N)$ (special unitary) gauge groups ($N = 3$ being the case relevant to QCD), so that $U_\mu^{-1}(n) = U_\mu^\dagger(n)$.

Let us consider the plaquette operator defined in equation (6.68) and see how it relates to continuum quantities in the $a \to 0$ limit, exploiting equation (6.65). Since the gauge group is non-Abelian, one has to make use of the Baker-Campbell-Hausdorff formula to properly treat the product of matrix exponentials:

$$\exp(A)\exp(B) = \exp\left(A + B + \frac{1}{2}[A, B] + \dots\right) \qquad (6.69)$$

where the dots stand for an infinite series of nested commutators, which in our case means a power series in a, since both A and B are linear in the lattice spacing, see equation (6.65). Let us divide the plaquette into two pieces and compute separately each of them:

$$U_\mu(n)\, U_\nu(n+\hat{\mu}) \quad \simeq \quad e^{igA_\mu(n)}e^{igA_\nu(n+\hat{\mu})} \qquad (6.70)$$

$$\simeq \quad e^{igaA_\mu(n)+igaA_\nu(n)+iga^2\partial_\mu A_\nu(n)-\frac{g^2a^2}{2}[A_\mu(n),A_\nu(n)]}$$

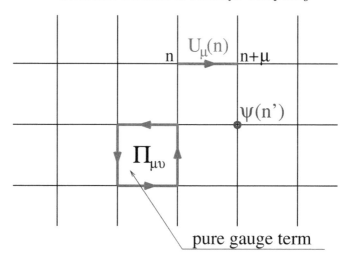

FIGURE 6.7: A pictorial representation of the main elements characterizing a lattice gauge theory.

$$U_\mu^\dagger(n+\hat{\nu})\,U_\nu^\dagger(n) \;\simeq\; e^{-igA_\mu(n+\hat{\nu})}e^{-igA_\nu(n)} \tag{6.71}$$

$$\simeq\; e^{-igaA_\mu(n)-iga^2\partial_\nu A_\mu(n)-igaA_\nu(n)-\frac{g^2a^2}{2}[A_\mu(n),A_\nu(n)]}$$

where we Taylor expanded the gauge field and applied equation (6.69) keeping only terms up to $O(a^2)$ in the exponentials. Putting the two pieces together, one easily finds:

$$\Pi_{\mu\nu}(n) = e^{iga^2(\partial_\mu A_\nu(n)-\partial_\nu A_\mu(n)+ig[A_\mu(n),A_\nu(n)])+O(a^3)} = e^{iga^2 F_{\mu\nu}(n)+O(a^3)}$$

meaning that the plaquette operator contains information, at the lowest order in a, about the flux of the field strength tensor across the plaquette itself. When considering its trace, which is gauge invariant, the first non-trivial contributions appear at the second order in the Taylor expansion of the exponential, because the exponent itself lives in the algebra (the plaquette operator is a group element); hence, it is traceless for $SU(N)$. We can then write

$$\mathrm{Tr}\left(\Pi_{\mu\nu}(n)\right) = N - \frac{g^2a^4}{2}\,\mathrm{Tr}\left(F_{\mu\nu}^2(n)\right) + O(a^5) \tag{6.72}$$

from which it follows that a discretized gauge invariant action having the correct continuum limit can be built by properly summing over all plaquettes of the lattice. Indeed, using the fact that $\sum_n a^4 \to \int d^4x$ as $a \to 0$, it is easy to check that

$$S_{plaq} \equiv \beta \sum_{n,\,\mu<\nu} \left(1 - \frac{1}{N}\mathrm{Re}\,\mathrm{Tr}\left(\Pi_{\mu\nu}(n)\right)\right) \to \int d^4x \sum_{\mu,\nu} \frac{1}{2}\mathrm{Tr}\left(F_{\mu\nu}^2\right) = S_G$$

$$\tag{6.73}$$

as $a \to 0$, provided that $\beta = 2N/g^2$, where β is the so-called inverse gauge coupling. Taking the real part is necessary, since the trace of the plaquette is in general a complex quantity, because of higher order contributions.

We can now state our final result, i.e. that the continuum path integral distribution of gauge fields, reported in equation (6.60) for a pure gauge theory, has a possible lattice counterpart as follows:

$$P[U]\,\mathcal{D}U \propto \prod_{n,\mu} dU_\mu(n) e^{-S_{plaq}[U]} \tag{6.74}$$

where $dU_\mu(n)$ represents the extension to the gauge group of the integration measure originally defined over the group algebra, which is nothing but the gauge invariant Haar measure and will be discussed in more detail in the explicit examples we are going to illustrate in the following.

As a final remark, before discussing the possible algorithms available for sampling the distribution in equation (6.74), we stress that the plaquette action is not the only possibility. Many other, actually infinite, possibilities are available, considering different kinds of loops (e.g. loops made up of 6 link variables, or even larger). The actual choice can be based on the attempt to reduce discretisation effects while maintaining the lattice action simple enough for efficient numerical simulations. We are not going to make further comments on this aspect and we will consider just the plaquette action in the following.

6.3.2 Numerical algorithms: the $SU(2)$ case

There are various reasons to start our illustration of the numerical treatment of lattice gauge theories from the $SU(2)$ case: it is simple enough to permit a clear view about all technical details; it is nevertheless general enough to illustrate all essential aspects of the problem; as a matter of fact, most algorithms developed for $SU(N)$ gauge theories include the $SU(2)$ ones as core components.

Let us start from the representation of $SU(2)$ variables and their integration measure. An element of $SU(2)$ in the fundamental representation is a 2×2 matrix U satisfying the general unitary relation $UU^\dagger = \text{Id}$, $U_{ih}U_{jh}^* = \delta_{ij}$ in index notation[4], plus the special condition $\det U = 1$. The unitary relations mean explicitly

$$|u_{11}|^2 + |u_{12}|^2 = 1 ; \quad |u_{21}|^2 + |u_{22}|^2 = 1 ; \quad u_{11}u_{21}^* = -u_{21}u_{22}^* \tag{6.75}$$

taking the squared modulus of the last and exploiting the first two one finds $|u_{21}| = |u_{12}|$ and $|u_{22}| = |u_{11}|$, so that one can write in general (the two phases in the second row must be equal again to satisfy the last condition)

$$U = \begin{pmatrix} a & b \\ -b^* e^{i\phi} & a^* e^{i\phi} \end{pmatrix} \tag{6.76}$$

---

[4]Here δ_{ij} stands for the Kronecher delta function, defined so that $\delta_{ij} = 1$ if $i = j$ and 0 otherwise. We also adopt the Einstein convention: an implicit sum is understood for indexes appearing twice in the expression.

with $|a^2| + |b^2| = 1$, while the condition $\det U = 1$ further fixes $\phi = 0$. A convenient parametrisation is given in terms of the Pauli matrices σ_i, indeed writing $a = u_0 + iu_3$ and $b = u_2 + iu_1$ with $u_0^2 + u_1^2 + u_2^2 + u_3^2 = 1$ one has

$$U = u_0 \mathrm{Id} + i\vec{u} \cdot \vec{\sigma}; \quad \sigma_1 \equiv \begin{pmatrix} 0 & 1 \\ 1 & 0 \end{pmatrix} \quad \sigma_2 \equiv \begin{pmatrix} 0 & -i \\ i & 0 \end{pmatrix} \quad \sigma_3 \equiv \begin{pmatrix} 1 & 0 \\ 0 & -1 \end{pmatrix} \quad (6.77)$$

where $\vec{u} \equiv (u_1, u_2, u_3)$ and the Pauli matrices σ_i satisfy the following commutation and anticommutation relations[5]:

$$[\sigma_i, \sigma_j] = 2i\,\epsilon_{ijk}\,\sigma_k \;; \quad \{\sigma_i, \sigma_j\} = 2\,\delta_{ij} \quad (6.78)$$

One reaches the same result starting from the infinitesimal form of an $SU(2)$ matrix. Indeed, writing $U = \mathrm{Id} + i\alpha\,\Omega$, where α is an infinitesimal parameter, the condition of special unitarity implies that Ω be a traceless hermitian matrix. Since the Pauli matrices are a basis for 2×2 traceless hermitian matrices, one can write $\Omega = \hat{n} \cdot \vec{\sigma}$, where $\hat{n}$ is a unit vector. The infinitesimal form can then be exponentiated and, exploiting the fact that the anticommutation relations in equation (6.78) implie $(\hat{n} \cdot \vec{\sigma})^2 = \mathrm{Id}$, one easily find the finite version

$$U = e^{i\alpha\,\hat{n}\cdot\vec{\sigma}} = \cos\alpha + i\,\sin\alpha\,\hat{n} \cdot \vec{\sigma} \quad (6.79)$$

which fixes $u_0 = \cos\alpha$ and $\vec{u} = \hat{n}\sin\alpha$.

The general expression reported in equations (6.77) and (6.79) shows that elements of $SU(2)$ are in one-to-one correspondence with the points of a four-dimensional spherical surface of unit radius. That implies that also the group integration measure, which is essential to define the probability distribution, can be explicited in that terms. The most convenient way to parameterise a point on the n spherical surface S^n is by $n-1$ spherical angles, in particular we can write

$$u_0 = \cos\alpha \;; \; u_1 = \sin\alpha\sin\theta\cos\phi \;; \; u_2 = \sin\alpha\sin\theta\sin\phi \;; \; u_3 = \sin\alpha\cos\theta \;;$$
$$\alpha, \theta \in [0, \pi] \;\; \phi \in [-\pi, \pi] \quad (6.80)$$

where θ and ϕ are the usual three-dimensional polar and azimuthal angles. The integration measure is then

$$dU = \sin^2\alpha\,d\alpha\,\sin\theta\,d\theta\,d\phi. \quad (6.81)$$

That this is indeed the correct group integration measure can be seen by checking that it reduces to the standard non-compact integration over the group algebra for infinitesimal group elements, indeed in that case $|\vec{u}| = \sin\alpha \sim \alpha \ll 1$, so that

$$dU = \sin^2\alpha\,d\alpha\,\sin\theta\,d\theta\,d\phi \simeq |\vec{u}|^2 d|\vec{u}|\,\sin\theta\,d\theta\,d\phi = du_1\,du_2\,du_3. \quad (6.82)$$

---

[5]Here ϵ_{ijk} is the Levi-Civita symbol, or completely antisymmetric tensor in three dimension, defined by $\epsilon_{123} = 1 \;; \epsilon_{jik} = \epsilon_{ikj} = -\epsilon_{ijk}$.

Before entering the details of the specific algorithms, let us analyze more closely the properties of the probability distribution. Notwithstanding the different nature of the stochastic variables and of their interaction, the theory is still local as for a standard spin system, i.e. each gauge link variable is coupled to a limited number of neighbouring gauge links. We can therefore easily define and compute a sort of mean field, which fixes the distribution of the single gauge links. This is why a local algorithm, which changes link variables one by one, is still the simplest, and generally effective, choice, as it happens for spin systems.

The part $S_\mu(n)$ of the discretized gauge action S_{plaq}, defined in equation (6.73), which involves the link variable $U_\mu(n)$ is made up of the six (in four dimensions) plaquettes containing the given link. It can be conveniently expressed in terms of the *force* $F_\mu(n)$, defined as the sum over all six plaquettes of the so-called *staples*, which are the three links paths connecting n to $n + \hat\mu$ going around a plaquette, in formulae:

$$F_\mu(n) = \sum_{\nu \neq \mu} U_\nu(n)U_\mu(n+\hat\nu)U_\nu^\dagger(n+\hat\mu)+U_\nu^\dagger(n-\hat\nu)U_\mu(n-\hat\nu)U_\nu(n-\hat\nu+\hat\mu) \quad (6.83)$$

so that one can write, up to an irrelevant constant term,

$$S_\mu(n) = -\frac{\beta}{N}\mathrm{ReTr}\left(U_\mu(n)\,F_\mu^\dagger(n)\right) . \quad (6.84)$$

The force $F_\mu(n)$, which in fact represents the mean-field, is the sum of $SU(2)$ elements (the staples), hence it is not a group element itself. However, a special property of $SU(2)$, stemming from the representation given in equation (6.77), is that the sum of group elements is proportional to a group element, indeed we can write in general

$$F_\mu(n) = F_0\mathrm{Id}+i\vec{F}\cdot\vec{\sigma} = F f_\mu(n) = F\left(f_0\mathrm{Id} + i\vec{f}\cdot\vec{\sigma}\right); \; F \equiv \sqrt{F_0^2 + |\vec{F}|^2} \quad (6.85)$$

where $f_\mu(n) \in SU(2)$. In this way, setting $U_\mu(n) = u_0\mathrm{Id} + i\vec{u}\cdot\vec{\sigma}$, the local action takes the simple form:

$$S_\mu(n) = -\frac{F\beta}{N}\mathrm{ReTr}[(u_0\mathrm{Id}+i\vec{u}\cdot\vec{\sigma})(f_0\mathrm{Id}-i\vec{f}\cdot\vec{\sigma})] = -\frac{2F\beta}{N}(u_0 f_0 + \vec{u}\cdot\vec{f}) \quad (6.86)$$

where we have used the fact that Pauli matrices are traceless and their properties in equation (6.78).

Equation (6.86) better clarifies the role of the force as a mean field, since the action is minimum (hence the probability is maximum) when the gauge link $U_\mu(n)$ is oriented along the same direction where the force points, just like it happens for a spin system. A difference with respect to spin systems, due to the different kind of local interaction, is that it is far less trivial to devise efficient non-local algorithms, like the cluster algorithm illustrated previously for the Ising and the $O(2)$ models.

A local Metropolis approach

It is relatively easy to devise a Metropolis-like algorithm which changes gauge links one by one, let us review its main structure:

---

1. Choose a lattice site n and a direction μ, with the purpose of updating the corresponding link variable. The choice can be random or follow a deterministic sweep of the whole lattice, detailed balance can be easily proved in the former case. Compute the force $F_\mu(n) = F f_\mu(n)$ associated with that link, see equations (6.83) and (6.85);

2. Make a tentative change of the current value of the link variable

$$U_\mu(n) \to U'_\mu(n)$$

   according to some reversible stochastic rule (so as to apply a straight Metropolis), i.e. having the same probability of bringing back to $U_\mu(n)$ starting from $U'_\mu(n)$ (how to implement that is discussed in more details below);

3. Compute the ratio of probabilities according to distribution in equation (6.74), since the action is local and since the group integration measure is invariant, this is simply given by

$$r = \exp\left(-S'_\mu(n) + S_\mu(n)\right) = \exp\left(\frac{\beta}{N}\mathrm{ReTr}[(U'_\mu(n) - U_\mu(n))F^\dagger_\mu(n)]\right)$$

4. Draw a uniform deviate $x \in [0, 1)$, if $x \leq r$ then $U'_\mu(n)$ will be the new link variable; otherwise, we will keep $U_\mu(n)$.

Note that it can be computationally convenient to repeat the steps above a certain number of consecutive times (hits) for the same link variable, since the costly step 1), involving the computation of the force, remains unchanged and needs to be performed only for the first hit; typically one would perform $O(10)$ hits for each link variable. It should also be noticed (and this applies also to other local algorithms illustrated below) that the force does not involve any link in the same direction sitting on nearest-neighbours sites, so that it is possible (and could be convenient for parallel computing purposes), to apply the Metropolis steps in parallel to all links in the same direction sitting on sites with the same parity according to a checkerboard scheme, since they do not enter the computation of each other force.

Let us now focus on step 2, i.e. how to move to a tentative new link $U'_\mu(n)$ in a symmetric (reversible) way. Moving around the group is achieved by multiplication by some other group element, so step 2 can be detailed in this way: choose stochastically an element of the gauge group g and set

$$U'_\mu(n) = g\, U_\mu(n) \tag{6.87}$$

then symmetry is trivially guaranteed if one has the same probability of drawing g or the inverse element g^{-1}. In addition, one would like g to be in a certain range around the identity (not too small, not too large), so as to guarantee a good efficiency, usually corresponding to an acceptance rate around 50%. The above requirements can be easily achieved in various ways; for instance, one could build a sufficiently large library of random matrices with a distribution peaked around the identity, taking care that if g is in the library, g^{-1} is there as well. We propose another possible practical implementation:

---

a. Choose $g_0 \in [0, 1]$ with any distribution peaked towards 1, for instance g_0 uniform in $[1 - \delta, 1]$ where δ is some parameter to be tuned in order to optimize the acceptance rate;

b. Fix $\vec{g} = \sqrt{1 - g_0^2}\, \hat{n}$ where $\hat{n}$ is uniformly distributed on the unit's three-dimensional spherical surface, in practice

$$n_1 = \sqrt{1 - z^2}\, \cos\phi\; ; \quad n_2 = \sqrt{1 - z^2}\, \sin\phi\; ; \quad n_3 = z$$

with z and ϕ drawn uniformly, respectively, in $[-1, 1]$ and $[-\pi, \pi]$.

c. Then

$$g = g_0\, \mathrm{Id} + i\vec{g}\cdot\vec{\sigma}$$

---

reversibility is trivially guaranteed, because

$$g^{-1} = g_0\, \mathrm{Id} - i\vec{g}\cdot\vec{\sigma}$$

and $\hat{n}$ and $-\hat{n}$ have the same probability to be drawn in step b.

Single link distribution and heat-bath: Creutz and Kennedy-Pendleton algorithms

Local heat-bath algorithms follow a scheme quite similar to the Metropolis illustrated above, at least for what concerns the selection of the link and the construction of the force. Then, the new link variable $U_\mu(n)$ is drawn from scratch according to the single link distribution:

$$P(U_\mu(n))\, dU_\mu(n) \propto \exp\left(\frac{F\beta}{N}\mathrm{Re\,Tr}(U_\mu(n)f^\dagger_\mu(n))\right) dU_\mu(n). \tag{6.88}$$

Since the probability distribution has a maximum for $U_\mu(n) = f_\mu(n)$, it is convenient to "translate" around the maximum, i.e. to rewrite $U_\mu(n) = U F_\mu(n)$, with $U = u_0 \mathrm{Id} + i\vec{u} \cdot \vec{\sigma}$. Since the group measure is invariant by left or right multiplication, U is distributed as (we make explicit reference to equations (6.80) and (6.81)):

$$\tilde{P}(U)\, dU \propto e^{2F\beta u_0/N}\, dU = e^{2F\beta u_0/N} \sin\alpha\, d(\cos\alpha)\, d(\cos\theta)\, d\phi \qquad (6.89)$$

meaning that one can set $\vec{u} = \sqrt{1 - u_0^2}\, \vec{n}$ where $\vec{n}$ is a uniformly distributed unit vector, while $u_0 = \cos\alpha$ is distributed according to

$$p(u_0) du_0 \propto \sqrt{1 - u_0^2}\, e^{F\beta u_0}\, du_0\,; \quad u_0 \in [-1, 1] \qquad (6.90)$$

where we have substituted $N = 2$. We have therefore reduced the problem to the sampling of a one-variable distribution, which resembles the exponential distribution apart from the correction factor $\sqrt{1 - u_0^2}$. Actually, this sampling problem has been already explicitly discussed in section 3.2.1, where we have proposed the so-called Creutz algorithm[6] based on the von Neumann method: one first samples according to the exponential, then corrects by the accept-reject step.

However, as we have shown on that occasion, the acceptance of the Creutz algorithm decreases as $1/\sqrt{\beta F}$ in the large β limit. As we will discuss later on, this limit is the one relevant to the continuum limit of lattice gauge theories: that means that the Creutz algorithm will perform worse and worse as the continuum limit is approached. Note that, contrary to a Metropolis algorithm, a low acceptance in this case does not simply mean a low efficiency with long autocorrelations times: we actually need a successful draw before moving to the next link variable, so that in case of extremely low acceptance the algorithm could get completely stuck.

This is the reason alternatives to the Creutz algorithm have been searched, like the one proposed by A.D. Kennedy and B.J. Pendleton in 1985 [81]. The idea is again an accept-reject algorithm, but the distribution is modified by a smart change of variable, so that the acceptance is much better in the large β limit, reaching 1 for $\beta \to \infty$. The change of variable is the following: $r = \sqrt{1 - u_0}$, i.e. $u_0 = 1 - r^2$ with $0 \leq r \leq \sqrt{2}$. It is easily checked that equation (6.90) implies the following distribution for r:

$$p_r(r)\, dr \propto r^2 \sqrt{1 - r^2/2}\, e^{-\beta F r^2}\, dr\,. \qquad (6.91)$$

The idea is to first draw r according to the tentative distribution $r^2 \exp(-\beta F r^2)$, which can be easily generated as shown below, and then correct for the factor $\sqrt{1 - r^2/2}$ by an accept–reject step, as in the Creutz algorithm. However, this time the tentative distribution has a maximum for $r = 1/\sqrt{\beta F}$, which approaches zero as $\beta \to \infty$, while the correction factor is

---

[6]The algorithm is named after Michael Creutz, because it was used by him in his pioneering simulations of lattice gauge theories [80].

significantly different from one only far from zero; therefore, the acceptance will approach one as $\beta \to \infty$. On the contrary, for low values of β (in particular for $\beta F \lesssim 1$) the Creutz algorithm performs better.

Let us now turn to the problem of drawing r according to $dr\,r^2\exp(-\beta Fr^2)$, which is solved as follows. Let $y \in (-\infty, \infty)$ be a Gaussian variable, drawn according to $dy\exp(-\beta Fy^2)$ following, e.g. the Box-Müller algorithm discussed in Chapter 1. Let $x \in [0, \infty)$ be a variable distributed according to $dx\,x\exp(-\beta Fx^2)$: that can be easily obtained by substitution, $x = \sqrt{z}$ with $z = -\log(1-w)/(\beta F)$ and w uniformly distributed in $[0,1)$, indeed $z = x^2$ is distributed as $dz\exp(-\beta Fz)$. We are now going to write the combined distribution of x and y, then changing variables as follows $x = r\cos\phi$, $y = r\sin\phi$, with $r \in [0, \infty)$ and $\phi \in [-\pi/2, \pi 2]$:

$$dxdy\,x\,e^{-\beta F(x^2+y^2)} = rdrd\phi\,r\cos\phi\,e^{-\beta Fr^2}\,.$$

The distribution for r is then obtained by integrating over ϕ, that however brings a factor 2 which is independent of r, so that $r = \sqrt{x^2 + y^2}$ is distributed proportionally to $r^2\exp(-\beta Fr^2)$.

To summarise, the Kennedy-Pendleton algorithm to sample u_0 according to equation (6.90) runs as follows:

---

1. Draw four independent uniform deviates w_1, w_1, w_1, w_4 in $[0, 1)$.

2. Set $r^2 = x^2 + y^2$, with

$$x^2 = -\log(1-w_1)/(\beta F)\,;\quad y^2 = -\log(1-w_2)/(\beta F)\,(\cos(2\pi w_3))^2\,.$$

3. If $r^2 < 2$ and $w_4^2 \le 1 - r^2/2$ then $u_0 = 1 - r^2$,
   otherwise go back to step 1.

Over-relaxation methods

Over-relaxation methods are inspired by an analogous technique used in iterative minimisation (optimisation) problems for systems with many degrees of freedom, known as *successive over-relaxation* (SOR), where local changes of the variables point towards a local minimum but do not stop right there, going instead a bit further: that reveals quite effective in order to accelerate the convergence towards the global minimum of the system. The idea has been exported to Monte Carlo simulations for the first time by Adler in the context of multiquadratic actions [82], then by Brown and Woch for lattice gauge theories [83].

The method can be placed within the general framework of Metropolis algorithms. We will illustrate only a special case, which, however, revealed to be quite simple and effective, in which the idea is to perform a reversible step which moves exactly to the other side of the minimum, without changing the energy (action in our case), so that the acceptance is one by default; for this reason such algorithms are also called *microcanonical*, since they move the system within a hypersurface of constant energy. They are usually ergodic, even if deterministic, within the given energy hypersurface, but are not able to change the energy itself, so they must be combined with some other algorithm which guarantees global ergodicity.

To see how the implementation works for $SU(2)$, let us consider again the single link distribution reported in equation (6.88). We know it has a maximum for $U_\mu(n) = f_\mu(n)$, however usually the current value of the link variable will be away from the maximum, in particular one can write

$$U_\mu(n) = g \, f_\mu(n) ; \quad g \equiv U_\mu(n) f_\mu^\dagger .$$

The idea is to change $U_\mu(n)$ into the group element $U'_\mu(n)$ which is equally displaced from the maximum of the distribution, but in the opposite direction, i.e.:

$$U'_\mu(n) = g^{-1} f_\mu(n) = f_\mu(n) U_\mu^\dagger(n) f_\mu(n) . \tag{6.92}$$

In other words, multiplication of $U_\mu(n)$ by g^{-1} would bring it to the local minimum, but we multiply it by g^{-2}, so as to bring it to the opposite side[7]. It is trivial to check that $\mathrm{ReTr}(U'_\mu(n) f_\mu^\dagger(n)) = \mathrm{ReTr}(U_\mu(n) f_\mu^\dagger(n))$, so that the action remains indeed unchanged. Moreover, the step is reversible: if applied to $U'_\mu(n)$, it brings back to $U_\mu(n)$.

A few hybrid variants have been proposed over the years, for instance, the *over-heat* algorithm of Ref. [84], where one performs a standard heat-bath to select how far from the minimum the new link will be (i.e. the value of u_0), then choosing the direction opposite to that of the previous link variable.

This closes our short overview of local algorithms available for the numerical simulation of the $SU(2)$ lattice gauge theory. The optimal combination is usually to alternate heat bath with over-relaxation, with the latter performed a few times (typically 4–5) more frequently than the former. It is, however, frequent to simulate modified models for which the single link distribution is more difficult to sample: in this case, Metropolis is a standard and safe backup choice.

6.3.3 Extension to SU(N) gauge theories

The extension to $SU(N)$ gauge theories is quite clear from the point of view of the action and of the statistical weight of gauge configurations, see equation (6.73) and (6.74) for the Wilson plaquette action. What is less trivial is

---

[7]In general over-relaxation schemes, one would multiply by $g^{-\omega}$ where ω is the over-relaxation parameter with $\omega \in [0, 2]$.

how to represent gauge link variables and write explicitly the correct integration measure over the gauge group.

$SU(N)$ elements in the fundamental representation are $N \times N$ complex matrices U (so, in principle, $2N^2$ real numbers), subject to the unitarity condition $(U^\dagger U)_{ij} = \delta_{ij}$ plus the special condition $\det U = 1$. Since $U^\dagger U$ is hermitian, the unitarity condition means $N + 2N(N-1)/2 = N^2$ constraints (one for each $i = j$, two for each $i < j$); the special condition adds one more constraint, since unitarity alone already imposes $|\det U| = 1$. We conclude that $SU(N)$ elements are described in terms of $N^2 - 1$ independent parameters. The same conclusion can be reached reasoning in terms of the algebra: considering the infinitesimal for $U = \mathrm{Id} + i\Omega$, the unitarity condition requires that Ω is hermitian (N^2 free parameters); however, the special condition imposes also that it is traceless, ending up with $N^2 - 1$ free parameters.

While it is quite easy to write down a complete representation of the algebra (think, for instance, of the Gell-Mann matrices in the case of $SU(3)$), it is far less trivial (already for $SU(3)$) to give an explicit non-redundant parameterisation of finite elements, as we did in the $SU(2)$ case. This is why in most numerical codes $SU(N)$ matrices are represented as $N \times N$ complex matrices, then imposing the group constraints by hand and from time to time during the Markov Chain evolution (in order to avoid the accumulation of machine precision errors). The only exceptions can be found in codes where memory saving issues become essential[8].

A somewhat irreducible problem is how to write down the correct integration measure over the gauge group. In order to do that, one needs an explicit non-redundant parameterisation and the exact representation of the group integration measure in terms of these parameters, as we did for $SU(2)$: this is already quite difficult for $SU(3)$.

Knowledge about the integration measure is however not essential for Metropolis-like algorithms, in which the new tentative gauge link variable is obtained by multiplying the old one by another group element, $U'_\mu(n) = gU_\mu(n)$, because the group measure is invariant by left or right multiplication, hence only the action difference enters the accept–reject step. A standard approach could be to select g randomly from a given collection of $SU(N)$ matrices distributed around the identity a containing both g and g^{-1} to ensure reversibility.

Another possibility is to try something close to an over-relaxation step [85, 86]. Given any $SU(N)$ element g_0, chosen independently of $U_\mu(n)$, a reflection around it, i.e.

$$U'_\mu(n) = (g_0\, U^\dagger_\mu(n))^2\, U_\mu(n) = g_0\, U^\dagger_\mu(n)g_0 \qquad (6.93)$$

defines a good (reversible) tentative step for a Metropolis algorithm, however

---

[8]Nowadays RAM availability is typically not an issue, but it may happen to face a limited memory bandwidth compared to the computational capability, as it happens in some Graphics Processing Units (GPUs). In those case, it is usual, for instance, to store and write/read at runtime only the first two rows of each $SU(3)$ matrix (saving $1/3$), since the third row can be reconstructed by the processor itself.

the acceptance in general will be low. However, if one manages to obtain a good approximation for the $SU(N)$ element which maximizes $\mathrm{ReTr}(g_0 F_\mu^\dagger(n))$, the tentative step will be almost microcanonical, leading to a good acceptance. However, previous possibilities could be not particularly efficient or not easy to implement. On the other hand, the implementation of heat-bath algorithms require the explicit knowledge of the integration measure: examples have been reported for $SU(3)$ [87], but are again not simple and particularly efficient. A smart alternative is the one proposed by Cabibbo and Marinari [88], which is based on the update of $SU(2)$ subgroups and is in fact the present standard in the numerical simulations of $SU(N)$ pure gauge theories.

The Cabibbo Marinari method

The general idea of local algorithms is to make a step in which only a portion of the configuration is modified; the step will be small in configuration space, but can be convenient anyway because of the easiness and cheapness of the implementation. Till now, we have considered cases in which the small part consists of a single gauge link variable. The idea of the Cabibbo-Marinari method is to consider something even smaller than that: an $SU(2)$ subgroup of the given gauge link variable.

Given the gauge link variable $U_\mu(n)$, let us consider the set of all $SU(N)$ elements which can be written as $U'_\mu(n) = u\,U_\mu(n)$, where $u \in$ some $SU(2)$ subgroup of $SU(N)$. The Cabibbo-Marinari idea is to perform a heat-bath update restricted to this set: the set itself is not a subgroup, however, since the group measure is invariant under left or right multiplication, the integration measure over the set is the same as that over the subgroup, which is the well known $SU(2)$ measure. One has then to repeat the same step for a sufficiently large number of different subgroups, so as to ensure ergodicity by covering the whole $SU(N)$ group.

A standard way to select a $SU(2)$ subgroup, labelled by a couple of indexes $i \neq j$, is to consider matrices u which are the identity but for a couple of columns and rows corresponding to i and j, i.e.

$$u_{kl} = \delta_{kl} \text{ if either } k \text{ or } l \notin \{i, j\}. \tag{6.94}$$

For instance, if $i = 1$ and $j = 2$, we will write $u = u_0\mathrm{Id} + i\vec{u}\cdot\vec{\sigma}$, with $u_0^2 + |\vec{u}|^2 = 1$ for the first 2×2 diagonal block, and the identity otherwise, as follows:

$$u = \begin{pmatrix} u_0 + iu_3 & u_2 + iu_1 & 0 & 0 & \dots \\ -u_2 + iu_1 & u_0 - iu_3 & 0 & 0 & \dots \\ 0 & 0 & 1 & 0 & \dots \\ 0 & 0 & 0 & 1 & \dots \\ \dots & \dots & \dots & \dots & 1 \end{pmatrix}. \tag{6.95}$$

There are $N(N-1)/2$ such diagonal subgroups, and $N-1$ if one considers just couples of consecutive rows and columns, i.e. $j = 1 + 1$.

The local action, restricted to the link n, μ, reads:

$$S_\mu(n) = -\frac{\beta}{N} \text{ReTr} \left(U'_\mu(n) F^\dagger_\mu(n) \right) - \frac{\beta}{N} \text{ReTr} \left(u\, U_\mu(n) F^\dagger_\mu(n) \right) . \tag{6.96}$$

$U_\mu(n) F^\dagger_\mu(n)$ is in general a $N \times N$ complex matrix, it can be easily checked that the only part relevant to the trace above, having a nontrivial dependence on u_0 and $\vec{u}$, is that restricted to rows and columns i and j. In other words, defining the 2×2 matrix

$$\begin{pmatrix} C_{11} & C_{12} \\ C_{21} & C_{22} \end{pmatrix} \equiv \begin{pmatrix} P_{ii} & P_{ij} \\ P_{ji} & P_{jj} \end{pmatrix} \tag{6.97}$$

where $P \equiv F_\mu(n) U^\dagger_\mu(n)$, and setting $u = u_0 \text{Id} + i\vec{u} \cdot \vec{\sigma}$ as a short for the whole matrix, we have

$$S_\mu(n) = \text{constant} - \frac{\beta}{N} \text{ReTr} \left(u\, C^\dagger \right) . \tag{6.98}$$

C is a general 2×2 complex matrix, which can be parameterised again in terms of the Pauli matrices, $C = C_0 \text{Id} + i\vec{C} \cdot \vec{\sigma}$, where $C_0 = A_0 + iB_0$ and $\vec{C} = \vec{A} + i\vec{B}$ are four complex numbers. However, it can be easily checked that, after taking the real part of the trace, only the real part of $C_0, \vec{C}$ survives, so that we can finally write:

$$S_\mu(n) = \text{constant} - \frac{A\beta}{N} \text{ReTr} \left(u\, a^\dagger \right) . \tag{6.99}$$

where $(A_0, \vec{A}) \equiv A (a_0, \vec{a})$ with $A \equiv \sqrt{A_0^2 + |\vec{A}|^2}$, so that $a \equiv a_0 \text{Id} + i\vec{a} \cdot \vec{\sigma}$ is a $SU(2)$ matrix.

At this point, we are ready to implement a standard $SU(2)$ heat-bath step in order to sample u, indeed its distribution is

$$p(u)du \propto e^{\frac{A\beta}{N} \text{ReTr}\left(u\, a^\dagger \right)} du \tag{6.100}$$

which is maximum for $u = a$, therefore we can rewrite $u = ga$, with $g = g_0 \text{Id} + i\vec{g} \cdot \vec{\sigma}$, and generate g_0 using the Creutz or Kennedy-Pendleton algorithm, and then $\vec{g}$ uniformly over the sphere of radius $\sqrt{1 - g_0^2}$. Alternatively, one can perform a microcanonical step by simply setting $u = a^2$.

The Cabibbo-Marinari algorithm is made ergodic by repeating it over different $SU(2)$ subgroups. A minimal choice, which guarantees that no part of the original gauge link variable remains unchanged, is to explore at least all $N-1$ $SU(2)$ subgroups with consecutive indexes, $j = i+1$. However it is usual, to allow for a better decorrelation of the gauge link, to consider all $N(N-1)/2$ diagonal subgroups. In doing so, one can alternate updating sweeps consisting of just microcanonical or heat-bath sweeps, with proportions similar to those suggested for $SU(2)$.

Let us summarize, for the sake of clarity, the main steps of the Cabibbo-Marinari heat-bath update of a given gauge link variable $U_\mu(n)$. The microcanonical case, as discussed above, is trivially similar.

---

1. Select a $SU(2)$ subgroup by choosing a pair of indexes $i, j \in [1, N]$ with $i < j$;

2. Construct the 2×2 matrix C from the current value of the gauge link and from the force as in equation (6.97), $C \equiv (F_\mu(n)U_\mu^\dagger(n))_{[i,j]}$;

3. Compute the four real numbers

$$A_0 + iA_3 = \frac{C_{11} + C_{22}^*}{2} \; ; \quad A_2 + iA_1 = \frac{C_{12} - C_{21}^*}{2}$$

and from them compute $A \equiv \sqrt{A_0^2 + |\vec{A}|^2}$ and the $SU(2)$ matrix $a \equiv (A_0 \, \mathrm{Id} + i\vec{A} \cdot \vec{\sigma})/A$;

4. Draw a real number g_0 according to the probability distribution

$$p(g_0)dg_0 \propto \sqrt{1 - g_0^2} \exp\left(\frac{2A\beta}{N} g_0\right) dg_0$$

using either the Creutz or the Kennedy-Pendleton algorithms illustrated for the $SU(2)$ case

5. Set $g_3 = \sqrt{1 - g_0^2}(1 - 2w_1)$; $g_1 = \sqrt{1 - g_0^2}\sqrt{1 - g_3^2} \cos(2\pi w_2)$
   $g_2 = \sqrt{1 - g_0^2}\sqrt{1 - g_3^2} \sin(2\pi w_2)$
   where w_1 and w_2 are uniform deviates in $[0, 1)$;

6. Set $u = ga$ and then, after embedding u back into $SU(N)$:

$$U_\mu'(n) = uU_\mu(n)$$

7. repeat from step 1. for a new pair i, j, running over a large enough set of $SU(2)$ subgroups before moving to the next gauge link.

6.3.4 Scaling, asymptotic freedom and dimensional transmutation

Till now, we have not considered the problem of actually taking the continuum limit of lattice gauge theories. We have computed the $a \to 0$ limit of the plaquette action, checking that it gives back the continuum Yang-Mills action, but this is usually called the naïve continuum limit. As we have discussed above, a continuum QFT can only be defined approaching a critical point of the corresponding statistical system, where dimensionless correlation lengths diverge.

In non-interacting theories, this is usually simple and well-defined right from the beginning. For instance, in the free scalar field theory introduced in section 6.2.2, the lattice model is defined in terms of a single parameter $\hat{m}$, which is related to the physical mass by the relation $\hat{m} = am$, without any further renormalisation involved. It is clear, on one hand, that the continuum limit requires $\hat{m} \to 0$; on the other hand, one actually finds that the dimensionless correlation length of the scalar field $\hat{\xi} \sim 1/\hat{m}$, so that $\hat{m} = 0$ is indeed a critical point.

For an interacting field theory, things can be less trivial, because of renormalisation. Let us analyse the $SU(N)$ lattice gauge theory: it has just one parameter, the inverse bare gauge coupling $\beta = 2N/g^2$, which is also dimensionless for 3+1 space-time dimensions, so that its relation with the lattice spacing cannot be inferred on simple dimensional grounds, as we did in fact for the free scalar field. There are therefore two main questions we should ask:

1. Is there any critical value g^* of the bare coupling, around which correlation lengths diverge?

2. Assuming one such point exists, how do we know that the critical behaviour around it correctly reproduces the physics of the continuum Yang-Mills theories?

Regarding the first question, the answer is yes and the critical point is found at $g^* = 0$, corresponding to $\beta \to \infty$. Actually, for $N > 4$, another phase transition is found, for finite values of g, which, however, is first order, hence not associable with any possible continuum theory. Close enough to $g = 0$, i.e. for large enough β, correlation lengths will start scaling proportionally to each other, according to the same critical index ν: in this region, which defines the so-called *scaling regime*, one can safely measure ratios of correlation lengths, since the ratios will remain almost unchanged moving closer and closer to the critical point, providing the ratios of the corresponding physical masses in the continuum QFT.

Regarding the second question, the answer comes, at least for $SU(N)$ gauge theories and QCD, from a property known as *asymptotic freedom* and discovered in 1973 by David Gross, Frank Wilczek, and David Politzer [89, 90], stating that the renormalized coupling of the theory becomes asymptotically small as the energy scale increases, i.e. as the corresponding length scale decreases. In practice, when one renormalizes the theory, the renormalized coupling turns out be dependent on the renormalisation energy scale μ, $g_R = g_R(\mu)$. This dependence is dictated by the so-called β function, $\beta(g_R) \equiv \mu dg_R(\mu)/d\mu$, which in perturbation theory turns out to be a power series of g_R, starting as

$$\mu \frac{dg_R}{d\mu} = -\beta_0 g_R^3 - \beta_1 g_R^5 + \dots \, ; \quad \beta_0 = \frac{1}{4\pi^2} \left(\frac{11N - 2N_f}{3} \right) \tag{6.101}$$

where N_f is the number of quark families, hence $N_f = 0$ in a pure gauge theory. The expression is valid just for small g_R, however one has $\beta_0 > 0$ for $N_f \leq 11N/2$, so that $g_R = 0$ is an attractive fixed point as $\mu \to \infty$. In principle, physical quantities are independent of the renormalisation scale μ, however in perturbation theory, where one expresses physical quantities as power series of g_R, the accuracy of the truncated series is best if μ matches the energy scale of the physical process. From that it follows that the interaction coupling actually *runs* with the energy scale and vanishes at high energies, while it becomes large at low energies, where perturbation theory fails and Monte Carlo simulations become the best available tool to investigate the theory.

How is that linked to the problem of approaching the continuum limit? The regularized theory is defined in terms of the bare coupling g, which defines the theory at some microscopic scale a, i.e. at the UV cut-off $\Lambda_{UV} = 1/a$. That induces a renormalized coupling g_R at some low energy scale μ, for instance, we could think ideally of measuring the interaction potential between two ideal probes at some distance $r = 1/\mu$. In principle, g_R is a function of g, a and μ, however only the dimensionless combination $a\mu$ can play a role; for instance, we will determine the interaction coupling at some distance r which is fixed in lattice units, i.e. $r = a\hat{r}$, and g_R will naturally be a function of just g and $\hat{r}$. So we can write $g_R = g_R(g, a\mu)$.

If a consistent continuum limit exists, we should be able to send $g \to g^*$ as we send $a \to 0$ so that g_R remains unchanged, i.e. we require

$$dg_R = \frac{\partial g_R}{\partial g} dg + \frac{\partial g_R}{\partial (a\mu)} \mu da = 0 \tag{6.102}$$

and this equation will set the implicit dependence between a and g, hence the scaling of dimensionless correlation lengths $\hat{\xi} \propto 1/a$ towards the continuum limit.

Now, the crucial point is that if the continuum limit we are approaching is indeed Yang-Mills, the dependence of g_R on $a\mu$ must be the same expected according to the β-function of that theory. Since the interesting critical point is perturbative, $g^* = 0$, and since also g_R is small for large enough μ, we can make use of perturbative results, at least close enough to the continuum limit. Putting these considerations and equations (6.101) and (6.102) all together, and exploiting the fact that g_R is also a power series of g, starting as $g_R = g + O(g^3)$, one easily arrives at the following result:

$$\beta_{LAT}(g) \equiv a\frac{\partial g}{\partial a} = \frac{\mu \, dg_R/d\mu}{\partial g_R/\partial g} = \beta_0 g^3 + \beta_1 g^5 + \dots \tag{6.103}$$

so the first two coefficients of the so-called *lattice β-function* are the same of the continuum β function. The fact that $\beta_0 > 0$ proves that indeed $g = 0$ is an attractive fixed point where the continuum limit can be taken, however equation (6.103) says much more than that. Indeed, by integrating the differential equation one finds the actual dependence of a on g (and vice versa) which, taking into account just β_0, is

$$a(g) = \frac{1}{\Lambda_{LAT}} \exp\left(-\frac{1}{2\beta_0 g^2}\right) \tag{6.104}$$

where Λ_{LAT} is an integration constant. The fact that a dimensionful quantity, like the lattice spacing, depends on a dimensionless quantity like the bare gauge coupling is usually called *dimensional transmutation*, and is a direct effect of non-trivial renormalisation effects, leading to the appearance of anomalous dimensions.

Turning back to the continuum limit, equation (6.104) fixes the asymptotic dependence of correlation lengths on the bare coupling, $\hat{\xi}(g) \sim 1/a(g)$, if the $g = 0$ critical point describes continuum Yang-Mills: this is called *asymptotic scaling*. Therefore, the actual check, by numerical simulations, that asymptotic scaling is reached for small enough g (large enough β) substantiates the fact that we are approaching the correct continuum theory, i.e. the same theory that correctly describes high-energy experimental results at a perturbative level. Of course, it is not necessary to reach the asymptotic scaling regime to get sensible physical information: it is sufficient to be in the scaling regime, so as to obtain reliable determinations of mass ratios.

6.4   Adding fermion fields

Quantum Chromodynamics describes strong interactions among quarks and gluons. Till now, we have illustrated the lattice discretisation and the numerical treatment of a pure gauge theory: in order to make reliable predictions about our world, we need to include quarks into the game. Quarks are particles with a non-integer intrinsic angular momentum (spin), which are described by fermionic fields. The introduction of fermionic fields in the path-integral lattice formulation is accompanied by a number of theoretical and technical problems, which we cannot discuss in much detail in the present context, and for which we refer the reader to specialized textbooks on the subject [73, 74, 75, 76]. However, we would like to sketch the main framework of the problem, especially regarding the applicability of Monte–Carlo methods and the available algorithms: this is indeed of quite general interest, as can be appreciated from the fact that HMC algorithms have been invented right in this context.

Let us start from a brief summary of the continuum theory. We will consider directly the Euclidean time formulation, since this is the one relevant to numerical simulations. The full QCD action reads as follows:

$$S_{QCD} = \int d^4x \, \mathcal{L}_{QCD} \tag{6.105}$$

$$\mathcal{L}_{QCD} = \sum_f \bar{\psi}_f \left(D_\mu \gamma_u + m_f \right) \psi_f + \frac{1}{2} \text{Tr}(F_{\mu\nu} F_{\mu\nu})$$

where f is the flavour index indicating the type of quark (f = up, down, charm, strange, top or bottom), and m_f is the associated mass. Each field ϕ_f is a function of the space-time point x and has two additional indexes which are not shown explicitly: the Dirac index $\alpha = 1, 2, 3, 4$, on which the 4×4 Dirac matrices act, and the colour index $i = 1, 2, 3$, on which the gauge fields contained in the covariant derivative D_μ act; unit matrices over both indexes are understood otherwise. The (Euclidean) Dirac matrices satisfy the Clifford algebra $\{\gamma_\mu, \gamma_\nu\} = 2\delta_{\mu\nu}$ and, in the so-called chiral representation, take the following block form:

$$\gamma_0 = \begin{pmatrix} 0 & \text{Id} \\ \text{Id} & 0 \end{pmatrix} ; \; \gamma_i = \begin{pmatrix} 0 & i\sigma_i \\ -i\sigma_i & 0 \end{pmatrix} ; \; \gamma_5 \equiv \gamma_0\gamma_1\gamma_2\gamma_3 = \begin{pmatrix} -\text{Id} & 0 \\ 0 & \text{Id} \end{pmatrix} \tag{6.106}$$

where the matrix γ_5 will be used later on and anticommutes with each Dirac matrix. Finally, $\bar{\psi}$ is the so-called Dirac adjoint of ψ, defined as $\bar{\psi} = \psi^\dagger \gamma_0$.

In the discretized theory, fermions live on lattice sites and are parallel transported from site to site by gauge link variables; it is also usual to rescale them by a power of the lattice spacing so as to work with dimensionless variables, in particular for fermion fields in 3+1 dimensions one defines $\hat{\psi}_{f,n} \equiv a^{3/2}\psi_f(an)$, where all other indexes are understood, and $\hat{m}_f \equiv am_f$. One can then proceed to the discretisation of the covariant derivative term, paying attention to the following detail: in the continuum, D_μ is an anti-hermitian operator, and this is essential to guarantee that the continuum QFT is well defined; the simplest discretisation of the derivative keeping the anti-hermiticity property is the symmetric one. Therefore, the simplest discretisation of the lattice action for the single flavour field reads:

$$\sum_n \left[\hat{m}_f \hat{\bar{\psi}}_{f,n} \hat{\psi}_{f,n} + \frac{1}{2} \sum_\mu \hat{\bar{\psi}}_{f,n} \gamma_\mu \left(U_\mu(n)\psi_{f,n+\hat{\mu}} - U_\mu^\dagger(n-\hat{\mu})\hat{\psi}_{f,n-\hat{\mu}} \right) \right] \tag{6.107}$$

which appears as a bilinear form in the quark field, hence it is usually rewritten as $\hat{\bar{\psi}}_{f,n} M^{(f)}_{n,m} \hat{\psi}_{f,m}$, where $M^{(f)}_{n,m}$ is the so-called fermion matrix:

$$M^{(f)}_{n,m} = \delta_{n,m}\hat{m}_f\text{Id} + \frac{1}{2}\sum_\mu \gamma_\mu \left(\delta_{m,n+\hat{\mu}}U_\mu(n) - \delta_{m,n-\hat{\mu}}\hat{U}_\mu^\dagger(n-\hat{\mu}) \right) \tag{6.108}$$

and Id stands for the identity over colour and Dirac indices. It is usual to write $M^f[U] = \hat{m}_f + K[U]$ where $K[U]$ defines the so-called *hopping matrix*, which connects nearest-neighbours sites.

The discretization in equation (6.107) is usually called *naïve*: the reason is that it reveals to describe, in the continuum limit, 16 fermion fields, instead of just one; this is known as the *fermion doubling problem*. The reason is subtle but understandable: the symmetric derivative measures variations between fields on sites with the same parity for each direction, hence fluctuations among sites with different parities cost zero action and it is in fact like having 16 independent fields. The solution to this problem was revealed to be non-trivial and requires some compromise, like the explicit breaking of chiral symmetry[9]. The reader can find extensive treatments of this subject in standard textbooks on lattice gauge theories [73, 74, 75, 76]. Since the problem is not fully relevant to a description of the numerical methods employed for the numerical simulation of QCD, we will go on assuming the naïve discretization, adding some comments, when needed, on how the possible alternatives change the picture.

Just writing the action is not sufficient to understand how to treat the problem numerically: one needs to write the full path-integral of the theory. Fermionic fields are represented by local operators which anti-commute (instead of commuting) with each other: this is why the computational basis cannot be simply represented by the field operator eigenstates and the corresponding eigenvalues, as in section 6.2.2 for the scalar field. Also in this case, the reader is referred to standard textbooks for a comprehensive discussion of the subject, and we will just state the relevant results.

The path-integral integration for fermionic fields is over anti-commuting set of variables are known as *Grassmann variables*. The full lattice QCD path integral therefore reads

$$\int \mathcal{D}U \prod_f \mathcal{D}\hat{\bar{\psi}}_f \mathcal{D}\hat{\psi}_f \exp\left(-\sum_f \hat{\bar{\psi}}_f M^{(f)}\hat{\psi}_f - S_{YM}\right). \qquad (6.109)$$

Among the various properties of Grassmann variables, the one which is most relevant to our purposes is that the functional integration of the exponential of a bilinear form, like that in equation (6.109), returns the determinant of the bilinear form matrix, i.e. integration over fermionic variables can be performed exactly and the expression in equation (6.109) can be rewritten as

$$\int \mathcal{D}U \prod_f \det\left(M^{(f)}[U]\right) e^{-S_{YM}[U]}. \qquad (6.110)$$

---

[9]Chiral symmetry is a property of QCD in the limit in which the quark masses vanish. Even if quark masses are non-vanishing, two of them (up and down) are particularly light compared to the typical energy scale of strong interactions, hence the symmetry is almost exact.

Equation (6.110) appears as an integral over gauge link variables only: however fermion fields have not disappeared from the game, their contribution is encoded in the determinants and will influence expectation values non-trivially. Two are the main questions when one considers a Monte Carlo evaluation of the integral in equation (6.110):

1. Is the determinant of the fermion matrix real and positive, so that the path-integral measure remains real and positive, thus allowing for Monte Carlo sampling?

2. What are the available algorithms in order to properly take into account the fermion determinant in the Monte Carlo sampling?

As we will illustrate, the answer to the first question is usually positive, but significant exceptions are found where it is not, i.e. the determinant can be negative or even complex, something which is usually known as the *sign problem*. Regarding the second issue, we will learn that the fermion determinant can be interpreted as a highly non-local interaction term among gauge-link variables, so that standard local algorithms become extremely inefficient: this is why this reveals the ideal field of application of the HMC algorithm illustrated in Chapter 5. More than that, this is actually the numerical problem where the HMC algorithm was born.

6.4.1 Positivity of the fermion determinant

The determinant of the fermion matrix can be expressed as the product of its (complex) eigenvalues, hence the question about its positivity is a question about the spectrum of the fermion operator. Let us analyse the issue first from the point of view of the continuum theory.

$M^{(f)}$ is a discretization of the operator $D_\mu \gamma_\mu + m_f$. whose eigenvectors are also eigenvectors of $D_\mu \gamma_\mu$. Since the Euclidean Dirac matrices are hermitian, while $D_\mu = \partial_\mu + ig A_\mu$ is anti-hermitian, $D_\mu \gamma_\mu$ is anti-hermitian overall, so its spectrum is purely imaginary. In addition, given that γ_5 anticommutes with all γ_μ matrices, if ψ is an eigenvector of $D_\mu \gamma_\mu$, then $\gamma_5 \psi$ is an eigenvector as well, with opposite eigenvalue. Putting all together, we conclude that the eigenvalues of $D_\mu \gamma_\mu + m_f$ come in pairs, $m_f \pm i\lambda = 0$, so their contribution to the determinant is

$$(m_f + i\lambda)(m_f - i\lambda) = m_f^2 + \lambda^2 > 0$$

the only exception being the case $\lambda = 0$, where one may have single eigenvalue, but in that case one has $m_f > 0$ anyway. Therefore, the continuum determinant is positive flavour by flavour.

However this nice property breaks, already in the continuum, as soon as one considers some variations, like the introduction of a chemical potential μ_f coupled to the quark number operator N_q. In the continuum theory that means adding a term

$$-\mu_f N_q = -\mu_f \int d^3x \psi_f^\dagger \psi_f = -\mu_f \int d^3x \bar{\psi}_f \gamma_0 \psi_f$$

to the Hamiltonian, and in turn the same term with an additional minus sign to the Lagrangian. For that reason, the temporal part of the fermion kernel gets modified in the following way

$$D_0\gamma_0 = (\partial_0 + igA_0)\gamma_0 \rightarrow (\partial_0 + ig(A_0 - i\mu_f/g))\gamma_0$$

which makes it lose its anti-hermiticity properties. Overall, the eigenvalues of $D_\mu\gamma_\mu$ are not guaranteed anymore to come into purely imaginary pairs, and the fermion determinant becomes complex.

This is the infamous *sign problem*, presently hindering our exploration by Monte Carlo methods of the part of the QCD phase diagram involving a baryon chemical potential, which includes cases of particular interest, like the interior structure of neutron stars. Note that this is not a problem of the theory itself, but just of the choice of the computational basis: the grand-canonical partition function is still written as the sum of positive terms when one works with Hamiltonian eigenstates. This is why it makes sense to find a solution to the problem, as better discussed in Chapter 7.

Let us now turn to the lattice theory. In the naïve discretization, the hopping operator $K[U]$ keeps the essential properties of its continuum counterpart: it is anti-hermtian and anticommutes with γ_5; hence, the spectrum of the fermion matrix still contains eigenvalue pairs like $\hat{m}_f \pm i\lambda$, leading to a positive determinant. Things may change as one tries to modify the fermion matrix in order to cure the fermion doubling problem.

For instance, a popular solution (Wilson fermions) is to add to the lagrangian a term $\bar{\psi} \,\Box\psi$, where $\Box$ is the four-dimensional Laplacian operator, which is irrelevant in the continuum limit but adds a coupling between nearest-neighbours sites, thus killing the doublers. The operator is hermitian and commutes with γ_5, hence it changes the overall structure of the fermion matrix. However, it preserves a property known as γ_5-hermiticity, $\gamma_5 M^{(f)}\gamma_5 = M^{(f)\dagger}$, which is sufficient to prove that the eigenvalues of $M^{(f)}$ are real, since

$$\det\left(M^{(f)} - \lambda\mathrm{Id}\right) \;=\; \det\left(\gamma_5(M^{(f)} - \lambda\mathrm{Id})\gamma_5\right) \tag{6.111}$$
$$=\; \det\left(M^{(f)\dagger} - \lambda\mathrm{Id}\right) = \left(\det\left(M^{(f)} - \lambda^*\mathrm{Id}\right)\right)^*$$

hence if λ is an eigenvalue, λ^* is an eigenvalue as well. However, that does not exclude the possibility of having a single eigenvalue which is real and negative, so that the determinant itself can be real and negative, leading to a genuine "sign" problem. This problem is usually solved in the Wilson fermions approach by considering only pairs of flavours with the same mass, leading to the appearance of $(\det M^{(f)})^2$ in the path-integral measure, which is positive if the determinant is real.

6.4.2 Non-locality of the fermion determinant

There are various ways to realise that the fermion determinant involves highly non-local interaction terms among gauge links, thus making a local algorithm approach extremely inefficient. The first way we illustrate is the so-called *loop expansion* of the determinant. It starts from the general formula $\det A = \exp(\operatorname{Tr} \log A)$, which is trivially proved once the matrix A is diagonalized. Then, rewriting $M^{(f)} = \hat{m}_f + K[U] = \hat{m}_f(\mathrm{Id} + K[U]/\hat{m}_f)$ and exploiting the Taylor expansion of the logarithm around the identity, we can write

$$\det M^{(f)} \propto \exp\left(\operatorname{Tr} \left(\frac{K}{\hat{m}_f} - \frac{K^2}{2\,\hat{m}_f^2} + \cdots + \frac{(-1)^n}{n} \frac{K^n}{\hat{m}_f^n} + \cdots \right) \right). \qquad (6.112)$$

Equation (6.112) is formal to some extent, since convergence of the series is guaranteed only for sufficiently high values of $\hat{m}_f$, for this reason it is also called a high-mass expansion. The trace is taken over all indexes, i.e. space-time, Dirac, and color ones, meaning that it is non-zero if the corresponding element of the expansion has non-zero diagonal (over all indexes) elements. Since K contains only elements connecting nearest-neighbours sites, its trace vanishes. The generic power K^n, instead, contains elements connecting pairs of lattice sites by the product of n gauge link variables along a lattice path connecting the two sites: hence diagonal elements are present only if n is even, and contain color traces of parallel transports along closed loops; the first non-trivial contribution comes from $n = 4$ and is proportional to a plaquette. That proves at the same time that the fermion determinant is gauge invariant, and that it contains contributions from loops of arbitrarily large size.

Another point of view is the following. We have obtained the determinant as the result of a Gaussian integral over Grassmann variables; a similar formula holds also for the integration over standard complex variables, but involves the inverse of the fermion matrix. We shall consider the case of two flavors of equal mass (which is relevant to many physical situations, for instance, when one considers just the up and down quarks, which have almost equal masses) and, in order to simplify the notation, will omit the flavour index in the following. Assuming the determinant is real, the squared determinant can be rewritten as follows:

$$(\det M)^2 = \det(M^\dagger M) = \int \mathcal{D}\Phi^\dagger \mathcal{D}\Phi \, e^{-\Phi^\dagger (MM^\dagger)^{-1}\Phi}. \qquad (6.113)$$

In the above expression, Φ has exactly the same degrees of freedom (space-time, Dirac and colour indexes) as the original fermion field, however, it is a standard complex field, like that associated with a bosonic particle, and it is indeed called the *pseudo-fermion* field. Equation (6.113) offers the possibility to take into account the fermion determinant by simply introducing a new set of integration variables in the path-integral, the pseudo-fermions, which can be treated, from a numerical point of view, as a new set of standard stochastic variables.

There is a price to pay, however. The pseudofermion fields interact through the inverse of the fermion matrix, which can be Taylor expanded similarly to what we did before:

$$\frac{1}{M} = \frac{1}{\hat{m} + K} = \frac{1}{\hat{m}} \sum_{n=0}^{\infty} \left(\frac{K}{\hat{m}}\right)^n \qquad (6.114)$$

showing that it connects fields at arbitrarily large distances.

6.4.3   Monte Carlo sampling and the HMC algorithm

We will focus on the theory with two quark flavours of equal mass. After the introduction of pseudofermion fields, the partition function reads

$$\int \mathcal{D}U \mathcal{D}\Phi^{\dagger} \mathcal{D}\Phi \, e^{-\Phi^{\dagger}(MM^{\dagger})^{-1}\Phi} e^{-S_{YM}} . \qquad (6.115)$$

Pseudofermion fields can be updated quite efficiently by a global heat-bath step performed at fixed values of the gauge links. Indeed, introducing the auxiliary field $\eta = (M^{\dagger})-1\Phi$, one has that η is Gaussian distributed, like $\mathcal{D}\eta^{\dagger} \mathcal{D}\eta \, e^{-\eta^{\dagger}\eta}$, so one can generate η from scratch by a standard Box-Müller algorithm and then compute $\Phi = M^{\dagger}\eta$.

The update of the gauge link variables at fixed values of the pseudo- fermion fields is less trivial, because of the non-local dependence of the pseudo-fermionic effective action on them. In the HMC approach, one first introduces the conjugate momenta of the gauge link variables. Since the latter are elements of the gauge group, the conjugate momenta live in the algebra of the group, i.e. for each lattice link n, μ one has a hermitian and traceless 3×3 matrix $H_\mu(n)$, with an associated kinetic term which is conventionally normalized as $\text{Tr}(H_\mu(n)H_\mu(n))/2$, so that the partition function reads

$$\int \mathcal{D}H \mathcal{D}U \mathcal{D}\Phi^{\dagger} \mathcal{D}\Phi \, e^{-\sum_{n,\mu} \text{Tr} H_\mu^2(n)/2} e^{-\Phi^{\dagger}(MM^{\dagger})^{-1}\Phi} e^{-S_{YM}} . \qquad (6.116)$$

The HMC algorithm then consists first of a heat-step where momenta are updated according to their Gaussian distribution; then, one performs a molecular dynamics trajectory, where the gauge link variables and the conjugate momenta are updated following the Hamiltonian equations of motion, followed by a heat-bath step. To summarise, the algorithm reads as follows:

---

1. Heat-bath generation of the pseudofermion field Φ at fixed values of the $U_\mu(n)$ and $H_\mu(n)$;

2. Heat-bath generation of the momenta. Expanding over a proper basis of generators, $H_\mu(n) = h_\mu^a(n)T^a$, it can be rephrased in terms of a gaussian distribution of the $h_\mu^a(n)$ real numbers.

3. Molecular dynamics trajectory in a fictitious time evolution. Hamilton equations are easily written and integrated for the gauge link variables:

$$U_\mu(n) \to e^{iH_\mu(n)\Delta t}U_\mu(n)$$

while the $\dot{H}_\mu(n)$ is less trivial but can written down systematically. (see discussion below).

4. Perform a Metropolis accept–reject step

This is not the end of the story, as new complications enter the game when one tries to deal with the fermion doubling problem, like the necessity of simulating fractional powers of the fermion determinant. However, we will not go further on that, given the introductory nature of the present treatment.

There is just one point that is worth further discussion. Most of the numerical effort during the HMC algorithm is spent in the computation of $\dot{H}_\mu(n)$: since this is obtained by differentiation with respect to the link variables of $\Phi^\dagger(MM^\dagger)^{-1}\Phi + S_{YM}$, the computation involves the application of the inverse fermion matrix M^{-1} over Φ, and that must be done separately at each step of the HMC trajectory, since the evolution of link variables modifies M after each step (while Φ is constant).

The inversion of a sparse matrix can be performed using efficient iterative algorithms (like the Conjugate Gradient algorithm or its variants), which typically return the solution $X = M^{-1}\Phi$ with some residual uncertainty which is fixed apriori (i.e. the algorithm is iterated until the norm of the residual $|MX - \Phi|$ goes below a predefined threshold). In principle, one could profit of the fact that M is slowly varying during the trajectory, in order to use $M^{-1}\Phi$ as a first guess for the iterative inversion process at the next step, so as to save time, however, that would violate the reversibility of the HMC step, thus ruining detailed balance, unless the inversion is done to machine precision. For this reason the inversion is repeated from scratch at each step, as a benefit one can be more relaxed with the inversion precision, which in this case affects only the Metropolis acceptance but not the validity of detailed balance.

As a final comment, we stress that the computational difficulty of simulating QCD is strongly enhanced by the particular physical properties of the theory.

Chiral symmetry, which is an almost exact symmetry of the theory, happens to be spontaneosly broken; this is revealed, for instance, by the presence of relatively light hadronic particles (the pions), corresponding to large correlations lengths, which would be the Goldstone massless bosons associated with spontaneous symmetry breaking, were the light quarks really massless. It can be proved that the spontaneous breaking of chiral symmetry is equivalent to an accumulation of eigenvalues of the Dirac operator around zero: that in turn, because of the small values of the light quark masses, implies very small eigenvalues for the fermion matrix itself. The appearance of very small eigenvalues in general means that the matrix is ill-conditioned, making the inversion process is much harder from a computational point of view.

6.5   Exercises

Exercise 6.1 *Consider the probability distribution for the discretized path-integral of the one-dimensional harmonic oscillator, as reported in equation (6.43).*

   a. *Devise a local Metropolis algorithm for sampling the distribution.*

   b. *How should one tune the step parameter δ of the Metropolis as $\eta \to 0$, in order to maintain a reasonable acceptance? Try a guess and then check by direct numerical experiment (remember to take $N\eta$ fixed to some value).*

Exercise 6.2 *Consider again the discretized harmonic oscillator and implement a local heat-bath and a local over-relaxation (microcanonical) algorithm for the same system (the local distribution is always Gaussian). Find what is the optimal mixture of the two algorithms, then compare its efficiency, for some value of η, with that of the optimal Metropolis developed in exercise 6.1 for the same η. (Suggestion: measure the efficiency in terms of autocorrelation times of some quantity, e.g. the average value of y^2 over each path; remember to take also the numerical cost of each step into consideration).*

Exercise 6.3 *According to Quantum Mechanics and to the correspondence $y_j = x(aj)\sqrt{m\omega/\hbar}$ leading to the distribution in equation (6.43), the distribution of each coordinate y_j should be proportional to e^{-y^2} when the harmonic oscillator is in the ground state.*

a. *Take a very low value of the temperature, e.g. $\beta\hbar\omega = N\eta = 20$, so that only ground state physics is relevant, perform a numerical simulation for different values of η, prepare an histogram for the distribution of y_j obtained during the simulation and compare it to the expected ground distribution mentioned above. How accurate is the matching if $\eta \sim 1$?*

b. *From the same set of numerical simulations, compute $\langle y^2 \rangle$ and test if it is compatible, within statistical errors, with the expected value $\langle y^2 \rangle = 0.5$ (you can average over different values of j, given the translational invariance in Euclidean time). Do you find agreement?*

c. *Now Ttry to take discretization effects into account by a best fit of results obtained for different values of η according to*

$$\langle y^2 \rangle_\eta = A + B\eta^2 .$$

Does the fit work well if the fitted range of η values is close enough to zero? Does the parameter A agree within errors with the expected $\langle y^2 \rangle_{continuum} = 0.5$?

d. *Prove that the observed distribution of y_j reflects, apart from the discretization errors, the thermal distribution of the coordinate for any temperature. Then consider the opposite limit of very high temperature and compare numerical results with the expected classical Boltzmann distribution for the harmonic oscillator.*

Exercise 6.4 *Consider again the study of ground state physics for the harmonic oscillator. Now we want to explore the behaviour of the "kinetic" term*

$$\langle \frac{1}{2}\frac{\Delta y^2}{\eta^2} \rangle ; \qquad \Delta y_j \equiv y_{j+1} - y_j .$$

a. *Perform numerical simulations for different values of η at fixed $N\eta = 20$, e.g. $\eta = 0.5, 0.2, 0.1, 0.05, 0.02, 0.01, 0.005$. Compute $\langle \Delta y^2 \rangle/(2\eta^2)$ for each η and show that results can be well described by a divergent behaviour $1/(2\eta)$.*

b. *Try to explain your results on the basis of the probability distribution function in equation (6.43).*

Exercise 6.5 *Show that, when properly taking into account the normalization factor in front of the discretized partition function, the normalized internal energy of the harmonic oscillator can be written as*

$$\frac{U}{\hbar\omega} = -\frac{1}{\hbar\omega}\frac{\partial}{\partial\beta}\log Z = \frac{1}{2}\langle y^2 \rangle + \frac{1}{2\eta} - \frac{1}{2}\frac{\langle \Delta y^2 \rangle}{\eta^2}$$

a. *Interpreting the last two terms as the correct kinetic contribution, show that it is expected to be positive on the basis of the probability distribution in equation (6.43).*

b. *Compare numerical results obtained for different values of T and η with the exact results, derived from the thermodynamics of the harmonic oscillator taking into account our change of variables*

$$\frac{U}{\hbar\omega} = \frac{1}{2} + \frac{1}{e^{1/(N\eta)} - 1}.$$

Exercise 6.6 *Consider again the study of the ground state physics for the harmonic oscillator, fixing, e.g. $N\eta = 20$.*

a. *Consider the two-point functions $\langle y_j y_{j+k} \rangle$ and $\langle y_j^2 y_{j+k}^2 \rangle$ and show that, for the harmonic oscillator and in the continuum limit, they are expected theoretically to be proportional to single exponentials (in particular, $e^{-\eta k}$, and $e^{-2\eta k}$) giving information about the energies of the first and the second excited state of the system.*

b. *Compute such two-point functions numerically for different values of η, try a best fit to single exponentials and compare with the expected results.*

c. *Repeat the exercise for the two-point function $\langle y_j^3 y_{j+k}^3 \rangle$, what do you expect in this case?*

Exercise 6.7 *Modify the algorithm for the harmonic oscillator in order to include a quartic corrections λx^4 to the potential.*

a. *Repeat in this case the numerical computations done above for the ground state wave function, for the internal energy and for the energies of the first and second excited states.*

b. *Compare your results, for small enough values of λ, with analytical predictions obtained in first-order perturbation theory (it is useless to go beyond first order, since the perturbative series does not converge).*

Exercise 6.8 *Consider the problem of a quantum particle, freely moving (i.e. without any external potential) around a circular path. A more detailed numerical discussion of this problem is reported in Ref. [91].*

a. Write down the continuum thermodynamical path-integral of the system and show that it divides into topological sectors, characterized by an integer winding number counting how many times the (continuous) Euclidan path winds around the circle.

b. Compute the expected, continuum distribution for the winding number.

c. Now consider the numerical simulation of the discretized system, in order to verify theoretical expectations. As you will experience, a simple local Metropolis algorithm badly fails to correctly update the winding number as the continuum limit is approached. Can you devise some different, non-local algorithm to solve the problem?

d. Now consider the same problem for a charged particle with a non-zero magnetic flux going through the circle. Show that now the thermodynamical path-integral becomes non treatable by Monte Carlo simulations, since paths with a non-zero winding number have a complex action.

Exercise 6.9 *Consider the path-integral formulation for a system made up of two bosonic (fermionic) particles moving in a one-dimensional potential., e.g. a harmonic one.*

a. Taking into account that position eigenstates must be properly symmetrized (antisymmetrized), show that the thermodynamical path-integral is made up of two classes of paths, those made up of two separate paths (one for each particle), those made up of a single path rounding twice around the Euclidean circle, which can also be interpreted as an "exchange" pair of paths.

b. Show that, for fermion particles, the exchange paths are weighted by a minus sign.

c. Set up the numerical simulation for the discretized path-integral of two bosons in a harmonic potential: you should devise a new Markov Chain step capable of moving from disconnected to exchange paths. Compare numerical results with theoretical expectations, e.g. for the ground state energy.

d. Can you perform similar numerical simulations for the two fermion system? Suggestion: you can make use of the code for bosons, and try to take the minus sign for the exchange paths into account by a reweighting procedure, i.e. by rewriting

$$\langle O \rangle_F = \frac{\langle O \; sign \rangle_B}{\langle sign \rangle_B},$$

where F/B stands for fermions/bosons. However, how this strategy works for low temperatures?

Exercise 6.10 *Consider the probability distribution for a discretized scalar field theory reported in equation (6.51).*

    a. *Devise a numerical algorithm for the Monte Carlo sampling of the system, considering a local Metropolis, heat-bath, over-relaxation or their possible combinations, comparing their efficiencies.*

    b. *Now we want to study the thermodynamics of the system. What do you obtain by deriving $\log Z$ with respect to $\hat{m}$? Is it proportional to the internal energy? How should the internal energy be computed?*

Exercise 6.11 *Consider a $U(1)$ pure gauge theory in two space-time dimensions.*

    a. *Show that, for periodic boundary conditions in both directions (i.e. on a torus), the Euclidean path-integral of the system separates into topological sectors, characterized by quantized values of the field strength flux across the torus (winding number).*

    b. *Consider now the discretized path integral, written in terms of elementary parallel transports. Write down an expression for the action and for the winding number. How the continuum limit should be taken in this case?*

    c. *Devise a local algorithm (e.g. Metropolis) for the Monte Carlo sampling of the discretized system. Do any problems emerge for the decorrelation of the winding number as the continuum limit is approached?*

Chapter 7

Current challenges in Monte Carlo Simulations

In this chapter, we will explain how standard Monte Carlo techniques can spectacularly fail in practical applications. For example, the so-called sign problem can be particularly challenging in quantum Monte Carlo simulations. It can restrict the types of quantum systems that can be accurately studied, limiting our understanding of fundamental physical phenomena. Despite of more than 50 years of efforts, it still implies that cold and dense quantum matter can only simulated for very simple systems and is elusive for theories such as quantum chromodynamics, the theory of strong interactions.

7.1   Sign and overlap problems

Sign and overlap problems severely restrict the possibility of using standard stochastic simulations. In this chapter, we give an introduction to those problems and explain selected techniques to alleviate the issue.

7.1.1   What is the issue?

Monte Carlo simulations are based upon the (semi-)positivity of the integrand that allows to interpret the integrand as a probability density function (after proper normalisation). However, this positivity is not granted in some impor-

DOI: 10.1201/9781315156156-7

tant cases. High-profile examples are for instance Fourier transforms of many variables or quantum field theories at finite densities of matter. In the latter case, the mapping of the quantum field theory onto a statistical system of many variables generically produces a complex (or a real, but in parts negative) integrand. Losing the probability interpretation of the integrand has the immediate consequence that the legacy of extremely successful Monte Carlo techniques cannot be applied. A prominent example is the theory of the strong interactions, quantum chromodynamics, at finite matter density. Here, the regime of cold but dense matter, prevailing in compact stars, evades a first principle study by computer simulations since more than thirty years, and coined the name "sign problem".

Let us illustrate the "sign problem" using the Fourier transform for one variable only:

$$ Z = \int_{-\infty}^{\infty} dx \, \exp\{-x^2/2\} \, \exp\{i\, p\, x\} \, . \tag{7.1} $$

Although there is not a probability interpretation of the integrand anymore, the generalised expectation values such as

$$ \langle x^2 \rangle = \frac{1}{Z} \int_{-\infty}^{\infty} dx \, x^2 \, \exp\{-x^2/2\} \, \exp\{i\, p\, x\} \, , \tag{7.2} $$

are very useful objects in the quantum physics context. Note that there is no guarantee anymore that $\langle x^2 \rangle$ is positive due to the complex nature of the integrand. In fact, the integral can be carried out analytically with the result:

$$ Z = 2\sqrt{\pi} \, \exp\{-p^2/2\} \, , \qquad \langle x^2 \rangle = 1 - p^2 \, . \tag{7.3} $$

Indeed, $\langle x^2 \rangle$ is negative for $|p| > 1$.

We can already make an important observation: assume that we generate samples for x in a Monte Carlo approach from a cleverly designed "effective" probability measure and then use these samples to estimate, in the Monte Carlo sense, $\langle x^2 \rangle$. This approach would necessarily produce a *positive* estimate, and, hence, the approach must fail at least for a large p. Any "effective theory" approach exploiting Monte Carlo simulation must also change the observable ($\langle x^2 \rangle$ is our example) in order to be successful.

The so-called *reweighting* formally solves the sign problem and was discussed in the late eighties as a potential remedy for QCD simulations at finite densities. The central idea is to define a positive probability density and rewrite partition function and expectation values as

$$ Z_{\text{mod}} = \int_{-\infty}^{\infty} dx \, \exp\{-x^2/2\} \, , $$

$$ \langle f(x) \rangle_{\text{mod}} = \frac{1}{Z_{\text{mod}}} \int_{-\infty}^{\infty} dx \, f(x) \, \exp\{-x^2/2\} \, , \tag{7.4} $$

$$ \langle x^2 \rangle = \frac{\langle x^2 e^{ipx} \rangle_{\text{mod}}}{\langle e^{ipx} \rangle_{\text{mod}}} \, . \tag{7.5} $$

The expectation values $\langle\ldots\rangle_{\mathrm{mod}}$ are sometimes called the *phase-quenched* expectation values. Note that these expectation values can be estimated using standard Monte Carlo simulations. To this aim, we generate a sequence of N values $x_1 \ldots x_N$ that are distributed in accordance with Z_{mod} (Gaussian for our example). We then estimate as usual

$$[f] \; = \; \frac{1}{N} \sum_{k=i}^{N} f(x_k) \,, \quad \langle f(x) \rangle_{\mathrm{mod}} = M[f] + \epsilon_N \,, \quad \epsilon_N = \mathcal{O}\left(1/\sqrt{N}\right) . \quad (7.6)$$

The so-called *phase factor expectation value* $\langle \mathrm{e}^{ipx} \rangle_{\mathrm{mod}}$ is of central importance. We find, for example

$$\langle x^2 \mathrm{e}^{ipx} \rangle_{\mathrm{mod}} \; = \; - \frac{d^2}{dp^2} \langle \mathrm{e}^{ipx} \rangle_{\mathrm{mod}} . \quad (7.7)$$

A direct calculation leads to

$$\langle \mathrm{e}^{ipx} \rangle_{\mathrm{mod}} \; = \; \exp\{-p^2/2\} \,, \quad (7.8)$$

confirming

$$\langle x^2 \mathrm{e}^{ipx} \rangle_{\mathrm{mod}} \; = \; - \frac{d^2}{dp^2} \exp\{-p^2/2\} \; = \; (1 - p^2) \exp\{-p^2/2\} .$$

Now let us return to the object of interest in (7.5):

$$\langle x^2 \rangle \; = \; \frac{\langle x^2 \mathrm{e}^{ipx} \rangle_{\mathrm{mod}}}{\langle \mathrm{e}^{ipx} \rangle_{\mathrm{mod}}} = \frac{(1 - p^2) \exp\{-p^2/2\}}{\exp\{-p^2/2\}} . \quad (7.9)$$

For large p, only in an analytical calculation, the small factors $\exp\{-p^2/2\}$ cancel. If we replace the involved expectation values by Monte Carlo estimators, we find instead

$$\langle x^2 \rangle \; = \; \frac{\langle x^2 \mathrm{e}^{ipx} \rangle_{\mathrm{mod}} + \epsilon_1}{\langle \mathrm{e}^{ipx} \rangle_{\mathrm{mod}} + \epsilon_2} = \frac{(1 - p^2) \exp\{-p^2/2\} + \epsilon_1}{\exp\{-p^2/2\} + \epsilon_2} . \quad (7.10)$$

We must have at least

$$\langle \mathrm{e}^{ipx} \rangle_{\mathrm{mod}} > \epsilon_2 \quad (7.11)$$

to have a significant signal-to-noise ratio. In our example, we demand that

$$\exp\{-p^2/2\} > \epsilon_{1,2} \,,$$

and find approximately

$$\langle x^2 \rangle \; \approx \; 1 - p^2 + \mathcal{O}\left(\mathrm{e}^{p^2/2}\epsilon_1\right) + \mathcal{O}\left(\mathrm{e}^{p^2/2}\epsilon_2\right) . \quad (7.12)$$

We find that, although the result can be of order one, the absolute errors can be very large:

$$\text{error} \approx \frac{\epsilon_{\mathrm{MC}}}{\langle \mathrm{e}^{ipx} \rangle_{\mathrm{mod}}} . \quad (7.13)$$

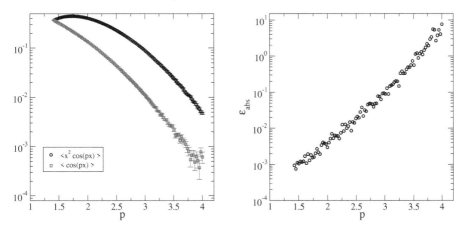

FIGURE 7.1: For a fixed $N = 10^7$, estimates for (7.14) as a function of p (left). The absolute error for the estimate for $\langle x^2 \rangle$ (7.15).

In practice, this condition usually limits the range of p to small numbers:

$$\frac{\epsilon_{MC}}{\langle e^{ipx} \rangle_{\mathrm{mod}}} \ll 1 \quad \Rightarrow \quad e^{-p^2/2} \gg 1/\sqrt{N} \quad \Rightarrow \quad p^2 \ll \ln(N) \,.$$

The rise of the absolute error with $\langle e^{ipx} \rangle_{\mathrm{mod}}^{-1}$ is called *overlap* problem. Although (7.5) makes the theory accessible by standard Monte Carlo simulations by avoiding the sign-problem, in practice, however, the *overlap* problem limits the applicability to a small parameter range.

Let us study the overlap problem for the current example in a computer experiment. We have generated a set of $N = 10^7$ values x_k and have estimated

$$\langle x^2 e^{ipx} \rangle_{\mathrm{mod}} \approx [x^2 e^{ipx}] \,, \qquad \langle e^{ipx} \rangle_{\mathrm{mod}} \approx [e^{ipx}] \,. \tag{7.14}$$

We observe that both numerator and denominator for $\langle x^2 \rangle$ (7.5) rapidly decrease with increasing p while the statistical error for fixed N remains constant. We define the absolute error by

$$\epsilon_{\mathrm{abs}}(p) := \left| \frac{[x^2 \cos(px)]}{[\cos(px)]} - (1 - p^2) \right| \,. \tag{7.15}$$

Our numerical findings are shown in Figure 7.1, right panel. Note the logarithmic scale on the y-axis: we observe an exponential increase of the absolute error with increasing p. This increase severely limits the applicability of the reweighting technique to the range of small p.

7.1.2 Theories with sign problems

A well-known starting point for quantum field theories is the partition function Z formulated as a Hilbert space expectation value of an operator-valued Gibbs factor:

$$\langle A \rangle = \frac{1}{Z} \, \mathrm{Tr} \left[A \, \exp(-H/T) \right], \qquad Z = \mathrm{Tr} \, \exp(-H/T), \qquad (7.16)$$

where H is the Hamiltonian and T is the temperature. A is an operator representing an observable and as such commutes with the Hamiltonian, i.e. $[H, A] = 0$.

Key ingredient to make quantum field theories accessible to computer simulations, is to map (7.16) to a *statistical field theory*. Such mapping has been already illustrated in Chapter 6; however, let us here only touch upon the main idea and then illustrate the rise of a sign-problem for a large class of quantum field theories. Assume that $|c\rangle$ is a basis in Hilbert space with the usual properties:

$$1 = \sum_c |c\rangle \langle c| , \qquad \mathrm{Tr} A = \sum_c \langle c| A |c\rangle .$$

Using n sets of complete sets of vectors $c_1 \ldots c_n$, we can write [92]

$$Z = \sum_{c_1 \ldots c_n} \langle c_1 | e^{-aH} | c_2 \rangle \ldots \langle c_{n-1} | e^{-aH} | c_1 \rangle =: \sum_{\{c\}} P(c) \quad \text{(classical repr.)} \quad (7.17)$$

$P(c)$ is referred to as the *probabilistic weight* of the (classical) *configurations* $c = c_1 \ldots c_n$, and the " lattice spacing" $a = \frac{1}{nT}$ is the regulator we introduce. Similarly

$$\mathrm{Tr} \left[A \, \exp(-H/T) \right] = \sum_{\{c\}} A(c) \, P(c) . \qquad (7.18)$$

If $P(c)$ turns out to be positive, the latter formulation can be used to set-up a Monte Carlo approach. This is indeed the case for many scenarios. There is, however, an important class of quantum field theories for which $P(c)$ evades positivity. This is the large class quantum field theories at finite densities of its constituents, such as electrons in metal or hadrons in compact stars. In this case, it is a $U(1)$ symmetry that is associated with particle number conversation. Because of this symmetry, the particle number operator N commutes with the Hamilton operator. In order to study systems with finite particle number, the mathematical approach uses a generalised Hamiltonian,

$$H = H_0 + \mu N .$$

with a so-called Lagrange multiplicator μ, which enjoys the physical interpretation of a *chemical potential*.

Let us illustrate the rise of a sign problem using a simple quantum mechanical model. The Hamiltonian H_0 has a translation invariant interaction, implying that the momentum operator $\hat{p}$ commutes with H_0. Rather than a system with finite particle number, we here consider a theory with a finite average of the momentum operator and define

$$H = H_0 + \mu \hat{p} .$$

Re-deriving the steps that map the partition function in operator language to a statistical field theory, it generically turns out that the "weights" $P(c)$ are complex numbers. For example consider, the quantum mechanical system

$$H_0 = \frac{1}{2} \hat{p}^2 + V(x),$$

where $\hat{p}$ is the momentum operator. Using momentum $|p\rangle$ and coordinate $|x\rangle$ eigenstates, we find

$$
\begin{aligned}
\langle x_n | e^{-aH} | x_{n+1} \rangle &= \int_{p_n} \langle x_n | e^{-aV(x_n)} | p_n \rangle \langle p_n | x_{n+1} \rangle \exp\left(-\frac{a}{2} p_n^2 - a\mu\, p_n \right) \\
&= \int_{p_n} e^{-aV(x_n)}\, e^{-\frac{a}{2} p_n^2 - a\mu\, p_n}\, e^{ip_n(x_{n+1} - x_n)} \\
&\propto \exp\left(-\frac{a}{2} \left[\frac{x_{n+1} - x_n}{a} + i\mu \right]^2 - a\, V(x_n) \right).
\end{aligned}
$$

In the limit $a \to 0$, the partition function can be written as an ensemble average over closed "world lines" parameterised by $x(\tau)$ with $x(0) = x(L)$, $L = 1/T$:

$$Z \propto \int \mathcal{D}x \, \exp\left(-\int_0^L d\tau \left[\frac{1}{2}(\dot{x} + i\mu)^2 - V(x(\tau)) \right] \right). \tag{7.19}$$

Here, $P(c)$ is a complex function of the configurations c specified by closed paths. While the operator formulation tells us that the partition function is real and positive, its statistical field theory representation (7.19) is only real and positive after taking the ensemble average. The formulation evades simulations with direct stochastic methods. We encounter a sign-problem. Note that the sign problem is somewhat artificial and hinges on the choice of the set of eigenstates. Different choices lead to a different "severeness" of the sign problem. If we choose eigenstates e of the hamiltonian

$$\left(H_0 + \mu N \right) |e_a\rangle = E_a \, |e_a\rangle ,$$

we find that every contribution to the partition function is real and positive:

$$Z = \sum_{\{e\}} \exp\{-E_a/T\} .$$

However, if we find the set of eigenstates to the Hamiltonian, it means that we have solved the quantum field theory in operator language a fact that is only feasible in some special cases.

7.2 Introduction to overlap problems

Overlap problems occur when the importance sampling algorithm suggests stochastically important Monte Carlo configurations that, however, are not relevant for a particular observable of interest. If that occurs in an application, often advanced simulation techniques are needed.

7.2.1 Prototype issue

Monte Carlo simulations must be ergodic so that a stochastic estimate converges to the integral when the size N_s of the random number sample grows over all boundaries. While we have the mathematical proof that, e.g. the Metropolis Algorithm is ergodic and exact, the algorithm still might fail for cartoon practical applications. The reason is that the N_s is necessarily finite in actual computer simulation. In fact, we rely on the feature that even for finite N_s the algorithm samples *all regions* of configuration space that are *important* for the observable at hand. There are certain types of problems, even classes of problems that are highly interesting for physics applications, where this assumption does not come true. In this case, we encounter a "practical" ergodicity problem.

Let us illustrate this with a very simple model of one variable x. The probability distribution is bimodal and given by

$$P(x) \; \propto \; \exp\left\{-\beta x^2\right\} \; + \; \exp\left\{-\beta(x-2)^2\right\} \; . \tag{7.20}$$

Such bimodal distributions occur naturally. For example, in statistical physics, β would be the inverse temperature, and the marginal x the total energy of systems. In real cases, the integral is typically high-dimensional and over the microscoping variables such as positions and speeds of water molecules, the degrees of freedom of the theory. At the so-called critical temperature, a first-order phase transition might occur where, e.g. water and ice coexist. The (total) energy distribution becomes bimodal: the peak at high energies, say x, emerges from the watery region of space, and the peak at low energy from the icy regions. Two states of matter are *equally* important. Let us denote the maximum value of the probability distribution P_{max} and the local minimum at $x = 1$ as P_{min}. The flat region around $x \approx 1$ is formed by states of matter with interfaces between ice and water regions. A particular object of interest is the *interface tension* σ. If we consider the statistical system as part of a cube with side length L, the interface tension can be calculated from:

$$P_{min}/P_{max} \; = \; \exp\left\{-\sigma L^2\right\} \; .$$

For our toy model, we would have (for sufficiently large β):

$$P_{min}/P_{max} \; \approx \; 2 \; \exp\{-\beta\} \; .$$

The above ratio is notoriously difficult to estimate by standard methods. Those would enlist a random sequence $x_1, \ldots, x_{N_s}$ and generate a histogram for the estimate of the probability distribution.

The issue emerges as follows: An important sampling algorithm would be attracted by one of the regions of high probability density. Local update methods such as Metropolis-Hasting algorithm only propose incremental changes of the variable x. It is therefore very unlikely that a sequence that started around $x = 0$ will "discover" the other region of high probability density around $x \approx 2$. In particular, histogram entries around $x = 1$, which inform $P_{\min}$, are ill conceived. The algorithm with finite N_s fails to be practically ergodic for large β.

In the following, we will explore this observation for the toy model at hand. We use a Metropolis Hasting to generate the random number sequence $x_1, \ldots, x_{N_s}$:

1. Generate a random spatial increment Δx from the normal distribution with zero mean and a standard deviation of $\delta/\sqrt{2}$.

2. Set $x_{n+1} = x_n + \Delta x$ with probability $min(1, P(x+\Delta x)/P(x)$ or $x_{n+1} = x_n$ else. We start with $x_1 = 1$.

3. continue with item [1] until the sample size N_s has been reached.

For the simulation at $\beta = 4$, we have chosen $N_s = 4,000,000$ and $\delta = 0.7$, which gives about a 50% Monte Carlo acceptance rate. We have chosen a 100 bins for the histogram. The result is shown in Figure 7.2, left panel. We find an acceptable approximation of the exact probability density function. Note

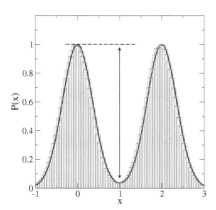
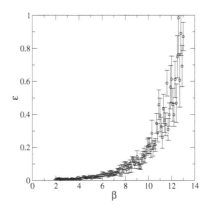

FIGURE 7.2: Left: A probability density function with a bimodal distribution for $\beta = 4$. Also shown is the histogram estimate from a Metropolis Hasting simulation.

that both graphs were normalised such that $P_{\max} = 1$. We also estimated the desirable ratio and found a good agreement:

$$P_{\min}/P_{\max} \approx 3.6653 \times 10^{-2}, \qquad \text{exact:} \quad P_{\min}/P_{\max} = 3.6631 \times 10^{-2}.$$

Apparently, the algorithm is still *practically ergodic* at $\beta = 4$. Let us study systematically what happens if we increase β. We will keep $N_s = 4,000,000$ fixed. Since we are interested in an estimate of the " interface tension", we study the relative error:

$$\epsilon = \text{abs}\left(1 - \left\langle \frac{P_{\min}}{P_{\max}} \right\rangle \exp\{\beta\}/2 \right).$$

Our result is shown in Figure 7.2, right panel. Error bars are obtained from 10 independent runs. We observe that for a fixed sample size, there is a quite sharp transition to a regime where the ϵ reaches order one and the simulation no longer produces valid results due to a lack of practical ergodicity.

7.2.2   Exponential Error Suppression with LLR

In order to solve "overlap problems", we need to estimate the probability distribution in the " rare event" regime with a good *relative error*. Generating events and estimating the density by histogramming generically leads to a roughly constant *absolute error*. Take the toy model of the previous subsection as an example: in the regime where the probability distribution is smallest, we might find only a few events and, thus, the estimate of $P_{\min}$ in relation to $P_{\max}$ is plagued by a large relative error. The LLR method solves this issue and usually produces a constant *relative error* over the domain of interest. This is called *exponential error suppression*.

In the following, we illustrate this phenomenon using a very simple model. Assume that we have random numbers x_k at our disposal that are Normal distributed with the probability density function $\exp\{-x^2/2\}$. Using the random numbers only, we will then find an analytical approximation of the probability density function for the variable x by

1. Calculating the LLR function $a(x)$ at a discrete set of points x_k, $k = 1 \ldots n_x$ using the LLR method.

2. Find an analytical approximation for the LLR function using a Taylor polynomial

$$a(x) = \sum_{\ell=0}^{N_{\text{tay}}} b_\ell \, x^{2\ell+1} \tag{7.21}$$

by fitting the Taylor coefficients b_ℓ to the numerical data. We have capitalised on the fact that the LLR function is an odd function in x, since the density of states $\rho(x)$ is even.

3. Reconstruct the density of states ρ by

$$\rho(x) = \exp\left\{ \int_0^x dy \, a(y) \right\} . \tag{7.22}$$

Up to the normalisation constant, the density of states ρ is an analytical function, i.e. differentiable to all orders. Any expectation value of an observable $f(x)$ is the estimated by the ratio of two integrals

$$\langle f \rangle = K_1/K_2 , \tag{7.23}$$

$$K_1 = \int_0^\infty dx \, x^2 \, \rho(x) , \qquad K_2 = \int_0^\infty dx \, \rho(x) .$$

Note that $\rho(x)$ is analytically known (though depending on the Taylor coefficient b_ℓ) so that the integrals can be evaluated using some high-precision numerical integration method.

Let us now look into some details of the calculation. At the heart of the LLR algorithm are the expectation values

$$\langle\langle F \rangle\rangle = \frac{1}{\mathcal{N}} \int dx \, F(x) \, \exp\{a\,x\} \, P(x) \, W(x,\delta) , \tag{7.24}$$

$$\mathcal{N} = \int dx \, \exp\{a(x_0)\,x\} \, P(x) \, W(x - x_0, \delta) ,$$

where x_0 is position at which we would like to calculate the LLR coefficient $a(x_0)$, W is the "window function" and $P(x)$ is the probability density function, which is for our case $P(x) = \exp\{-x^2/2\}$. For ease of calculation, we choose a Gaussian window function

$$W(x - x_0, \delta) = \exp\left(-\frac{(x - x_0)^2}{\delta^2} \right) .$$

The LLR approach states the following: Choose $a(x_0; \delta)$ such that

$$\langle\langle x - x_0 \rangle\rangle = 0 \quad \Rightarrow \quad a(x_0) := \lim_{\delta \to 0} a(x_0; \delta) = -\frac{d}{dx} \ln P(x) \Big|_{x = x_0} . \tag{7.25}$$

Once we have estimated the LLR coefficient, we can invert the last equation to obtain an estimate of the probability density function up to a normalisation function:

$$P(x) = \exp\left\{ -\int_0^x a(y) \, dy \right\} .$$

Note that, in practice, we only have access to sequences of random numbers rather than the probability density function of the marginal that we wish to estimate. For the one-variable scenario here and for the simplicity of the example, we can actually verify the LLR statement by a direct calculation.

All involved integrals are Gaussian, and we find by completing the square and direct integration:

$$\langle\langle x - x_0 \rangle\rangle \;-\; \frac{\delta^2}{2 + \delta^2} \left[a - x_0 \right] \;=\; 0 \,.$$

Solving the equation yields

$$a(x_0; \delta) \;=\; x_0 \qquad\qquad \rightarrow \qquad\qquad a(x_0) := \lim_{\delta \to 0} a(x_0; \delta) \;=\; x_0 \,.$$

Note that the solution $a(x_0; \delta)$ does not depend on δ. This is generically *not* the case and due to the Gaussian nature of the integrals involved. We indeed find:

$$-\frac{d}{dx} \ln P(x) \Big|_{x=x_0} \;=\; \frac{d}{dx} (x^2/2) \Big|_{x=x_0} \;=\; x_0 \,.$$

Numerical stochastic simulations are only delivering estimates and that only for a limited parameter range, e.g. $\delta > 0$ although we are interested in $\delta \to 0$. We will only have noise-afflicted estimators for the expectation value in (7.25). The purpose of the numerical study is now to find out how these inaccuracies affect our final answer for (7.23).

To this aim, we are going to estimate $\langle\langle x - x_0 \rangle\rangle$ by a Monte Carlo simulation. We need to generate random numbers from the probability density distribution

$$\exp\{a\,x\} \; P(x) \; W(x, \delta) \;=\; C \, \exp\left\{ -A \, (x - B)^2 \right\}$$

with

$$A \;=\; \frac{1}{2} + \frac{1}{\delta^2} \,, \qquad\qquad B \;=\; \frac{2x_0 + a\,\delta^2}{2 + \delta^2} \,.$$

C is an irrelevant constant. Hence, we can generate those random numbers very efficiently by using the Box-Muller algorithm without the need of invoking a Markov Chain.

The estimate of $a(x_0; \delta)$ arises from the Robbins-Monro iteration:

$$a_{n+1} \;=\; a_n \;-\; \frac{2}{\delta^2\,n} \left[x - x_0 \right] \qquad (n \geq 1) \,, \tag{7.26}$$

where $[x - x_0]$ is the stochastic estimator for $\langle\langle x - x_0 \rangle\rangle$. Naturally, we need to stop the iteration at some $n = n_{\mathrm{RM}}$. We then end up with a noisy representative $a_{n_{\mathrm{RM}}}$ from a perhaps not fully converged sequence. If we restart the experiment, these final values for a form a random sequence. The seminal paper from Robbins and Monro guarantees that these random values are normal distributed around the true value $a(x_0; \delta)$. This paves the way to standard error analysis including propagation.

A disadvantage of the LLR approach is the many parameters of the simulation that needs controlling. All the parameters needed for an estimate (7.23)

TABLE 7.1: Parameters needed for the LLR estimate of the expectation value (7.23)

δ	width of the LLR window function
n_{ita}	number of random numbers used to estimate the MC expectation value $\langle\!\langle\ldots\rangle\!\rangle$
n_{RM}	number of Robbins-Monro iterations for finding $a(x,\delta)$
n_x	number of points x_0 for which the LLR coefficient a is estimated
N_{tay}	number of Taylor coefficients used to approximate the function $a(x_0)$.
N_{rept}	number of repeats used for estimating the stochastic error of the final result.
Λ	cutoff for the numerical integrations over the density of states.

are listed in Table 7.1. It plays to our advantage that the algorithm is robust against the choice of those and that exact error estimates of the result are obtained. This means in practice that a less optimal choice makes the algorithm less efficient but still delivers a valid result with error bars. We have tested some of the parameters in Figure 7.3. In particular, we observe that the error for the estimator for $a(x_0;\delta)$ with the number n_{RM} of Robbins-Monro iterations decreases like $1/\sqrt{n_{\mathrm{RM}}}$. Without much fine tuning of parameters, we have reached an optimal situation.

The next step of the programme is to fit point data for the LLR coefficient a to a Taylor polynom of degree $2N_{\mathrm{tay}} + 1$, see (7.21). A stable solution is to use Householder rotations for that. Once the Taylor coefficients are available, it is straightforward to integrate the polynomial and reconstruct the density of states:

$$\ln\rho(x) \;=\; -\sum_{\ell=0}^{N_{\mathrm{tay}}} \frac{b_\ell}{2\ell + 2}\, x^{2\ell+2}\,. \tag{7.27}$$

We have chosen the normalisation constant such that $\rho(x = 0) = 1$.

Let us test this approach in practice. We use the parameters (see Table 7.1):

$$n_{\mathrm{ita}} = 200, \qquad n_{\mathrm{RM}} = 1000, \qquad N_{\mathrm{tay}} = 8, \qquad N_{\mathrm{rept}} = 100,$$

$$\delta = 0.1, \qquad n_x = 100\,.$$

The approach is as follows:

- Generate estimators for $a(x_k;\delta)$ for $x_k = k\,\delta$ from *one* Robbins-Monroe iteration.

- Find the Taylor coefficients.

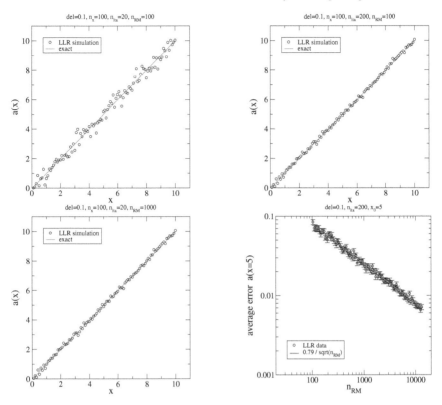

FIGURE 7.3: Estimates for the LLR coefficient $a(x; \delta)$ from a single Robbins-Monro iteration. (upper left) low statistics $n_{\text{ita}} = 20$, $n_{\text{RM}} = 100$; (upper right) more statistics for the MC estimator $n_{\text{ita}} = 20 \to 200$; (lower right) more iterations $n_{\text{RM}} = 100 \to 1000$; (lower right) stochastic error for a at $x_0 = 5$ as a function of the number of Robbins-Monro iterations.

- Reconstruct the density of states from (7.27) at the points x_k. Note that at this stage we have an analytical approximation for ρ and could choose any other set of points.

- Repeat this $N_{\text{rept}} = 100$ times and calculate the average for $\rho(x_k)$ and its *standard deviation*.

Figure 7.4 shows the estimate for the density of states $\rho(x)$ as a function of x. Note that the size of the error bars approximately stays constant over the x domain that we explored, although the graphs stretch over many orders of magnitude on a logarithmically scale. This observation is called *exponential error suppression*.

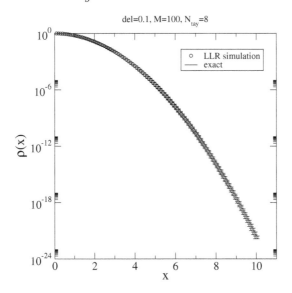

FIGURE 7.4: Estimates for the density of states over about 20 orders of magnitude. The red line shows the exact result.

7.2.3 Solution to overlap problems using LLR

Let us now return to the example in subsection 7.2.1 and explore whether the feature of Exponential Error Suppression let us reliably estimate P_{min}/P_{max} – a task for which standard Importance Sampling MC simulations fail.

The probability density is now given by (7.20) (up to a normalisation constant):

$$P(x) = \exp\left\{-\beta x^2\right\} + \exp\left\{-\beta(x-2)^2\right\} . \tag{7.28}$$

The LLR method attempts to reconstruct the density of states ρ of the marginal[1] x:

$$P(x) \propto \rho(x) = \exp\left\{-\int_0^x a(y)\, dy\right\} . \tag{7.29}$$

The LLR coefficient at position x_0 emerges from the solution of a non-linear stochastic equation:

$$\langle\langle x - x_0 \rangle\rangle(a, \delta) = 0 ,$$

where the double-bracket expectation values are defined in (7.24).

A few words about our implementation are in order. We again use a Gaussian Window function

$$W(x - x_0, \delta) = \exp\left(-\frac{(x - x_0)^2}{\delta^2}\right) . \tag{7.30}$$

---

[1]The marginal equals the actual distribution for the case of one variable.

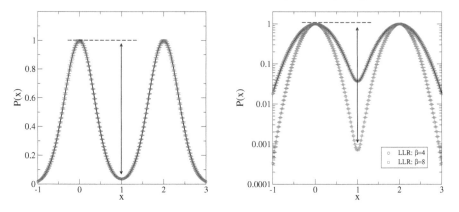

FIGURE 7.5: Density of States estimate as a function of x also for two β on logarithmic scale (right).

We found it efficient to generate a random number sequence $x_1, \ldots, x_{n_{ita}}$ with respect to the distribution (7.30) and calculate the double-bracket expectation value by re-weighting:

$$\langle\!\langle F \rangle\!\rangle \;=\; \frac{\Big\langle F(x)\, \exp\{a\,x\}\, P(x)\, W(x - x_0, \delta) \Big\rangle_W}{\Big\langle \exp\{a\,x\}\, P(x)\, W(x - x_0, \delta) \Big\rangle_W}\,. \qquad (7.31)$$

We use the following parameters for the simulation (see Table 7.1 for a reference):

$$n_x = 100, \quad \delta = 0.02, \quad n_{ita} = 1000, \quad n_{RM} = 1000, \quad N_{rept} = 6.$$

Figure 7.5 shows our numerical estimate in comparison with the exact probability density. We find a very good agreement with small error bars for $\beta = 4$ (left panel). We also show the results for $\beta = 4$ and $\beta = 8$ on a logarithimc scale (right panel). Note that the error bars seem of roughly the same size. For a logarithmic scale, this means that the *relative error* is (approximately) constant. This is what we called *exponential error suppresion* .

We now have a perfect algorithm for estimating the "interface tension", or more precisely P_{min}/P_{max}, even for large β, where this ratio becomes exponentially small with increasing β. The result is shown in Figure 7.6 again on a logarithmic scale. We observe a very good match with the exact result over six orders of magnitude.

7.2.4   The Ising model showcase

Let us now move away from illustrative toy models and study a "real life" problem. To this aim, we are coming back to the Ising model in two dimensions. We studied the Ising model at length in section 4.4.

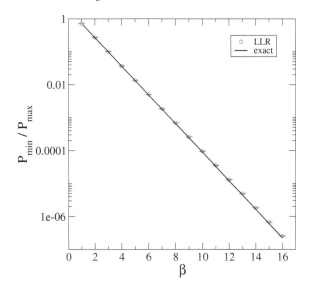

FIGURE 7.6: $P_{\min}/P_{\max}$ as a function of β: LLR estimate (symbols) and exact result (solid line).

For a brief recap, we note that the degrees of freedom are Ising spins $\sigma_x \in \{-1, +1\}$ located at a $L \times L$ square lattice. We consider periodic boundary condition only:

$$\sigma_{x+L,y} = \sigma_{x,y} , \qquad \sigma_{x,y+L} = \sigma_{x,y} .$$

The Hamiltonian and the partition function are given by

$$\mathcal{H} = -\sum_{\langle xy \rangle} \sigma_x \sigma_y , \qquad Z = \sum_{\{\sigma\}} \exp\{-\beta \mathcal{H}\} , \tag{7.32}$$

where $\langle xy \rangle$ denotes that x and y are nearest neighbours on the square lattice. The notation

$$\sum_{\langle xy \rangle}$$

means that we sum over all $2L^2$ links of the lattice. We will be studying the Ising model in symmetry broken phase with $\beta = 0.48 > \beta_c = 0.44 \ldots$. The Hamiltonian is invariant under the transformation $\sigma_x \to -\sigma_x \; \forall x$.

The "indicator light" for symmetry breaking is the total magnetisation

$$M = \sum_x \sigma_x . \tag{7.33}$$

A good quantity to study the properties of the model is the marginal distribution $P(M)$, which is the probability to find the total magnetisation of the system (either in a Monte Carlo simulation or in real life by measuring the magnetisation of a ferromagnet). Note that, for any finite lattice, we have $\langle M \rangle = 0$ by virtue of the reflection symmetry $\sigma_x \to -\sigma_x$. How this vanishing expectation value is realised, however, depends on the temperature. In any case, symmetry dictates

$$P(M) = P(-M),$$

but, for $\beta < \beta_c$, $P(M)$ is a convex function, which peaks at $M = 0$ while, for $\beta > \beta_c$, $P(M)$ is maximal at $M = \pm M_{\max}$ with $M_{\max} > 0$. The most likely states have a non-vanishing magnetisation at low temperatures.

In the broken phase ($\beta < \beta_c$), Ising configurations with $M \approx 0$ show a characteristic pattern: the Ising configuration is made of large regions with $\sigma_x = 1$ neighbouring equally large regions with $\sigma_x = -1$ separated by an *interface*. We already said that such a configuration with $M \approx 0$ is not the most likely one. In physics, we associate an energy cost with the presence of the interfaces. Since the size of the interfaces scales with L (the interfaces are one dimensional), this energy is given by $E_{\text{inter}} = 2\sigma L$, where σ is called the *ordered-ordered interface tension*. The factor 2 is convention and is attributed to the fact that on a lattice with periodic boundary conditions, it is impossible to have just one interface dividing the lattice into two halves. The state's minimal energy is one with two interfaces, which are viewed as a strip going through the lattice.

The interface tension is our objective here. It can be extracted from

$$P_{\max}/P_{\min} = \exp\{-\sigma L\}, \qquad (7.34)$$

where $P_{\max}$ and $P_{\min}$ are the maximum and the minimum of the marginal distribution $P(M)$. Hence, the task is to estimate $P(M)$, and we will use LLR techniques for that (see the previous subsection).

To this aim, we add the reweighting term

$$a \sum_x \sigma x$$

to the Hamiltonian, where a is the LLR coefficient. For the window function, we use a step function

$$W(M - M_0; \delta) = \begin{cases} 1 & \text{for } M \in \{M_0 - \delta, M_0 + \delta\} \\ 0 & \text{else} \end{cases}.$$

We estimate the LLR double expectation values $\langle\langle \ldots \rangle\rangle$ using a heatbath method. We first attempt the update of a single spin σ_x according the probability

$$\exp\{(\beta b + a)\sigma_x\}, \qquad b = \sum_{y \in \langle xy \rangle} \sigma_y.$$

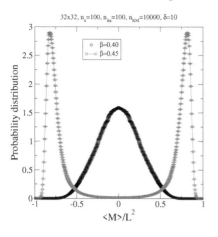
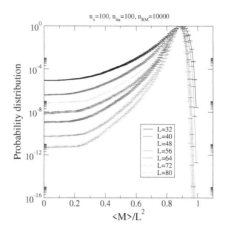

FIGURE 7.7: (left) Estimated marginal distribution $P(M)$ for the Ising (total) magnetisation for a $L = 32$ lattice: $\beta = 0.4$ (symmetric phase) and $\beta = 0.45$ (broken phase). (Right) $P(M)$ for fixed $\beta = 0.48$ and several lattice sizes L.

We then update the magnetisation $M \to M'$. We then reject the spin update if $M' \notin \{M_0 - \delta, M_0 + \delta\}$.

The next steps follow closely the procedure outlined in subsection 7.2.3. In summary, we solve the nonlinear stochastic equation

$$\langle\langle M - M_0 \rangle\rangle(a) = 0$$

for the LLR coefficient a. Using the solution $a(M_0)$, we reconstruct the density of states

$$\rho(M) = \exp\left\{-\sum_{m=0} a(m)\right\},$$

which is proportional to the desirable marginal distribution $P(M)$. Figure 7.7 shows the estimated marginal distribution. Simulation parameters are

$$L = 32, \qquad \beta = 0.4,\ 0.45, \qquad n_{\text{ita}} = 100, \qquad n_{\text{RM}} = 10,000,$$

$$N_{\text{rept}} = 20, \qquad \delta = 10.$$

The estimates are normalised to unity $\sum_M P(M) = 1$. We clearly observe a single peak in the symmetric phase and a double peak at $\pm M_0$, $M_0 > 0$ in the low-temperature phase with $\beta = 0.45$.

Let us now venture deeply into the broken phase at $\beta = 0.48$. We have reconstructed the density of states as a function of the normalised magnetisation $M/L^2 \in [-1, 1]$ for several lattice sizes L. The result is shown in Figure 7.7. The curves are normalised such that $P_{\max} = 1$. The graphs should be suppressed as $\exp\{-2\sigma L\} = P_{\min}$ at $M = 0$. We indeed observe a linear spacing

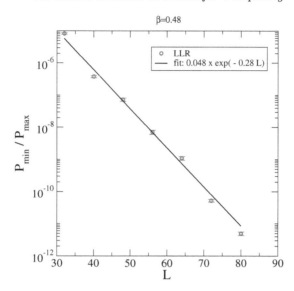

FIGURE 7.8: $P_{\min}/P_{\max}$ as a function of the lattice size L on a logarithmic scale. Also shown is a line that fits the data for large L.

between the curves for different lattice sizes L. This is quantified in Figure 7.8, which shows $P_{\min}/P_{\max}$ as a function of L. For sufficiently large L, we observe an exponential decrease with L. The line fit in Figure 7.8 corresponds to an interface tension of $\sigma \approx 0.14$.

7.3   Estimating probability density functions

Calculating the normalisations of probability density functions for systems with many degrees of freedoms has long been characterised as an "intractable problem". With modern advances, most notably *exponential error suppression* of certain stochastic methods, the situation has changed over the recent years. We will here map the task of normalising probability density functions to an overlap problem and then will discuss how to solve it using the worm algorithm.

Let us first illustrate the context. In Statistical Physics, the normalisation of probability density function is called *partition function*. For example, for the Ising model, the partition function is given by:

$$Z(\beta) \;=\; \sum_{\{\sigma=\pm 1\}} \exp\!\Big\{-\beta\, S[\sigma]\Big\}, \qquad S[\sigma] \;=\; -\sum_{\langle xy \rangle} \sigma_x \sigma_y\,. \tag{7.35}$$

The partition function usually involves very large numbers. First note that it is a monotonically rising function since:

$$\frac{1}{Z(\beta)} \frac{\partial}{\partial \beta} Z(\beta) = \langle -S(\beta) \rangle > 0, \qquad Z(\beta), \ \beta > 0. \qquad (7.36)$$

For a $L \times L$ square lattice, we find even for $\beta = 0$ that

$$Z(0) = 2^{L^2}.$$

Frequently, we use the so-called *Free Energy* instead of Z itself:

$$\beta F(\beta) = -\ln Z(\beta), \qquad \beta F(0) = -L^2 \ln 2. \qquad (7.37)$$

There are good reasons that for many statistical systems the Free Energy predominantly scales with the volume of the system. We then also define the *Free Energy Density* by

$$f(\beta) = F(\beta)/L^2, \qquad \beta f(0) = -\ln 2. \qquad (7.38)$$

The Free Energy for $\beta > 0$ cannot be calculated analytically. The aim here is to estimate it using stochastic methods.

Before we step into simulation possibilities, we can derive further information. First note that, because of (7.36),

$$\frac{\partial}{\partial \beta}(-\beta F(\beta)) = \frac{1}{Z(\beta)} \frac{\partial}{\partial \beta} Z(\beta) > 0, \qquad \text{for } \beta > 0. \qquad (7.39)$$

It is also a convex function:

$$\frac{\partial^2}{\partial \beta^2}(-\beta F(\beta)) = \frac{\partial}{\partial \beta} \left[\frac{1}{Z(\beta)} \frac{\partial}{\partial \beta} Z(\beta) \right] \qquad (7.40)$$

$$= \frac{1}{Z(\beta)} \frac{\partial^2}{\partial \beta^2} Z(\beta) - \left[\frac{1}{Z(\beta)} \frac{\partial}{\partial \beta} Z(\beta) \right]^2$$

$$= \langle S^2 \rangle - \langle S \rangle^2 = \left\langle (S - \langle S \rangle)^2 \right\rangle > 0. \qquad (7.41)$$

Since often analytical results are available for specific parameters (e.g. here for $\beta = 0$), ratio's of partition functions are often good enough to consider:

$$Z(\beta)/Z(0).$$

7.3.1 The worm algorithm

The worm algorithm approach dates back to the early seventies [93] and starts with the observation that the ratio of two partition functions can be written as a Monte Carlo expectation value:

$$\frac{Z(\beta_2)}{Z(\beta_1)} = \frac{1}{Z(\beta_1)} \sum_{\{\sigma=+1\}} \exp\left\{-(\beta_2 - \beta_1)\, S[\sigma]\right\} \exp\left\{-\beta_1\, S[\sigma]\right\} \quad (7.42)$$

$$= \left\langle \exp\left\{-(\beta_2 - \beta_1)\, S[\sigma]\right\}\right\rangle_{\beta_1}.$$

If we would apply this formula to estimate $Z(\beta)$, we would encounter an overlap problem:

$$\frac{Z(\beta)}{Z(0)} = \left\langle \exp\left\{-\beta\, S[\sigma]\right\}\right\rangle_{\beta=0} = \exp\left\{-\beta L^2\, (f(\beta) - f(0))\right\} \gg 0.$$

We would generate configurations at $\beta = 0$ and try to estimate the above expectation value. At $\beta = 0$, the spins are independent. Thus, we randomly generate spin configurations and try to infer information of the Ising model, e.g. in the broken phase $\beta > \beta_c$. This is clearly going to fail, this failure triggered the advent of *importance sampling*.

The worm algorithm breaks down the desirable ratio as follows:

$$\frac{Z(\beta)}{Z(0)} = \frac{Z(\Delta\beta)}{Z(0)} \frac{Z(2\Delta\beta)}{Z(\Delta\beta)} \frac{Z(3\Delta\beta)}{Z(2\Delta\beta)} \cdots \frac{Z(n\Delta\beta)}{Z((n-1)\Delta\beta)} \quad (7.43)$$

where we choose $\beta = n\,\Delta\beta$. Defining $\beta_k = k\Delta\beta$ and

$$A(\beta_k) = \frac{Z(\beta_k)}{Z(\beta_{k-1})} = \left\langle \exp\left\{-\Delta\beta\, S[\sigma]\right\}\right\rangle_{\beta_{k-1}}, \quad (7.44)$$

we obtain

$$-\beta(F(\beta) - F(0)) = \ln \frac{Z(\beta)}{Z(0)} = \sum_{k=1}^{n} \ln A(\beta_k). \quad (7.45)$$

Note that the later formula is exact for any choice of n. For large enough n, the exponentials in (7.44) are close enough to unity. In other words, the theory for β_{k-1} has enough overlap with the theory of β_k meaning that the expectation values (7.44) can be reliably estimated by reweighing. The disadvantage is that for large volumes L^2, n needs to be very large such that the expectation values stay close to unity.

Let us study the approach with a numerical experiment. We used the heatbath algorithm to estimate the expectation values $A(\beta_k)$. After $1,000$ lattice sweeps for thermalisation, each expectation value is estimated using 100 independent configurations. We have repeated each experiment 10 times to get the estimate with error bars. The result is shown in Figure 7.9, left panel, for lattice size $L = 16$ and $L = 32$. $\Delta\beta$ has been 0.01 (n=50) and 0.005 (n=100) to confirm that our numerical findings are indeed independent of the size of $\Delta\beta$. Also shown in the figure, right panel, is the *Free Energy Density*. We find that is largely independent of the lattice size in accordance with expectations. We aslo fitted a parabola to the low β regime just to illustrate that $-\beta f(\beta)$ is not just a simple parabola – the most simple example of a convex function.

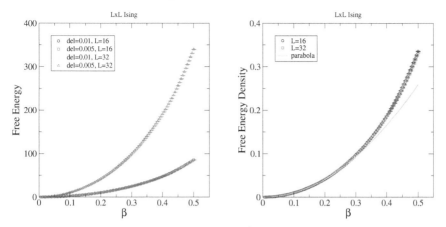

FIGURE 7.9: Worm algorithm: (left) Estimate of the Ising model Free Energy for two lattice sizes. (right) The Free Energy Density as a function of β. The plotted quantities are actually $-\beta F(\beta)$ and $-\beta f(\beta)$.

Chapter 8

Data Analytics and Statistical Systems

In a nutshell, data analytics aims at data and derive conclusions. The internet is increasingly used to serve as an exchange of information between automated units, which could be computers but also your refrigerator or your smart home heating system. This application of the web is therefore also frequently called the Internet of Things. The amount of data, which is produced on a daily basis, is steadily increasing as is the wish to put this information to good use.

There are four main types of data analytics:

Descriptive analytics has been around for a while and condenses datastes into a few outputs to describe the data to stakeholders. Nowadays, businesses and other institutions including Universities develop a set of key performance indicators (KPIs) that can help to track the performance of a process or even

DOI: 10.1201/9781315156156-8

the institution itself. It involves the collection of relevant data, processing of the data, data analysis and data visualisation.

Diagnostic analytics aims to "understand" data and often appears as an extension of descriptive analytics. It aims at identifying anomalies in data, perhaps collate more data connected to the anomalies and using statistical techniques to develop a model that explains these anomalies.

Predictive analytics comes into play when we want to analise the time dependence of data. Using historical data of a given context, we try to establish what is likely going to predict how the data will evolve in the same context. Thereby, it is not only important to predict the most likely outcome but also to quantify the uncertainty for it. There are no limits to the imagination, with prototype applications being time series such as, e.g. stock market indices. Predictive analytics uses a variety of statistical and machine learning techniques, such as neural networks, decision trees and regression.

Prescriptive analytics uses insights from the predictive analytics and aides data-driven decision making. For instance, during a pandemic outbreak, Prescriptive Analytics decides, on the basis of analytic results, which counter measures (face masks, lock-down) should be rolled out at which stage of the disease spread.

In modern data analytics, a central role is played by machine learning. Machine learning is a branch of artificial intelligence that uses mathematical algorithms to learn from data and make predictions. It is based on the idea that machines can learn from data, identify patterns, and make decisions with minimal human intervention. Machine learning algorithms can be used to solve complex problems, such as image recognition, natural language processing, and predictive analytics.

Machine learning is a powerful tool for mathematicians, as it can be used to solve complex problems that are difficult to solve with traditional mathematical methods. By using machine learning algorithms, mathematicians can quickly and accurately identify patterns in data and make predictions. This can help them to develop new theories and solve complex problems more efficiently.

On the ground of our knowledge of the data, machine learning is divided into " supervised" and " unsupervised". In supervised learning, the properties of a subset of data are known in full, and are used to drive the construction of a model that agrees with the available information. A model-independent approach is "unsupervised learning". It takes a stack of information (say, images) and identifies significant changes (variations). In this chapter, we are going to explore sample applications of both unsupervised and supervised machine learning methods in statistical mechanics and quantum field theory, often taking the Ising model as a target in order to explain the method in action.

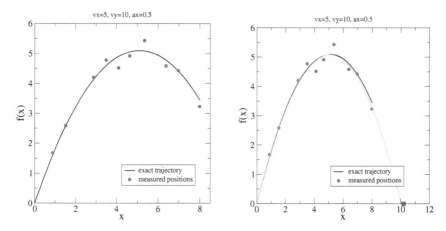

FIGURE 8.1: Left: Exact trajectory (solid) of an object and 10 "measured" positions (red symbols). Right: in green, is the optimal trajectory fitted to the measure values. The symbol on the x axis marks the exact landing point of the object.

8.1   Model regression – L2 norm

Let us consider a simple exercise in *predictive analytics*. Assume an object moves in the Earth gravitational field without its own propulsion. If $\vec{v} = (v_x, v_y)^T$ is the initial velocity, the trajectory is given by

$$y(x) = f(x) = \frac{v_y}{v_x} x - \frac{g}{2v_x^2} x^2 . \tag{8.1}$$

While the gravitational acceleration $g = 9.81 \, \mathrm{m/s}$ is given, the initial velocity is unknown. An observer undertakes $n = 10$ measurements of the position of the object: $(x_i, y_i)^T$, $i = 1 \ldots n$. Due to uncertainty in the measuring process, the apparent position $(x_i, y_i)^T$ is randomly displaced from the exact position, $(x_i = ih, y(x_i))^T$, with a random vector η, which is distributed according to

$$\exp\{-\eta^2/\delta^2\}.$$

We have thus generated *artificial data*, which we are going to use to analyse different predictive analytics techniques. Below, we will assume that all knowledge on how these data were generated is lost. Figure 8.1 shows these artificial data for $v_x = 5$, $v_y = 10$, $\delta = 0.5$, $h = 1$ and $n = 10$. The measured positions of the object are

x	0.87	1.51	2.87	3.49	4.1	4.63	5.34	6.38	6.96	7.98
y	1.68	2.59	4.21	4.78	4.52	4.92	5.44	4.59	4.43	3.24

Given the theoretical description (8.1) and the measured data, can we predict where the object hits the ground?

Our theory prediction f in (8.1) has two unknown parameters:

$$a_1 = \frac{v_y}{v_x}, \qquad a_2 = -\frac{g}{2v_x^2}.$$

Knowledge of both lets us reconstruct the velocity vector $(v_x, v_y)^T$. Note that the theory prediction for the point x_i would be $f(x_i; a_1, a_2)$. However, due to noise in the measurement, we obtain y_i instead. Using an L_2 norm, we can quantify the discrepancy between theory and measurement:

$$E(a_1, a_2) = \sum_{i=1}^{n} \left[y_i - f(x_i; a_1, a_2) \right]^2. \tag{8.2}$$

There are two important observations:

(i) $E(a_1, a_2) \geq 0$,

(ii) $E(a_1, a_2)$ only vanishes if there is no uncertainty in measured data (perfect alignment with the theory) and, secondly, if we found the correct parameter a_1 and a_2.

Both observations provide us with a strategy to obtain the "best" theory curve representing the data. We choose the parameters a_1 and a_2 such that

$$a_{1/2}: \qquad E(a_1, a_2) \rightarrow \min. \tag{8.3}$$

For our present scenario, we obtain a *linear regression model*, i.e.

$$\sum_{i=1}^{n} \left[y_i - a_1 x_i - a_2 x_i^2 \right]^2 \rightarrow \min,$$

which has a unique and analytic solution. A straightforward calculation in calculus gives a linear system to obtain the unknown parameters:

$$\left[x^2 \right] a_1 + \left[x^3 \right] a_2 = \left[yx \right],$$

$$\left[x^3 \right] a_1 + \left[x^4 \right] a_2 = \left[yx^2 \right], \qquad [A] := \frac{1}{n} \sum_{i=1}^{n} A_i.$$

The solution of the linear system is given by:

$$a_1 = \frac{[yx]\,[x^4] - [yx^2]\,[x^3]}{[x^2]\,[x^4] - [x^3]^2}, \qquad a_2 = \frac{[yx]\,[x^3] - [yx^2]\,[x^2]}{-[x^2]\,[x^4] + [x^3]^2}.$$

For the particular data in the table above, we find three decimal places:

$$a_1 = 2.075, \qquad a_2 = 0.208.$$

We now can predict the trajectory using (8.1). The result is shown in Figure 8.1. We also can answer the question for the position x_* where the object is most likely to land:

$$f(x_*) = 0 \qquad x_* = -\frac{a_1}{a_2} = 9.974 \quad (2\mathrm{dp}).$$

This needs to be compared with the exact position x_e for the landing site:

$$x_e = \frac{2v_x v_y}{g} = 10.194 \quad (2\mathrm{dp}).$$

An inspection of the fitted trajectory (green) in Figure 8.1 shows that this curve quite well represents the measured data. Our estimate x_* is reasonable.

Our findings stress an important point: the current approach does not give us the possibility to estimate how reliable our estimate is. An idea about the confidence interval for our finding is, however, essential for robust decision making. We hence need to extend the approach in the next sections below.

8.2 Gaussian Process

Gaussian processes (GPs) are a powerful and versatile tool in the field of machine learning and statistics. They are used, e.g. for:

Non-parametric Modeling: GPs provide a non-parametric framework for modeling data. Unlike many other machine learning models that have a fixed number of parameters, GPs can model complex, flexible functions without specifying a particular functional form. This makes them well-suited for problems where the underlying data structure is unknown or difficult to describe using traditional parametric models.

Uncertainty Estimation: GPs not only provide predictions but also offer estimates of uncertainty associated with those predictions. This is crucial in applications such as regression, classification, and optimization, where knowing the uncertainty allows for better decision-making. In cases where data are noisy or sparse, GPs can provide more reliable predictions by quantifying uncertainty.

Interpolation and Extrapolation: GPs excel at interpolation, i.e. making predictions within the range of the observed data. However, they are also capable of extrapolation, making predictions outside the observed data range. This is valuable in applications where understanding how a system behaves beyond the available data is essential, such as time series forecasting, geospatial modeling, and physics simulations.

Everyday life applications are many. They are used in Weather Forecasting by assisting predictions by capturing the complex spatial and temporal relationships in weather data. In E-commerce, Online shopping platforms use GP and your browsing and purchasing history to suggest products or services that you might be interested.

8.2.1 Stochastic generation of smooth curves

There are many ways to produce a complete set of functions of solutions of an eigenvalue problem of differential equations. There is a long history of using those sets to represent data. A typical example is the following:

$$A = -\frac{d^2}{dx^2} \,, \qquad x \in [0, L] \,, \qquad \psi(x + L) = \psi(x) \,.$$

The corresponding eigenvalue problem is

$$-\frac{d^2 \psi_n(x)}{dx^2} = k^2 \, \psi_n(x) \,, \qquad \psi_n(x) = \exp\left\{ i \, \frac{2\pi n}{L} x \right\} \,.$$

This leads to the well-known Fourier expansion of functions which are periodic on the interval $[0, L]$.

Here, we will follow a different path, and discuss sets of functions that are generated by a random process but that will be nevertheless *smooth* . The latter condition is a tall order. Consider a set of points x_i, $i = 1..n$ and a corresponding function values $f(x_i)$. Assume that each function value is generated by a random variable that is distributed by a Gaussian process:

$$f_i := f(x_i) : \qquad \frac{dP}{df} = \frac{1}{\sqrt{2\pi}\sigma_i} \exp\left\{ -\frac{f_i^2}{2\sigma_i^2} \right\} \,, \qquad \sigma_i = \sigma(x_i). \quad (8.4)$$

Note we allow a change of the variance $\sigma(x)$ with the position x. This provides some structure to the such generated function f, but the function is by no mean smooth. For example, choosing

$$\sigma(x) = 1 + x^2 \,,$$

we observe that the standard deviation is smallest around the origin and then rapidly rises. Figure 8.2 shows two curves, generated this way, in the interval $[-5, 5]$. Obviously, these curves are different and noisy.

The reason why those curves are noisy is that the functions values f_i, f_k, $i \neq k$ are generated with a Gaussian probability distribution and are, hence, uncorrelated. The key to make random curves smooth is to introduce correlations:

$$\langle f(x_i) \, f(x_k) \rangle = K(x_i, x_k) \,. \qquad (8.5)$$

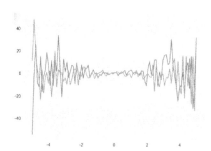 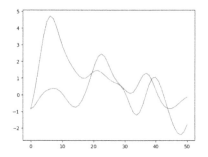

FIGURE 8.2: Left: Two random noisy functions generated by the process (8.4) with x dependent variance. Right: Two curves (i, f_i) generated with the Gaussian process (8.7). .

Thereby, $K(x, y)$ is called the *kernel* of the Gaussian process. We would like that function values, which are close, i.e. $d := |x - y| \sim 0$ are strongly correlated, while function values far apart should not be correlated very much. Selected popular choices for the kernel are

$K(x, y) = \sigma^2\, \delta(x, y)$	white Gaussian noise
$K(x, y) = \kappa\, \exp\{-d^2/2\ell^2\}$	Squared exponential
$K(x, y) = \kappa\, \exp\{-d^2/2\ell^2\} + \kappa\, \delta(x, y)$	Squared exponential with noise
$K(x, y) = \kappa\, \exp\{-d/\ell\}$	Ornstein-Uhlenbeck
$K(x, y) = \kappa\, \exp\{-2\, \sin^2(d/2)/\ell^2\}$	periodic

We stress that every set of the so-called *hyperparameters* such as (κ, ℓ) generates a different class of functions. If we will later use the smooth, random functions for data regression and employ optimisation for those parameters to find the best function system for the task.

We still need to specify the random process, among those leading to the formally written correlations in (8.5). A particular choice, which simplifies the explicit calculations enormously, is the *multivariate normal distribution*. This choice lends the name *Gaussian process* to the method.

Definition: For a given set of random variables indexed by a continuous variable (such as space), a Gaussian process is a stochastic process, such that every finite subset of those random variables has a multivariate normal distribution, i.e. every finite linear combination of them is normally distributed.

Consider a finite subset of function values f_i, $i = 1..n$, and making the choice that

$$\mu = \left\langle \sum_{i=1}^{n} n f_i \right\rangle = 0 \,,$$

the multivariate normal distribution is determined by the positive, symmetric $n \times n$ matrix M:

$$P(f_1..f_n) = \pi^{-n/2} \det^{1/2} M \, \exp\left\{-\vec{f}^T \, M \, f\right\} . \tag{8.6}$$

What remains to do now is to choose M in such a way that we find the desired correlation (8.5). The calculation can be performed in a closed form. We start by considering the generating function

$$Z[j] = \ln \int df_1 \ldots df_n \, \exp\left\{-\sum_{i,k} f_i \, M_{ik} \, f_k + \sum_k j_k f_k\right\} .$$

We leave it to the reader to check that the correlation is given by

$$\frac{\partial^2 \ln Z[j]}{\partial j_u \partial j_v} = \langle f_u \, f_v \rangle .$$

On the other hand, the multidimensional integral can be calculated analytically:

$$\ln Z[j] = \frac{1}{4} \sum_{i,k} j_i \left(M^{-1}\right)_{ik} j_k + \text{constant} .$$

Defining the kernel matrix by, $K_{ik} = \langle f(x_i) f(x_k) \rangle$, we obtain:

$$\frac{1}{2} \left(M^{-1}\right)_{uv} = \langle f_u \, f_v \rangle = K_{uv} .$$

In summary, the multivariate normal distribution

$$P(f_1..f_n | \Theta) = (2\pi)^{-n/2} \det^{-1/2} K \, \exp\left\{-\frac{1}{2} \vec{f}^T \left(K^{-1}\right) \vec{f}\right\} , \tag{8.7}$$

where Θ is the set of parameters of the kernel K, has a correlation matrix:

$$\langle f_u \, f_v \rangle = K_{uv} .$$

Let us see the Gaussian process in action: we choose a *squared exponential* with $\kappa = 1$ and $\ell = 3.5$. Note the function values are most correlated when they are close in x. ℓ dictates how much this correlation decreases with distance. We choose $n = 50$ and generate the 50 correlated function values:

$$f_i = f(x_i) , \qquad i = 1 \ldots n .$$

Two such curves are shown in Figure 8.2, right panel.

The conditions under which such generated curves are bounded and continuous have been the subject of great mathematical interest. Details of this reasoning are beyond the scope of this book. We instead point to the texts by Dudley [94] and by Marcus and Shepp [95].

8.2.2 Choosing hyperparamters

The random smooth curves that we have generated in the previous subsection have no bearing on any data. Yet, we would like them to represent data such as those of the "throw" experiment, red symbols in Figure 8.2. There is no doubt that human intervention and experience are needed to make an appropriate choice for the correlation kernel K. But if this choice has been made, how do we find the optimal parameters of this kernel?

As a starting point, we assume that those data have been generated by a Gaussian process with exactly the same kernel, say the "Squared Exponential with noise" kernel. Hence, we have a probability distribution for each function value f,

$$P(f_1 \ldots f_n | \kappa, \kappa_n, \ell) .$$

For representing the data, we thrive that the expectation values $\langle f_k \rangle$ are close to the measured values y_k, $k = 1 \ldots n$. To make the notation easier, we have assumed that $\langle f \rangle = 0$. We now need to make sure that our data also have a vanishing expectation value. This is easily achieved by

$$av := \frac{1}{n} \sum_{i=1}^{n} y_i , \qquad y_k \to y_k - av , \quad \forall k = 1 \ldots n .$$

Once we have modelled the data, we just need to keep in mind that we need to offset our findings by adding av. We will discuss this in greater detail below in subsection 8.2.4.

For finding the parameters, we are interested in the "inverse" problem, i.e. in the conditional probability distribution

$$P(\kappa, \kappa_n, \ell | y_1 \ldots y_n).$$

Both probability densities are connected by Bayes theorem:

$$P(\kappa, \kappa_n, \ell | y_1 \ldots y_n) = \frac{P(f_1 \ldots f_n | \kappa, \kappa_n, \ell) \, P(\kappa, \kappa_n, \ell)}{P(y_1 \ldots y_n)},$$

where $P(\kappa, \ell)$ is a prior distribution for the parameters. Note that the normalisation, i.e.

$$P(y_1 \ldots y_n) = \int P(y_1 \ldots y_n | \kappa, \kappa_n \ell) \, P(\kappa, \kappa_n, \ell) \, d\kappa \, d\kappa_n \, d\ell ,$$

is naturally independent from the parameter. We are trying to find the most likely parameter set that has generated the data set. Thus, we are going to minimise

$$- \ln P(\kappa, \kappa_n, \ell | y_1 \ldots y_n)$$

with respect to the parameters. Using (8.7), we are led to

$$-\ln P(\kappa, \kappa_n, \ell | y_1 \dots y_n) = \tag{8.8}$$

$$\frac{1}{2} \vec{y}^T (K^{-1}) \vec{y} + \frac{1}{2} \ln \det K + \text{constant} \xrightarrow{\kappa, \kappa_n, \ell} \min.$$

Let's $\theta \in \{\kappa, \kappa_n, \ell\}$ denote one of the parameters of this kernel. Minimising the above function with respect to θ is a demanding task especially if we have more than two parameters to play with. The standard approach is " steepest descent".

This approach leaves us with the task to calculate the gradient. The θ dependence is in the Kernel K. We observe

$$K^{-1}(\theta) K(\theta) = 1 \quad \Rightarrow \quad \left[\frac{\partial}{\partial \theta} K^{-1}(\theta)\right] K(\theta) + K^{-1}(\theta) \left[\frac{\partial}{\partial \theta} K(\theta)\right] = 0 .$$

We multiply with the inverse (and take care not to swap the order of any of the matrices) and find

$$\left[\frac{\partial}{\partial \theta} K^{-1}(\theta)\right] = - K^{-1}(\theta) \left[\frac{\partial}{\partial \theta} K(\theta)\right] K^{-1}(\theta) .$$

For the terms containing the determinant, we point out that the Kernel K is symmetric and positive. It admits an eigenvalue decomposition:

$$K = ODO^T , \qquad OO^T = 1 , \quad \det O = 1 , \qquad D = \text{diag}(\lambda_1 \dots \lambda_n) .$$

We then find

$$\det K = \det\left(ODO^T\right) = \det O \det D \det O^T = \det D = \prod_{i=1}^{n} \lambda_i .$$

We need to find out how the eigenvalues change when we change the kernel parameters. We firstly observe

$$\frac{\partial}{\partial \theta} K(\theta) = \left(\frac{\partial}{\partial \theta} O\right) DO^T + O\left(\frac{\partial}{\partial \theta} D\right) O^T + OD\left(\frac{\partial}{\partial \theta} O^T\right) .$$

Noting that

$$K^{-1}(\theta) = OD^{-1}O^T ,$$

we calculate

$$\text{tr}\left\{K^{-1}\frac{\partial}{\partial \theta} K\right\} = \text{tr}\left\{OD^{-1}O^T \left(\frac{\partial}{\partial \theta} O\right) DO^T\right\} + \text{tr}\left\{OD^{-1} \left(\frac{\partial}{\partial \theta} D\right) O^T\right\}$$
$$+ \text{tr}\left\{O\left(\frac{\partial}{\partial \theta} O^T\right)\right\} .$$

Using the cyclic invariance of the trace, i.e. $\mathrm{tr}ABC = \mathrm{tr}BCA$, we simplify

$$\mathrm{tr}\left\{K^{-1}\frac{\partial}{\partial\theta}K\right\} = \mathrm{tr}\left\{\left(\frac{\partial}{\partial\theta}O\right)O^T + \left(O\frac{\partial}{\partial\theta}O^T\right)\right\} + \mathrm{tr}\left\{D^{-1}\left(\frac{\partial}{\partial\theta}D\right)\right\}.$$

We can simplify further by noting

$$OO^T = 1 \quad\Rightarrow\quad \frac{\partial}{\partial\theta}[OO^T] = \left(\frac{\partial}{\partial\theta}O\right)O^T + \left(O\frac{\partial}{\partial\theta}O^T\right) = 0.$$

We find therefore

$$\mathrm{tr}\left\{K^{-1}\frac{\partial}{\partial\theta}K\right\} = \mathrm{tr}\left\{D^{-1}\left(\frac{\partial}{\partial\theta}D\right)\right\}.$$

Let us now consider the $\ln\det K$ term (with $\ln(ab) = \ln a + \ln b$):

$$\frac{\partial}{\partial\theta}\ln\prod_{i=1}^{n}\lambda_i = \frac{\partial}{\partial\theta}\sum_{i=1}^{n}\ln\lambda_i = \sum_{i=1}^{n}\frac{1}{\lambda_i}\frac{\partial\lambda_i}{\partial\theta} = \mathrm{tr}\left\{D^{-1}\frac{\partial}{\partial\theta}D\right\}.$$

Hence, we obtain altogether

$$\frac{\partial}{\partial\theta}\ln\det K = \mathrm{tr}\left\{K^{-1}\frac{\partial}{\partial\theta}K\right\}.$$

Our final result of this subsection is the gradient for the condition probability density function in (8.8):

$$-\frac{\partial}{\partial\theta}\ln P(\theta|y_1\ldots y_n) = \tag{8.9}$$
$$-\frac{1}{2}\vec{y}^T K^{-1}(\theta)\left[\frac{\partial}{\partial\theta}K(\theta)\right]K^{-1}(\theta)\vec{y} + \frac{1}{2}\mathrm{tr}\left\{K^{-1}\frac{\partial}{\partial\theta}K\right\}.$$

Let us study the method in action and reconsider the data set of a "throw" on page 328. Even with just eight pairs of data the gradient g can contain very large numbers. The means if we update the parameters, we need to rescale the gradients by some factor s:

$$\Theta_\nu \longrightarrow \Theta_{\nu+1} = \Theta_\nu - s\,g(\Theta_\nu).$$

in the present example, we choose

$$\Theta_0 = (\kappa, \kappa_n, \ell)^T = (1, 1, 3)$$

which yields (3 decimal places)

$$-\ln P(\theta|y_1\ldots y_n) = 10.693$$
$$g_\kappa = -0.598, \qquad g_\ell = 1.286, \qquad g_{\kappa n} = 3.7365.$$

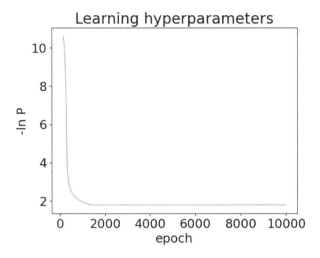

FIGURE 8.3: Learning the hyperparameters κ, ℓ by minimising the conditional probability $-\ln P(\theta|y_1 \ldots y_n)$.

We start the iteration with a small scaling factor $s = 10^{-7}$. Every 20 iterations we then doubled s until we reached 200 iterations. To avoid instabilities, we only allow $sc \leq 1$. After that, we kept s constant. In the Data Science context, finding the hyperparameters is called learning, and the iteration number is also known as "epoch". The number for $-\ln P$ rapidly comes down, followed by slow improvements towards the end of the iteration. We used $10,000$ epochs for learning and the evolution is shown in Figure 8.3. We thus obtain the optimised parameters and probability:

$$\kappa \;=\; 2.274 \qquad \kappa_n \;=\; 0.071, \qquad \ell \;=\; 1.870 \qquad -\ln p \;=\; 1.790. \tag{8.10}$$

We present a word of caution at the end of this subsection: while the convergence rate of a " steepest descent" iteration is quite good, it is by no means guaranteed that the method finds the *global* optimum. Depending on the non-linear function, the iteration might stop in local optimum. A standard technique which tries to circumvent this issue is Simulated Annealing, which is based upon a Markov Chain Monte Carlo simulation. However, despite the remarkable improvements over naive approaches, also with this setting, finding the *global* optimum is not always achieved.

8.2.3 Conditional Gaussian distributions

There is something special about multivariate Gaussian distribution what makes exceptionally useful in practical calculations. Assume that we have multivariate distribution of N variables $f_1 \ldots f_N$ with a given correlation kernel K:

$$\langle f_i f_k \rangle = K_{ik} \, , \qquad P(f_1 \ldots f_N) \propto \exp\left\{ -\frac{1}{2} \sum_{i,k} f_i \, (K^{-1})_{ik} \, f_k \right\} . \qquad (8.11)$$

If we choose $n < N$ variables $\vec{u} = (f_1 \ldots f_n)^T$ as given, what is the conditional probability of the remaining $m = N - n$ variables $\vec{v} = (f_{n+1} \ldots f_N)^T$? We can always split the bilinear form into terms containing variables from each category:

$$\sum_{i,k} f_i \, (K^{-1})_{ik} \, f_k = \vec{u}^T A \vec{u} \, + \, \vec{u}^T B \vec{v} \, + \, \vec{v}^T B^T \vec{u} \, + \, \vec{v}^T C^{-1} \vec{v} \, .$$

We then can "complete the square" and find:

$$\left[\vec{v} + C \, B^T \vec{u} \right]^T \, C^{-1} \, \left[\vec{v} + C \, B^T \vec{u} \right] \, .$$

Noting that C is a symmetric matrix, $C = C^T$, and so is its inverse, we prove the above equation by expanding the brackets. We find the important result that the conditional probability distribution for the remaining variables $\vec{v}$ is also multivariate Gaussian. The question then arises: what are the parameters, mean and standard deviation matrix, in terms of K?

To answer this question, it is convenient to introduce a matrix notation:

$$\vec{f} = \begin{pmatrix} \vec{u} \\ \vec{v} \end{pmatrix} , \qquad K = \begin{pmatrix} K_{uu} & K_{uv} \\ (K_{uv})^T & K_{vv} \end{pmatrix} , \qquad K^{-1} = \begin{pmatrix} A & B \\ B^T & C^{-1} \end{pmatrix} .$$

We particularly look for determining C and B from the equation

$$K \, K^{-1} = 1 \qquad \begin{pmatrix} K_{uu} & K_{uv} \\ (K_{uv})^T & K_{vv} \end{pmatrix} \begin{pmatrix} A & B \\ B^T & C^{-1} \end{pmatrix} = \begin{pmatrix} 1 & 0 \\ 0 & 1 \end{pmatrix} ,$$

since those inform the standard deviation and mean for the conditional distribution. We point out that only K_{uu}, K_{vv} and A, C are symmetric matrices and have an inverse. In fact, K_{uv} and B are $n \times m$ matrices. From the above matrix equation, we observe

$$\begin{aligned} K_{uu} B + K_{uv} C^{-1} &= 0 \quad \Rightarrow \quad B = -(K_{uu})^{-1} K_{uv} \, C^{-1} \, , \\ K_{uv}^T B + K_{vv} C^{-1} &= 1 \, . \end{aligned}$$

Inserting the former equation into the latter, we find

$$C = K_{vv} - (K_{uv})^T \, (K_{uu})^{-1} \, K_{uv} \, ,$$

For the "shift", i.e. the mean, we are interested in $-C \, B^T \vec{u}$. We find from above

$$BC = -(K_{uu})^{-1} K_{uv}, \quad \Rightarrow \quad -C B^T \vec{u} = (K_{uv})^T \, (K_{uu})^{-1} \vec{u} \, .$$

We point out as a reminder $(BC)^T = C^T B^T = CB^T$. We are in the position to formulate the main result of this subsection.

Assume that the variables $\vec{f} = (\vec{u}, \vec{v})^T$ have a multivariate Gaussian distribution with a correlation matrix

$$K = \begin{pmatrix} K_{uu} & K_{uv} \\ (K_{uv})^T & K_{vv} \end{pmatrix} .$$

The conditional probability for the variables $\vec{v}$ then also has a multivariate Gaussian distribution with the mean vector $\vec{\mu}$ and correlation matrix C with

$$P(\vec{v}|\vec{u}) \quad \propto \quad \exp\left\{ -\frac{1}{2} (\vec{v} - \vec{\mu})^T \; C^{-1} \; (\vec{v} - \vec{\mu}) \right\} , \qquad (8.12)$$

$$\vec{\mu} \quad = \quad (K_{uv})^T (K_{uu})^{-1} \vec{u} , \qquad C \; = \; K_{vv} \; - \; (K_{uv})^T (K_{uu})^{-1} K_{uv} .$$

8.2.4 Regression and prediction

Let us add now an additional point x to the sequence of points $x_1 \ldots x_n$ and ask ourselves: what is the most likely value y and what is the corresponding uncertainty e? If x lies inside the interval $[x_1, x_n]$, we speak about interpolation or regression. If the point lies outside, say $x > x_n$, the process is often called prediction, having a time series in mind.

To answer these questions, we need to study the condition probability:

$$P(y; x|y_1 \ldots y_n, x_1 \ldots x_n, \theta).$$

But this probability density is exactly what we have calculated in the previous subsection ($m = 1$, see (8.12)). The most likely value and the error bar for a 95% confidence level are

$$y \; = \; (K_{uv})^T (K_{uu})^{-1} \vec{y} , \qquad e \; = \; 2 \left(K_{vv} \; - \; (K_{uv})^T (K_{uu})^{-1} K_{uv} \right) .$$

We use the optimised hyperparameters from (8.20). The result is shown in Figure 8.4. We chose $1,200$ equally spread values x in the interval $[0, 12]$. We find that the Gaussian process remarkably well tracks the exact trajectory (orange line) within a 95% confidence error band except perhaps for the end point, where the error bar is underestimated. If it comes to *predicting* values

for about $x > 8$, we observe that error bar rapidly increases in such a way that any sensible prediction of the object hitting the ground is impossible. It even seems that the *most likely trajectory* rises again for $x > 0$, which we know is physically impossible since it would require some propulsion system of the particle.

And here lies the crux: the unsupervised Gaussian process doesn't have built-in information that we are dealing with a free-falling object. The measurements

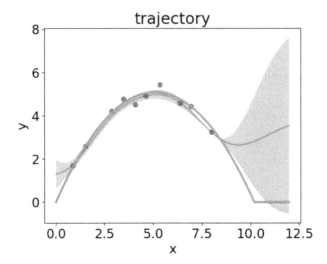

FIGURE 8.4: Measured trajectory (red symbols), exact trajectory (orange line), most likely position reconstructed from an unsupervised Gaussian process (blue line), error band: 95% confidence level.

could well relate to a drone with a propulsion system, where the prediction could well make sense. The only constraint we impose is that the curve we draw somehow comes close to the points (red symbols) and otherwise is smooth.

Unsupervised learning can give good results for interpolating data between measurements. However, if we want to increase the predictive power of the data modelling, we need to add some information from the underlying physics. This brings us to the topic of supervised *guided learning*.

8.2.5 Guided learning

We have seen in the previous subsection that the uncertainty grows very rapidly if we try to extrapolate data. This is due to the very reason that no behaviour is excluded for the continuation of the curves as long as the curve remains smooth. To get a better extrapolation, we would like to include knowledge, say on the free-fall trajectories, in the Gaussian process.

Here, we reconsider the example from section 8.1. The theory of the free fall has two parameters, $\theta \in \{a_1, a_2\}$, and the height of the object (without noise) is given by

$$f(x; \theta) = a_1 x + a_2 x^2 .$$

We model the noise, i.e. the deviation between the exact curve at point x_i from the measured value y_i due to disturbances or errors in measurement, by a Gaussian process:

$$P(\vec{y}, \vec{x}; \theta) = \frac{1}{\sqrt{2\pi \det K}} \exp\left\{ -\frac{1}{2} \sum_{i,k=1}^{n} (y_i - f(x_i; \theta)) \, K_{ik}^{-1} \, (y_k - f(x_k; \theta)) \right\} .$$

$$(8.13)$$

This means that the most likely curve of the process is a free-fall trajectory f in any case. This is the kind of information we wanted to include. An underpinning assumption is that there is some systematics in the correlation of the errors:

$$\epsilon_i = y_i - f(x_i; \theta) ,$$

which is then specified by the Gaussian kernel $K(x, y)$. To choose, we make the assumption that the errors are independent and distributed with a constant variance. Independence is a natural choice if these errors are made during measurements. Constance is a stark assumption: for instance, it could be that it is harder to measure precisely if the height y_i is large. With these assumptions, we choose the "white Gaussian noise" kernel:

$$K(x, y) = \sigma^2 \, \delta(x, y) .$$

For this choice, the probability distribution drastically simplifies to

$$P(\vec{y}, \vec{x}; \theta) = \frac{1}{\sqrt{2\pi} \, \sigma^n} \exp\left\{ -\frac{1}{2\sigma^2} \sum_{i=1}^{n} (y_i - f(x_i; \theta))^2 \right\} . \qquad (8.14)$$

To determine the optimal parameters, $\theta = (a_1, a_2)$ and σ, we invoke again Bayes theorem:

$$P(\theta, \sigma; \vec{y}, \vec{x}) = \frac{P(\vec{y}, \vec{x}; \theta, \sigma) \, P(\theta, \sigma)}{P(\vec{y})} .$$

We are choosing constant priors $P(\theta, \sigma)$. Maximising the above probability then implies minimising the "action":

$$S = \frac{1}{2\sigma^2} \sum_{i=1}^{n} (y_i - f(x_i; \theta))^2 + n \ln \sigma \xrightarrow{\theta, \sigma} \min.$$

Minimising with respect to $\theta = (a_1, a_2)$ then just produces the familiar equation from the *least square fit* with L2 norm in subsection 8.1. In addition, minimising with respect to σ gives us the desirable confidence interval of the fit:

$$-\frac{1}{\sigma^3} \sum_{i=1}^{n} (y_i - f(x_i; \theta))^2 + \frac{n}{\sigma} = 0 ,$$

implying

$$\sigma^2 = \frac{1}{n} \sum_{i=1}^{n} (y_i - f(x_i; \theta))^2 . \qquad (8.15)$$

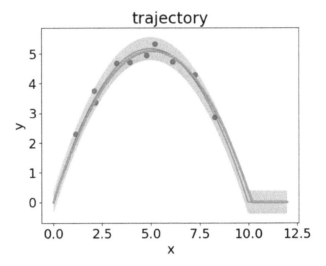

FIGURE 8.5: Measured trajectory (red symbols), exact trajectory (orange line), most likely position reconstructed from a *guided learning* Gaussian process (blue line), error band: 95% confidence level.

Let us now add a new point to the set of n pairs and estimate the probability distribution of the corresponding height y. Because of the nature of the kernel K, this distribution is independent of the other values:

$$P(y|x;\theta,\sigma) \;=\; \frac{1}{\sqrt{2\pi}\,\sigma}\,\exp\left\{-\frac{1}{2\sigma^2}\left(y - f(x;\theta)\right)^2\right\}.$$

Figure 8.5 shows the result with a 2σ, i.e. 95%, confidence error band. We observe that the shaded area covers the exact result even in the extrapolation regime $x > 8$.

8.3   Machine learning with graphs

Graph theory is of significant importance in the field of machine learning because it provides a powerful framework for modeling and solving a wide range of real-world problems. Graphs, which consist of nodes (vertices) and edges (connections between nodes), are a versatile way to represent and analyse complex relationships and structures in data. We here can only provide a glimpse of the powerful Mathematics underpinning graphs and point to the book by Karin Soub [96] for a good introduction to the matter.

8.3.1   Graph theory basics for machine learning

We here review the basic elements of Graph Theory and functions of random fields over graphs.

The basic elements of graph theory include

- **Vertex** (or node): A vertex is a fundamental unit in a graph and represents an object or entity. Nodes are usually visualised by points in an image of a graph, but bear in mind these points can present things such as cities in a road map or numbers in field theory.

- **Edge:** An edge is a connection between two vertices. Edges can be directed (arrows pointing from one vertex to another) or undirected (no arrow indicating direction). Edges can also be weighted to represent the strength or cost of the connection between two vertices.

- **Path:** A path is a sequence of edges connecting two vertices. The length of a path is the number of edges in the sequence.

- **Cycle:** A cycle is a path that starts and ends at the same vertex. Connected graph: A connected graph is a graph in which there is a path between any two vertices.

- **Degree of a vertex:** The degree of a vertex is the number of edges incident to it. For an undirected graph, the degree of a vertex is simply the number of edges that connect to it. For a directed graph, the degree of a vertex is the number of incoming and outgoing edges.

- **Adjacency matrix:** An adjacency matrix is a square matrix used to represent a graph. The entries in the matrix indicate whether there is an edge between two vertices.

Definition: A **undirected simple graph** is an ordered pair $G = (V, E)$ where V is the set of vertices and E is the set of edges.

Figure 8.6 shows an example with 7 edges and 10 edges. The corresponding adjacency matrix is

$$
\begin{bmatrix}
 & 1 & 2 & 3 & 4 & 5 & 6 & 7 \\
1 &   & 1 & 1 & 1 &   &   &   \\
2 & 1 &   &   & 1 & 1 &   &   \\
3 & 1 & 1 &   &   & 1 &   &   \\
4 & 1 & 1 & 1 &   &   & 1 &   \\
5 &   &   &   &   & 1 &   & 1 & 1 \\
6 &   &   &   &   & 1 &   &   & 1 \\
7 &   &   &   &   &   & 1 & 1 &
\end{bmatrix}
$$

It is often useful to have a concept of locality and neighbourhood. Neighbours could be friends in a social network and interacting fields on lattice for Statistical Field theory.

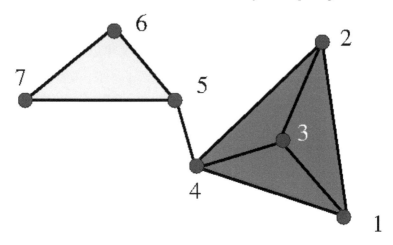

FIGURE 8.6: A graph with 7 vertices and 10 edges.

The mathematical concept of "neighbourhood" rises from the following definitions:

- A **clique** is a subset of vertices of an undirected graph such that every two distinct vertices of the clique are adjacent.

- A **maximal clique** is a clique that cannot be extended by including one more adjacent vertex, that is, a clique which does not exist exclusively within the vertex set of a larger clique.

- A **maximum clique** of a graph G is a clique, such that there is no clique with more vertices. The corresponding number $\omega(G)$ of vertices is called *clique number*.

- The **clique cover** of a Graph is the set of cliques with the smallest number whose union covers the set of vertices V of the graph. Finding a minimum clique cover is NP hard in computational complexity theory and is an exceptionally difficult algorithmic challenge.

Let us return to Figure 8.6 to familiarise ourselves with the above notations. It has 7 1-vertex clicks (the vertices itself) and 10 2-vertex clicks (the edges). Let us list the 5 3-vertex cliques: (123), (234), (134), (143), (567). The graph has only one 4-vertex clique: (1234), which is a maximal clique and the only maximum clique. The 3-vertex clique (567) is maximal, but (234) is not since the inclusion of vertex 1 can make the clique bigger.

8.3.2 Random fields on graphs and the Hammersley-Clifford theorem

We introduced the basic elements of a graph in the previous subsection purely on graphical grounds: edges have been lines in figures connecting points. We now extend the use of the formalism to consider random fields on undirected simple graphs.

To this aim, we consider a graph $G = (V, E)$ with n vertices, addressed by co-ordinates x_k, and introduce the random field ϕ which takes the values $\phi_1 \ldots \phi_n$ at the vertices. Those values are random variables distributed according to the multivariate distribution $p(\phi_1 \ldots \phi_n) > 0$. We call a given set of ϕ values a *configuration*. We use the probability distribution to define the edges E of the graph. Consider the conditional probability for ϕ_i when all the other variables are "frozen". We then say two vertices $\{x_i, x_k\}$ are a part of an edge if

$$p\Big(\phi_i | \{\phi_1 \cdots \phi_n\} \backslash \{\phi_i\}\Big) \neq p\Big(\phi_i | \{\phi_1 \cdots \phi_n\} \backslash \{\phi_i, \phi_k\}\Big) .$$

Here, we used the set theory symbol $\backslash$, say "without", for example, for a set with three elements: $\{a, b, c\} \backslash \{b\} = \{a, c\}$. The above definition of E connects the mathematical structure of the multivariate distribution p with the geometry of the graph. We then define the *neighbours* of the vertex x_i by

$$\mathcal{N}(x_i) = \Big\{ x_k \in V | \{x_i, x_k\} \in E \Big\} .$$

In addition, we define cliques in the usual way:

$$C \subseteq V \quad \text{is a clique if} \quad C \subseteq \{x, \mathcal{N}(x)\} \; \forall x \in C .$$

Finally, we say that a probability distribution *factorises* if there exists a set of non-negative functions ψ_C, called potential functions, with the property:

$$\forall \phi, \phi' : \quad \text{For} \;\; (\phi_i)_{i \in C} = (\phi'_k)_{k \in C} \quad \Rightarrow \quad \psi_C\Big(\phi\Big) = \psi\Big(\phi'\Big).$$

Or in other words, if two different field configurations ϕ and ϕ' happen to coincide doe a clique on the graph, the corresponding potential function is the same independent of the other random variables.

With these definitions at hand, we are now prepared to answer some of the most fundamental questions in stochastic simulations: What is the most general form of the conditional probability functions for interacting random variables such that they define a meaningful joint distribution (see, e.g. [97])?

The breakthrough came with the

Hammersley-Clifford theorem: For strictly positive distributions $p(\phi_1 \ldots \phi_n) > 0$, the following properties are equivalent:

- **Factorisation:** The above defined potential functions $\psi_C(\phi)$ exist, and we have

$$p(\phi_1 \ldots \phi_n) \;=\; \frac{1}{Z} \prod_{C \in \mathrm{cl}(G)} \psi_C\Big(\{\phi_k | x_k \in C, \forall k\}\Big),$$

where cl(G) is the set of maximal cliques of the graph, and Z is called *partition function*.

- **Local Markov Property:** The conditional probability of the random field ϕ_k depends only on its neighbours:

$$p\Big(\phi_k | \{\phi_1 \ldots \phi_n\} \backslash \{\phi_k\}\Big) \;=\; p\Big(\phi_k | \{\phi_\ell | x_\ell \in \mathcal{N}(x_k), \forall \ell\}\Big).$$

- **Global Markov Property:** For any disjoint vertex subsets A, B, and S in V such that S separates A and B (i.e. every path between a vertex in A and a vertex in B passes through a vertex in S), the random variables $\phi_{x \in A}$ are conditionally independent of $\phi_{x \in B}$ given $\phi_{x \in S}$:

$$p\Big(\{\phi_k | x_k \in A\} \,|\, \{\phi_i | x_i \in B \cup S\}\Big) \;=\; p\Big(\{\phi_\ell | x_\ell \in S\}\Big).$$

Equipped with the above structure, the fields ϕ are called *Markov Random Fields*.

It is nowadays very intuitive to combine statistical results with an underlying geometry. Examples include geographic information systems (GIS) and climate modeling, where the probability of rainfall overlays a map of a region. Other applications are Spatial and Temporal Modeling or Markov Points, where, e.g. crime statistics are analysed with a density function, quantifying the number of crimes around a point of interest and a certain radius. If this density is Markovian, the conditional probability of being affected only depends on the density function of the neighbourhood.

8.3.3 The ising model as a Markov random field

Let us explore in this section how the familiar ising model (see sections 4.4 and 5.2.1) fit into the framework of Markow Random fields. We start by study the graph of the 2d ising model, which is usually a square lattice. We consider two cases: (i) open and (ii) periodic boundary conditions (see Figure 8.7).

If we ask what are the sets of vertices in which all vertices are connected to each other, i.e., what are the cliques of the grid, we find that all cliques (apart from trivial the 1-vertex cliques) are 2-vertex cliques. This is true for vertices in the bulk of the lattice as well as for vertices on the edge boundary or corner (open boundary conditions). Hence, the *clique number* of the graph is 2 in both cases.

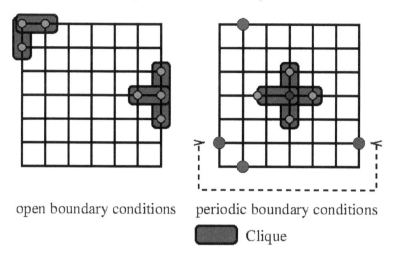

open boundary conditions   periodic boundary conditions

▭ Clique

FIGURE 8.7: Graph for the Ising model with open (left) and periodic boundary conditions (right).

Both graphs in Figure 8.7 differ by some of the degrees of vertices. Vertices in the bulk have always 4 neighbours and thus degree four. Since there is no boundary for periodic conditions, this is true out of the graph in this case. For open boundary conditions (left figure), the degree changes to 3 for vertices on the boundary edge and to 2 for vertices in the corners of the lattice. This affects how often a particular vertex is part of a clique, which is always 4 for periodic boundary conditions; for open boundary conditions, it is still 4 in the bulk, but it is 3 (edge vertices) or 2 (corner vertices) for vertices on the boundary, see Figure 8.7.

Let us now consider a field $\phi = z \in \{-1, +1\}$ on this graph. According to the Hammersley-Clifford theorem, any (positive) probability distribution can be written in terms of the positive, potential functions ψ. If we introduce the coordinates x and y for the lattice to address the vertices, then *all* cliques of the graph are the links $\langle xy \rangle$. We are thus led to that any distribution of an Ising random field on this graph necessarily can be written as:

$$p\left(z\right) = \frac{1}{Z} \prod_{\langle xy \rangle} \psi_k \left(z_x z_y\right). \qquad (8.16)$$

The original Ising model makes the choice $\psi_k(x) = \exp(x)$ for all cliques (i.e., links) k, with which recover the familiar Ising probability distribution:

$$p\left(z\right) = \frac{1}{Z} \exp\left\{ \sum_{\langle xy \rangle} z_x z_y \right\}. \qquad (8.17)$$

8.4   Emulation of statistical systems with Machine Learning

Monte Carlo simulations of statistical field theories aim to generate a sequence of independent field configurations $\{z\}$ distributed according the Gibbs factor

$$\exp\left\{-\beta\, S[z]\right\} \, / \, Z(\beta) \,,$$

where the normalisation Z is called partition function and S action. We spent a great deal of time in section 5 to discuss methods to generate those configurations. Emulation of a Statistical Field Theory aims to develop, using Machine Learning techniques, a stochastic system that produces *similar* field configurations $\{z\}$ at much less computational costs.

8.4.1   Restricted Boltzmann machines

Boltzmann machines were originally proposed by Sherrington and Kirkpatrick in 1975 to describe a spin-glass [98]. It is an Ising model where each spin $z_i = 2s_i - 1$, $s_i \in \{0, 1\}$ can interact with any other spin s_k, $k \neq i$ on the lattice with a random interaction strength w_{ik}, $w_{ii} = 0$, $\forall i$. Boltzmann machines nowadays have many applications in machine learning and the emulation of complex systems.

The starting point is the probability distribution:

$$P \;=\; \exp\left(\sum_{i<k} w_{ik} s_i s_k \;+\; \sum_i \theta_i s_i\right) \, / \, Z(w, \theta) \,, \qquad (8.18)$$

where Z ensures proper normalisation. θ_i acts like an external magnetic field, but is also known as *activation energy* in a machine learning context.

A standard method to generate configurations $\{s\}$ is the heat-bath method (see section 3), which we briefly discuss in the present context. We are seeking to update the spin s_ℓ while we consider all other spins as "frozen". We factorise the probability distribution in parts c, which do *not* depend on s_ℓ, and factors that do:

$$P \;=\; c \, \exp\left(\sum_{k>\ell} w_{\ell k} s_\ell s_k \;+\; \sum_{i<\ell} w_{i\ell} s_i s_\ell \;+\; \theta_\ell s_\ell\right)$$

The probability for $s_\ell = 0$ is then given by $p_0 = c$, while the probability for $s_\ell = 1$ is given by:

$$p_1 \;=\; c \, \exp\left(\sum_{k>\ell} w_{\ell k} s_k \;+\; \sum_{i<\ell} w_{i\ell} s_i \;+\; \theta_\ell\right).$$

Since the conditional probabilities need to add up to 1, i.e., $p_0 + p_1 = 1$, we find the probability with which we need to choose s_ℓ to be 1:

$$p_1 = \text{sig}(\Delta E_\ell), \qquad \Delta E_\ell := \sum_{k \neq \ell} w_{\ell k} s_k + \theta_\ell, \qquad (8.19)$$

where we have used that the matrix w_{ik} is symmetric. The function $\text{sig}(x)$ is well known in machine learning contexts and is called sigmoid function:

$$\text{sig}(x) = \frac{1}{1 + \exp(-x)}.$$

Given the parameters (w_{ik}, θ_i), we are now in the position to generate configurations $\{s\}$ and estimate observables. But how do we choose those parameters so that they represent a specific environment and emulate a given underlying system?

The basic idea to divide the spins or units s_i into two sets: the *visible* units, say σ_k, and the *hidden* units h_u:

$$\{s\} = \{\sigma\} \cup \{h\}.$$

The visible units are those related to the outside world. They could be a stack of photographs or Ising configurations generated by a Monte Carlo simulation. If we stick to the Ising model, we introduce the *observed*, i.e. exact, distribution of Ising spins σ_i by $p_e(\sigma)$. Introducing the conditional probability $p(\sigma|h)$, the Boltzmann Machine "emulates" the marginal distribution:

$$p_B(\sigma) = \sum_h p(\sigma|h). \qquad (8.20)$$

The task is now to make the emulated distribution $p_B(\sigma)$ as close as possible to the *observed* distribution $p_e(\sigma)$. To this aim, we need a measure that quantifies how different two probability distributions are. A widely used measure is the Kullback–Leibler divergence [99]:

$$D(p_B, p_e) = \sum_\sigma p_e(\sigma) \ln\left(\frac{p_e(\sigma)}{p_B(\sigma)}\right). \qquad (8.21)$$

Given that

$$\sum_\sigma p_B(\sigma) = 1, \qquad \sum_\sigma p_e(\sigma) = 1,$$

it turns out that D has the right "measure" properties:

$$D(p_B, p_e) \geq 0, \qquad D(p_B, p_e) = 0 \text{ only if } p_B(\sigma) = p_e(\sigma) \; \forall \sigma.$$

Proof can be found in the original publication from 1959 [99]. We briefly sketch the main elements. Consider:

$$F = \sum_i f_i \ln \frac{f_i}{g_i}, \qquad f_i, g_i > 0 \; \forall i, \qquad \sum_i f_i = \sum_i g_i = 1.$$

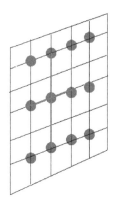 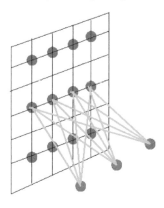

FIGURE 8.8: Left: The Ising model with 12 sites and next to nearest neighbours interaction. Right: Restricted Boltzmann Machine with 12 visible nodes and 3 hidden ones. All (blue) sites are connected with all hidden sites. For illustration purposes, we do not show all of these connections.

We then find:

$$F = \sum_i g_i \, \phi(t) \,, \qquad t := f_i/g_i \in (0, \infty) \,, \qquad \phi(t) = t \, \ln t \,.$$

Using Taylor's theorem, we have

$$\phi(t) = \phi(1) + \phi'(1)(t-1) + \frac{1}{2}\phi''(\bar{t})(t-1)^2 \,, \qquad 0 < \bar{t} < 1 \,.$$

Elementary calculus then gives

$$\phi(1) = 0 \,, \qquad \phi'(1) = 1 \,, \qquad \phi''(\bar{t}) = 1/\bar{t} \,,$$

leaving us with

$$F = \sum_i g_i \left[\frac{f_i}{g_i} - 1\right] + \sum_i g_i \frac{1}{2\bar{t}_i} \left[\frac{f_i}{g_i} - 1\right]^2 = \sum_i g_i \frac{1}{2\bar{t}_i} \left[\frac{f_i}{g_i} - 1\right]^2 \,,$$

where we have used that g_i and f_i both sum to 1. We conclude that $F \geq 0$, and since $g_i, \bar{t}_i > 0$ the only way F equals zero is that $f_i = g_i$ for all i.

Given that we established good measure properties for D in (8.21), finding the optimal parameter (w_{ik}, θ_l) for the Boltzmann Machine (8.18) means that we need to minimise (8.21) with respect to those parameters – a task that we will be discussed in detail in the next subsection.

A standard Boltzmann Machine, which connects every site with any other, is not much of practical use for emulating a statistical system for the reason that there are too many parameters available. This is illustrated in Figure 8.8:

the interaction network of the Ising model with nearest neighbour interactions is indicated by the red lines in the left panel. The left panel shows a Boltzmann Machine for which the Ising model constitutes a subset of the nodes – the so-called visible layer. If this Boltzmann Machine would have connections between *all* sites, there is nothing to optimise since we just choose the interaction matrix w_{ik} as of nearest neighbour type between all nodes of the visible layer. All other entries of w_{ik} are set to zero. This Machine is identical to the Ising model and trivially produces exact answers.

To make the Boltzmann Machine useful for our purposes, we need to restrict the connections between the sites. A popular choice is that there are *no connections* between the nodes of the visible layer, but that those nodes connect to all nodes of the hidden layer. The idea is to have more and more nodes in the visible layer than in the hidden one. But then the question arises: how well can such a Boltzmann machine represent the full statistical system?

8.4.2   Training a Boltzmann machine

Restricted Boltzmann Machines (RMBs) have a single hidden layer and a bipartite connection between the sites: every site in the visible layer is connected to every site in the hidden layer but the neurons in the same layer are not connected to each other. For further consideration, it is convenient to make these restrictions explicit in the probability distribution (8.18) of the BM. From the N sites in total, we attribute $1 \ldots n_v$ to label the visible nodes and $n_v + 1 \ldots n_v + n_h$ to the hidden layer nodes: of course, we have $N = n_v + n_h$. We split the set of units as well:

$$ s = (\sigma_i, h_k) , \quad \theta = (b_i, c_k) , \qquad i = 1 \ldots n_v, \quad h_k, k = 1 \ldots n_h. $$

We then can write for the probability distribution (8.18) recasting $(w, \theta) \to (W, b, c)$:

$$ P = \exp \left(\sum_{i=1}^{n_h} \sum_{k=1}^{n_v} W_{ik} \sigma_k h_i + \sum_{k=1}^{n_v} b_k \sigma_k + \sum_{i=1}^{n_h} c_i h_i \right) / Z(W, b, c) . \quad (8.22) $$

Our aim will be to minimise the measure (8.21), and, to this aim, we need to calculate the marginal distribution:

$$ p_B(\sigma) = \sum_h p(\sigma | h) . $$

The bipartite nature of the connection is helpful here, as we shall see

$$ p_B(\sigma) = \frac{\exp \left(\sum_{k=1}^{n_v} b_k \sigma_k \right)}{Z(W, b, c)} \sum_{\{h\}} \exp \left(\sum_{i=1}^{n_h} \sum_{k=1}^{n_v} W_{ik} \sigma_k h_i + \sum_{i=1}^{n_h} c_i h_i \right) $$

$$ = \frac{\exp \left(\sum_{k=1}^{n_v} b_k \sigma_k \right)}{Z(W, b, c)} \sum_{\{h\}} \exp \left(\sum_{i=1}^{n_h} h_i B_i \right) , $$

$$ B_i(\sigma) = \sum_{k=1}^{n_v} W_{ik} \sigma_k + c_i . \quad (8.23) $$

The latter sum factorises and the sum over all fields h_i can be carried out in closed form

$$p_B(\sigma) = \frac{\exp\left(\sum_{k=1}^{n_v} b_k \sigma_k\right)}{Z(W, b, c)} \prod_{i=1}^{n_h} \left(1 + e^{B_i(\sigma)}\right) . \tag{8.24}$$

The "exact" probability distribution for the Ising model is

$$p_e(\sigma) = \exp\left\{\beta \sum_{\langle xy \rangle} z_x z_y\right\} / Z(\beta) , \qquad z_x = 2\sigma_x - 1 . \tag{8.25}$$

The Kullback–Leibler measure (8.21) then becomes

$$
\begin{aligned}
D &= \sum_{\{\sigma\}} p_e(\sigma) \left[\beta \sum_{\langle xy \rangle} z_x z_y - \sum_{k=1}^{n_v} b_k \sigma_k - \sum_{k=1}^{n_h} \ln\left(1 + e^{B_k}\right)\right] \\
&+ \ln Z(W, b, c) - \ln Z(\beta) .
\end{aligned}
\tag{8.26}
$$

From a mathematically point of view, the task is easy enough formulated: one needs to make the artificially generated data by the Boltzmann machine as close as possible to that generated from the target Ising model. This is achieved by making both probability distributions as close as possible, where "closeness" is defined by the Kullback–Leibler measure. The issues lie with the technical implementation: (1) the Kullback–Leibler measure is a real value (bounded from below by zero) of many variables. It generically possesses many *local* minima. (2) the measure D for the current application contains the logarithm of a partition function, i.e. $\ln Z(W, b, c)$. Estimating those only recently became available by using techniques such as LLR (see Chapter 7). Even if feasible, to estimate a partition function for the purpose – say – one Monte Carlo step in minimising a function seems to be prohibitively costly. Because of the importance of machine learning and AI, a vast number of papers have appeared over the recent couple of years that offer ad hoc procedures, which are successfull for certain scenarios. Trying to give an overview of those publications would be beyond the scope of this paper. We refer instead to the two well-known publications by Hinton [100] and Bottou [101].

A standard approach to minimise complicated functions of many variables is the *steepest descent*. Since it only involves gradients, it also avoids to calculate a partition function for each minimisation step. This is not too unfamiliar to us: The metropolis update for Monte Carlo simulations was designed to avoid the explicit calculation of the normalisation, i.e. partition function (see section 3.2).

Our starting point is now to minimise D with respect to the set $\theta := (W, b, c)$ of parameters. Usually, one resorts to "steepest descend":

$$\theta_{n+1} = \theta_n - r \left.\frac{\partial D}{\partial \theta}\right|_{\theta = \theta_n} , \tag{8.27}$$

where r is the *scale factor*. For example, we find, e.g. for the gradient with respect to W_{ik}:

$$\frac{\partial D}{\partial W_{ik}} = -\sum_{\{\sigma\}} p_e(\sigma) \frac{\sigma_k}{1 + \exp\{-B_i\}} + \frac{\partial}{\partial W_{ik}} \ln Z(W, b, c) .$$

Both terms can be interpreted as Monte Carlo averages. We denote averages by the (exact) Ising model by $\langle \ldots \rangle_e$ and write $\langle \ldots \rangle_B$ for those rising from the Boltzmann machine with distribution (8.22). With these notations, the gradients are

$$\frac{\partial D}{\partial W_{ik}} = \left\langle h_i \sigma_k \right\rangle_B - \left\langle \frac{\sigma_k}{1 + \exp\{-B_i\}} \right\rangle_e , \tag{8.28}$$

$$\frac{\partial D}{\partial b_k} = \left\langle \sigma_k \right\rangle_B - \left\langle \sigma_k \right\rangle_e , \tag{8.29}$$

$$\frac{\partial D}{\partial c_i} = \left\langle h_i \right\rangle_B - \left\langle \frac{1}{1 + \exp\{-B_i\}} \right\rangle_e . \tag{8.30}$$

In (8.29), we could the reflection symmetry of the Ising model implying $\langle z_k \rangle_e = 0$ and, hence, $\langle \sigma_k \rangle_e = 1/2$.

As we can see from the last set of equations, we have avoided the explicit calculation of a partition function. If we we have *exact* expectation values at our fingertips, the use of those for the iteration (8.29) would result in a standard steepest descent method.

The Machine Learning and AI community has developed their own nomenclature alongside the long-standing mathematical language. We believe it is helpful to provide a brief mapping of those definitions:

- **Learning** refers to the **iteration**, such as the one in (8.27).

- **Learning rate** is the **scale factor** r of the gradient in (8.27).

- **Epochs** are labelled by n in (8.27) and refer to one complete update of all elements of the parameter space θ.

- **Stochastic Steepest Descent** – unless in very special cases, we are not in the position to have available exact expectation values. We will need to approximate those using the law of large numbers (see section 1.2):

$$\langle A \rangle \approx \frac{1}{N_B} \sum_{k=1}^{N_B} A_k$$

Using approximations in lieu of exact expectation values adds a stochastic noise to the steepest descent procedure, hence, the name " Stochastic Steepest Descent".

- **Minibatch size** is the number N_B independent random variables A_k used in estimating the expectation values.

- **Contrastive Steepest Descent**: please note that the gradient, e.g. in (8.29) is the difference between two expectation values of the same observables but calculated from the exact theory and the Boltzmann machine. If both produce the same result, the gradient is zero and the " steepest descent" stops. On the other hand, if the data is vastly different from what the Boltzmann machine produces, gradients are large, and the iteration moves quickly through parameter space [100].

Using a " steepest descent" method for a function with presumably many local minima is less than ideal: if the learning rate is small, it is very likely that the iteration converges to a local minimum for which the Kullback–Leibler measure is significantly larger than zero. This means that the exact probability distribution is poorly represented by the Boltzmann machine. Finding the global minimum of a multivariate function is certainly an area of high importance that deserves more research.

8.4.3   The 2d Ising model showcase

Let us now study the performance of Learning for the 2D – Ising model of size 8×8:

$$\exp\left\{\beta \sum_{\langle ik \rangle} z_i z_k\right\} /Z(\beta) , \qquad z_k \in \{-1,1\} \quad \forall k.$$

We use periodic boundary conditions.

The visible layer hence has $n_v = 64$ units $\sigma_x \in \{0,1\}$ which are connected to data by:

$$z_x = 2\,\sigma_x - 1 .$$

We study several sizes n_h of the hidden layer as well as several values for β. To test the training, we choose the *action per site*:

$$A = \frac{1}{2L^2} \sum_{\langle ik \rangle} z_i z_k .$$

Learning – Robbins Monro method

Learning means to find the parameters θ of the effective theory – the Boltzmann machine in our case – so that produces probability distribution for the spins z_k for effective and true theory are as close as possible. As explained in section 8.4.2, we minimise the Kullback–Leibler measure D (8.26) using steepest descent (8.27).

Unfortunately, the best we can do is to estimate the gradient $\partial D / \partial \theta$ using Markov chain–Monte Carlo (MCMC). However, this implies that we won't eb able to obtain the optimal parameters θ to arbitrary precision due to stochastic noise.

To alleviate this issue, we will generalise to Learning rate r to one that depends on the iteration number n:

$$\theta_{n+1} = \theta_n - r_n \left. \frac{\partial D}{\partial \theta} \right|_{\theta=\theta_n} , \tag{8.31}$$

In fact, if the gradients are known to arbitrary precision, a clever choice (involving the second derivative of D) will improve the convergence rate to second order in the deviation from the exact solution, and the approach is the well-known Newton-Raphson method. Since gradients are tainted with stochastic noise, a second–order approach is of little help.

We can reinterpret (8.31) as an iteration that solves the stochastic equation

$$\frac{\partial D}{\partial \theta} = 0 .$$

We have encountered such a mathematical task before in section 5.5 where we needed to solve a stochastic equation for the LLR approach. As discussed the solution dates back to 1951 and the seminal work by Robbins and Monro [60]. In effect, decreasing the learning rate with increasing n eliminates the noise. Furthermore, it is guaranteed that the iteration converges to the true solution if:

$$\sum_{n=1}^{\infty} r_n \to \infty , \qquad \sum_{n=1}^{\infty} r_n^2 \to \text{finite} . \tag{8.32}$$

A kind of optimal choice is $r_n \propto 1/n$.

At the beginning of the iteration when the parameters θ are far from optimal, we do not want to under-relax to ensure rapid learning. Thus, we choose $r_n = \lambda$ for $n < n_{RM}$. Over the next n_{ur} iterations, we would like to suppress the learning rate by a factor. We choose

$$r_n = \frac{A}{1 + B(n - n_{RM})} , \qquad \text{for } n \geq n_{RM} .$$

With the choice that we made, we find

$$r_{n_{RM}} = 1 \qquad \Rightarrow \qquad A = \lambda$$

and

$$\frac{A}{1 + B n_{ur}} = \frac{\lambda}{f} \qquad \Rightarrow \qquad B = \frac{f - 1}{n_{ur}} ,$$

leaving us with

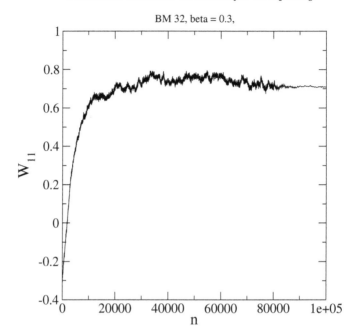

FIGURE 8.9: Training the Boltzmann machine ($n_h = 32$) for emulating a 8×8 Ising model at $\beta = 0.4$. The figure shows the parameter W_{11} as a function of the Learning step n.

$$
r_n = \begin{cases} \lambda & \text{for} \quad n < n_{RM}\ , \\ \frac{\lambda}{1+(f-1)\,(n-n_{RM})/n_{ur}} & \text{for} \quad n \geq n_{RM}\ . \end{cases} \tag{8.33}
$$

Figure 8.9 shows the evolution of the parameter W_{11} as a function of the learning index n. We have chosen

$$\lambda = 10^{-2}\ , \qquad n_{RM} = 80000\ , \qquad n_{ur} = 20000\ , \qquad f = 20\ .$$

Figure 8.9 shows the evolution of the parameter W_{11} under learning. We observe that, initially, there is a significant drift but then around $n \approx 50,000$ the parameter starts scattering around what appears an average value. We point out that the noise is significantly reduced for $n > 80,000$ when the Robbins-Monro under-relaxation sets in. We also have calculated the average Ising action per site for the true theory, A, and for the Boltzmann emulation, A_{BM}. We find

$$A = 0.361(3)\ , \qquad A_{BM} = 0.305(4)\ .$$

Learning – Hinton's approach

Standard Learning employs a steepest descent strategy (see (8.34) to minimise the Kullback–Leibler measure (8.21) to find the optimal parameter set

θ. The evaluation of the gradient involves the calculation of many convoluted functions using the chain rule, which lends the calculation its name: **backward error propagation**. The evolution of the machine for a given set of parameters and input is called **forward propagation**.

The iteration usually starts by choosing a random set $\theta^{(0)}$ from some prior distribution, e.g. Gaussian. As with the optimisation of many non-linear systems, the issue is that the iteration is attracted to a *local optimum* yielding a parameter set θ that does not represent the original theory well.

Methods for improving on this issue have been proliferating in the 80s in the context of statistical physics. One such method is *parallel tempering* also known as *replica exchange MCMC sampling* [102, 103, 104]. The central idea is to run the optimisation for several starting sets $\theta^{(1)} \ldots \theta^{(n)}$ in parallel and to come up with a strategy to use the information stored in all these simulations and to agree on the choice of the final set by some method.

An adaptation of this strategy in the context of machine Learning was put forward by Geoffrey Hinton in 2002 [100]. Hinton proposes to combine multiple probabilistic models of the same data by multiplying distributions together and then normalising. Then approach means that we have n replicas of the AI machine, each of them operating with their own parameter set $\theta^{(i)}$, $i = 1 \ldots n$.

There are many applications of Hinton's method freely available. Rather than applying it to the standard RBM for the Ising model, we here study the concept. Let us assume we have a target distribution for one random variable:

$$P_T(x) = \frac{1}{\sqrt{2\pi}\sigma} \exp\left\{-\frac{(x-\mu)^2}{2\sigma^2}\right\}. \tag{8.34}$$

We are seeking to represent this distribution by a "learning" model, which is a single Gaussian $P(x|\theta_1, \theta_2)$ with yet unknown means θ_1 and standard deviation θ_2. Rather than using the L_2 norm for the fitting process, we employ the Kullback–Leibler measure (8.21):

$$
\begin{aligned}
D &= \int dx \, P_T(x) \ln\left(\frac{P_T(x)}{P(x)}\right) = \int dx \, P_T(x) \left[\frac{(x-\theta_1)^2}{2\theta_2^2} - \frac{(x-\mu)^2}{2\sigma^2}\right] \\
&+ \ln\left(\frac{\theta_2}{\sigma}\right) = \frac{(\mu-\theta_1)^2 + \sigma^2 - \theta_2^2}{2\theta_2^2} + \ln\left(\frac{\theta_2}{\sigma}\right).
\end{aligned}
$$

If we take again the steepest descent method, we need to calculate the gradient:

$$\frac{\partial D}{\partial \theta_1} = \frac{\theta_1 - \mu}{\theta_2^2},$$

$$\frac{\partial D}{\partial \theta_2} = \frac{\theta_2^2 - \sigma^2 - (\mu-\theta_1)^2}{\theta_2^3}.$$

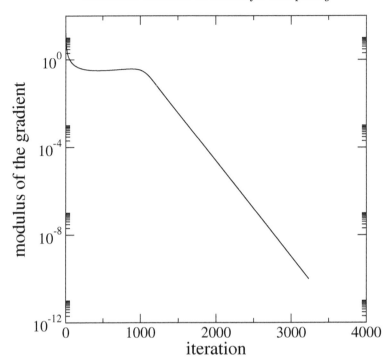

FIGURE 8.10: Convergence of "learning" for 1-Gaussian trial functions.

We can look for the parameters for which the gradient vanishes. The first equation necessarily yields $\theta_1 = \mu$ upon which the second equation uniquely provides $\theta_2 = \sigma$. We can also sthat if we use those values, the Kullback–Leibler measure vanishes, which means both distributions are equal.

The issue arises when the initial guess $P(x)$ is far off the true distribution $P_T(x)$. Let us consider $\sigma = 1$ and $\mu = 0$, and a starting distribution which has the correct width but is dislocated from the origin: $\theta_1 = 4$ and $\theta_2 = \sigma = 1$. Learning is an iterative process:

$$\theta_k^{(n+1)} = \theta_k^{(n)} - \lambda \frac{\partial D}{\partial \theta_k}, \qquad k = 1, 2.$$

Figure 8.10 shows the modulus of the gradient as a function of ten iteration "time" n for a fixed learning rate. The convergence is very slow until the θ_k is close enough to the solution when an exponential convergence sets in. This is a consequence of the Banach Fixed-Point theorem [105]. We here only outline the basic idea since it helps the understanding of convergence rates.

Assume that the function (of one variable x) is twice differentiable and has a minimum at $\bar{x}$. We have that at the minimum $f'' > 0$ and assume that there is an interval I around $\bar{x}$ with

$$0 < f''(\xi) < 2, \qquad \forall \xi \in I .$$

We then consider the iteration:

$$x_{n+1} = x_n - f'(x_n) .$$

Using Taylor's theorem, we find

$$f'(x_n) = f'(\bar{x}) + f''(\xi)(x_n - \bar{x}) = f''(\xi)(x_n - \bar{x}),$$

where we have used the minimum condition $f'(\bar{x}) = 0$.
We definded the error of the iteration as usual as $\epsilon_n := x_n - \bar{x}$ and find

$$\epsilon_{n+1} = \epsilon_n - f''(\xi)\,\epsilon_n = \left[1 - f''(\xi)\right]\epsilon_n$$

The inequality for the second derivative implies that there exists a $\kappa > 0$ with

$$|1 - f''(\xi)| \le \kappa < 1 ,$$

which leads us to

$$|\epsilon_{n+1}| = \left|1 - f''(\xi)\right| |\epsilon_n| \le \kappa |\epsilon_n|.$$

The solution of the latter equation is given by

$$|\epsilon_n| \le \kappa^n |\epsilon_0| ,$$

and, because of $\kappa < 1$, means an exponentially fast convergence to $\lim_{n \to \infty} \epsilon_n = 0$. Note the logarithmic scale of the y-axis in Figure 8.10 and "the linear appearance of the gradient for sufficiently large n when the iteration enters the interval I and exponential convergence sets in".

Let us now return to Hinton's original idea and explore whether we can represent bimodal data. A prototype case is a probability distribution that is an additive of two Gaussians:

$$P_T(x) = \frac{1}{\sqrt{2\pi}\sigma}\left[\exp\left\{-\frac{(x-\mu)^2}{2\sigma^2}\right\} + \exp\left\{-\frac{(x+\mu)^2}{2\sigma^2}\right\}\right]. \qquad (8.35)$$

The scattering plot of $10,000$ random variables generated with the above distribution, $\mu = 3$, $\sigma = 1$, is shown in Figure 8.11, black symbols. Hinton's idea is to use a product of the *same* probability distribution, yet with a different parameter set, to fit the scatter plot. If this distribution peaks in a certain domain, the product of those then can learn different elements of the data and has a chance for well representing the bimodal nature. Although intuitive, it

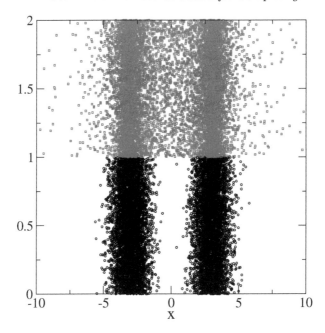

FIGURE 8.11: 10,000 events drawn from the bimodal distribution (8.35). The data are spread out along the y-axis in a strip between 0 and 1 for visualisation purposes. Also shown are events (red symbols) generated from an optimised product Lorentz distribution.

is challenging to make this idea work at a quantitative level. Let us choose the Gaussian distribution as an elementary ingredient:

$$p(x, \theta) = \frac{1}{\sqrt{2\pi}\sigma} \exp\left\{-\frac{(x-\mu)^2}{2\sigma^2}\right\} , \qquad \theta = (\mu, \sigma).$$

The product of two Gaussians, i.e.

$$p(x) = p(x, \theta_1)\, p(x, \theta_2) ,$$

is again a Gaussian, which fails to represent a bimodal distribution. In fact, the product distribution peaks at

$$x_{\max} = \frac{\sigma_1^2 \sigma_2^2}{\sigma_1^2 + \sigma_2^2} \left[\frac{\mu_1}{\sigma_1^2} + \frac{\mu_2}{\sigma_2^2}\right],$$

which is generically (for $\mu_1 \neq \mu_2$) is a region where neither constituting Gaussian has strong support. If we use instead of Gaussians a Lorentz distribution, i.e.

$$p(x, \theta) = \frac{1}{\pi b} \frac{1}{1 - (x-a)^2/b^2} , \qquad \theta = (a, b) ,$$

the product distribution is usually of bimodal type. Using optimisation and the Kullback–Leibler measure, we find the optimal parameter, representing 10,000 data points:

$$a_1 = 3.15(3), \qquad a_2 = -3.15(1), \qquad b_1 = 0.46(4), \qquad b_2 = 0.47(1) .$$

To check our results, we have then generated 10,000 data points using the product Lorentz distribution. The results are also shown in Figure 8.11 . We clearly observe a bimodal distribution. Note, however, that the Lorentz distribution has the tendency to spread further than the Gaussian. We say it has "heavy tails". This generates the effect that the events from the product Lorentz curve overestimate the abundance of points around $x = 0$ and for large $|x|$.

8.5   Categorisation in statistical physics: Naive Bayes

One of the primary tasks of Machine Learning (ML) is the categorisation of images: training a neural network with photos of buildings, the algorithm can "learn" to characterise new images according to classes such as "house", "factory" or "bungalow". *Naive Bayes* is probabilistic classification algorithm and falls into the class of so-called supervised learning algorithms. The algoritm creates a probability distribution for the classes. The class with with the highest probability distribution is most likely label for a given image.

In the case of *Naive Bayes*, it is assumed that the features of the image are independent: the contribution of a particular feature for predicting the class is not influenced by other features. The algorithm's cornerstone is the *conditional probability*: consider two events A and B; $P(A|B)$ is the probability that A happens under the condition that B has occurred. If we just looking at the outcome, and ask whether A and B took place, we have two ways to quantify the probability $P(A, B)$ for that

$$P(A, B) = P(A|B) \, P(B) \qquad \text{or} \qquad P(B|A) \, P(A) .$$

We are thus led to *Bayes theorem*:

$$P(A|B) = \frac{P(B|A) \, P(A)}{P(B)} . \qquad (8.36)$$

Let us consider a specific example; Figure 8.12 shows the images fo two Ising spin configurations on a $L = 64$ square lattice. We pretend that we do not know the parameter β with which those have been created. Can we infere the corresponding β value by just looking at the images?

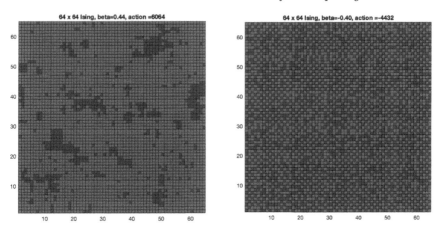

FIGURE 8.12: Two Ising configurations for a 64×64 lattice generated with a yet unknown β.

We are interested to estimate the conditional probability:

$$P\big(\beta|\text{image}\big)$$

is the desirable object and called the *posteriori*. $P(\beta)$ is called the *prior*, since it encodes our prior knowledge on the parameter set from which β came from. $P(image)$ is called *evidence*, since it is the probability fo what we are observing. Last but not least,

$$P(\{\sigma\}|\beta) \;=\; \frac{1}{Z(\beta)} \, \exp\Big\{-\beta \, S[\sigma]\Big\} \,, \qquad S[\sigma] \;=\; \sum_{\langle xy\rangle} \sigma_x \sigma_y \,, \qquad (8.37)$$

is the *likelihood*: it quantifies the probability that a particular image - represented by the set of spins $\{\sigma\}$ - is generated. In our case, the likelihood is provided by statistical physics and is the well known Gibbs factor. Note that $Z(\beta$ is the partition function and arises from the normalisation of the probability density:

$$\sum_{\{\sigma\}} P(\{\sigma\}|\beta) \;=\; 1 \,, \quad \Rightarrow \quad Z(\beta) \;=\; \sum_{\{\sigma\}} \exp\Big\{-\beta \, S[\sigma]\Big\} \,.$$

How can we apply this here? Someone has randomly chosen a parameter β from the distribution

$$p(\beta) \;=\; \frac{1}{2} \quad \text{for} \quad \beta \in [-1,1] \,, \qquad p(\beta) = 0 \ \text{else},$$

and used the heatbath algorithm to generate a sample Ising spin configuration, say the configuration at the left of figure 8.12. The prior would be in this case:

$$P(\beta) \;=\; p(\beta) \, d\beta \,.$$

Bayes theorem implies

$$P(\beta|\{\sigma\}) = \frac{P(\{\sigma\}|\beta)\, P(\beta)}{P(\text{image})} = q(\beta)\, d\beta\ , \qquad (8.38)$$

where $q(\beta)$ is the probability density of finding β in the interval $[\beta, \beta + d\beta]$ for a particular configuration $\{\sigma\}$. Proper normalisation leads us to

$$\sum_{\beta} P(\beta|\{\sigma\}) = 1\ , \qquad P(\text{image}) = \frac{1}{2}\int_{-1}^{1} d\beta\, \frac{1}{Z(\beta)}\, \exp\Big\{-\beta\, S[\sigma]\Big\}\ .$$

We observe that

$$q(\beta) = \frac{1}{2P(\text{image})}\, \frac{1}{Z(\beta)}\, \exp\Big\{-\beta\, S[\sigma]\Big\}\ , \qquad (8.39)$$

where, in practice, the normalisation constant can by always obtained by (numerically) calculating a 1-dimensional integral over β.

The biggest challenge in estimating $q(\beta)$ is the calculation of the partition function $Z(\beta)$. This is, however, a problem that we already solved in section 5.5 using the density of states $\rho(s)$ for the action s and the LLR algorithm with its remarkable property of exponential error suppression.

We proceed as follows:

1. Use the LLR algorithm to obtain an estimate for the LLR coefficients $a(s_k; \delta)$ for the set of actions $s_k = k\, \delta$.

2. Find a Taylor approximation for $a(s; \delta)$ (see subsection 7.2.2 for details).

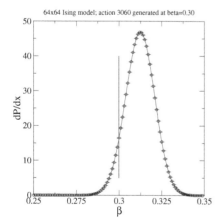 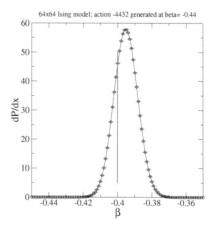

FIGURE 8.13: Naive Bayes: estimate of the posteriors as a function of β for the two Ising configurations in figure 8.12.

3. Reconstruct the density fo states by

$$\rho(s) = \exp\left\{\int_0^s ds' \, a(s'; \delta)\right\} .$$

4. Calculate the partition function (up to a β-independent constant by)

$$Z(\beta) = \text{const} \int ds \, \rho(s) \, \exp\{-\beta s\} .$$

5. Obtain the desirable density function $q(\beta)$ by

$$q(\beta) = \text{const} \, \frac{1}{Z(\beta)} \exp\left\{-\beta \, S[\sigma]\right\} .$$

6. Integrate $q(\beta)$ over β and fix the overall constant by $\int d\beta \, q(\beta) = 1$.

Simulation parameters have been (see table 7.1 for a reference)

$$n_{\text{ita}} = 20, \qquad n_{\text{RM}} = 10000, \qquad N_{\text{tay}} = 8, \qquad N_{\text{rept}} = 20,$$

$$\delta = 20, \qquad n_x = 255 .$$

We tested the approach for the two coinfigurations in figure 8.12. The left configuration has an action of $S[\sigma] = 3066$ while the right panel configuration $S[\sigma] = -4432$. The estimated probability distributions $q(\beta)$ are shown in figure 8.12. We observe a rather narrow distributions. Also shown are the actual β values (red line) with which the configurations have been generated.

8.6   Machine learning classification of phase transitions

A complex problem that lends itself to machine learning methods is the identification of phase transitions in statistical or quantum field theory. The traditional approach to this identification is "modelling": using the available knowledge of the system, the model tries to represent its key features – such as symmetries or interactions between degrees of freedom. Identifying the right degrees of freedom is of utmost importance. We then solve the theory at hand and eventually find out about phase transitions and other properties. On the plus side, if the model manages to reproduce observations to a great deal, the model delivers a wide spread "understanding" rather than only information about a phase transition. On the native side, we might not always be in a position to judge whether our model can grasp the essence of the model. If not, the model answers might produce flawed results without us noticing.

Unsupervised machine learning method offers a powerful complementary approach. Its main advantages are that in order to deploy unsupervised machine learning prior knowledge of the underpinning theory is not needed and that it can be applied to artificial as well as observed data. A disadvantage is that, even if the machine learning algorithms correctly identify features of the phase transition, they do not easily offer a further understanding of the underlying theory.

8.6.1 Principal Component Analysis – PCA

Principal component analysis (PCA) is a statistical technique used to identify patterns in data by transforming the original variables into a new set of variables called principal components. These new variables are a linear combination of the original variables, and they capture the maximum amount of variation present in the data.

PCA is often used to reduce the dimensionality of a large dataset. In other words, if you have a dataset with many variables, PCA can help you identify the most important variables that explain most of the variance in the data, and discard the least important ones.

The first principal component is the linear combination of variables that accounts for the largest amount of variance in the data. The second principal component is the linear combination of variables that accounts for the second-largest amount of variance, subject to the constraint that it is uncorrelated with the first principal component. Subsequent principal components are similarly defined, subject to the constraint that they are uncorrelated with all previous principal components.

PCA is often used in data visualisation and exploration, as it can help to identify relationships and trends in the data that may not be apparent when looking at the original variables. It can also be used as a preprocessing step for machine learning algorithms, as it can help to reduce the complexity of the data and improve the performance of the algorithm.

Let us consider N data points in an n-dimensional vector space, i.e. $\vec{x}_i \in \mathbb{R}^n$, $i = 1..N$. An example is shown in Figure 8.14. The data points are generated with the joint probability distribution

$$P(x, y) \propto \exp\left\{ -\frac{(x + y - 2)^2}{4} - \frac{(x - y + 1)^2}{1} \right\}. \tag{8.40}$$

The variables x and y are *not* independent. Indeed a scatter plot clearly shows a correlation between the data points. In order to understand better the distribution, we perform a rotation in variable space and define:

$$u = \frac{1}{\sqrt{2}}(x + y), \qquad v = \frac{1}{\sqrt{2}}(x - y).$$

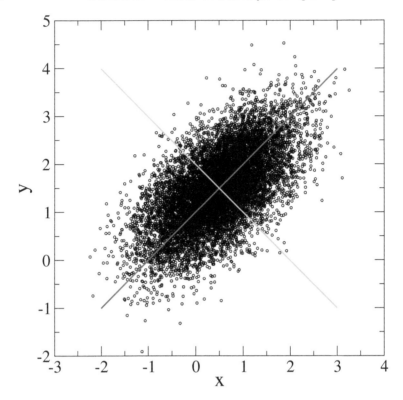

FIGURE 8.14: Scatter plot of $10,000$ Gaussian distributed points (x, y).

It turns out that u and v are independent variables with the (product) distribution:

$$P(u, v) \propto \exp\left\{-\frac{(u - 2\sqrt{2})^2}{2}\right\} \exp\left\{-\frac{(v + \sqrt{2})^2}{1/2}\right\}.$$

Both variables have the variance

$$\sigma_u^2 = 1, \qquad \sigma_v^2 = 1/4.$$

This means that u labels the direction of greatest variance. The u-axis is found by setting $v = 0$ leaving us with $y = x$. Hence, the direction of the first principal component is $(1, 1)^T$ with a variance of 1.

How do we find the principal components if we only have sample data at hand rather than access to the full probability distribution?

To answer this question, we are now looking for the direction $\vec{n}$, $\vec{n}^2 = 1$, for which data have the biggest average distance if projected onto his direction. Hence, data projected onto the line

$$\vec{x}(t) = \vec{n}t + \vec{b}, \qquad \vec{n}^2 = 1, \qquad \vec{n}, \vec{b} \in \mathbb{R}^n$$

have the biggest average spread. Let u_i be the distance of the projected point $\vec{x}_i$:

$$u_i = \left(\vec{x}_i - \vec{b}\right)\vec{n}.$$

The averaged distance D of all points is then derived from

$$D^2 = \frac{1}{N}\sum_{i=1}^{N} u_i^2$$

We are led to maximise

$$S := \frac{1}{N}\sum_{i=1}^{N}\left(\left(\vec{x}_i - \vec{b}\right)\cdot\vec{n}\right)^2 - \lambda\vec{n}^2 \longrightarrow \max,$$

where λ is a Lagrange multiplicator to ensure that $\vec{n}^2 = 1$. We find the equations that determine $\vec{b}$ and $\vec{n}$:

$$C\vec{n} = \lambda\vec{n}, \qquad C = \frac{1}{N}\sum_{i=1}^{N}\left(\vec{x}_i - \vec{b}\right)\left(\vec{x}_i - \vec{b}\right)^T, \tag{8.41}$$

$$\vec{n}\left(\vec{b} - \frac{1}{N}\sum_{i=1}^{N}\vec{x}_i\right) = 0. \tag{8.42}$$

The matrix C is a $n \times n$ symmetric matrix and therefore has n eigenvectors $\vec{e}_1 \ldots \vec{e}_n$ with eigenvalues $\lambda_1 \ldots \lambda_n$. We solve these equations by identifying $\vec{b}$ with the "centre of mass":

$$\vec{b} = \frac{1}{N}\sum_{i=1}^{N}\vec{x}_i, \tag{8.43}$$

which makes C the covariance matrix. We then choose a eigenvector for $\vec{n}$, say $\vec{n} = \vec{e}_k$. The corresponding distance is given by

$$D = \sqrt{\lambda_k}.$$

Hence, the first principal component is found by choosing the eigenvector with the largest eigenvalue.

Let us come back to the $n = 2$ example in Figure 8.14. We find for the $10,000$ points listed:

$$\vec{b} = \begin{pmatrix} 0.4970359 \\ 1.4951350 \end{pmatrix}, \qquad C = \begin{pmatrix} 0.6210747 & 0.3748418 \\ 0.3748418 & 0.61980520 \end{pmatrix}.$$

For a simple Gaussian distribution such as (8.40), the covariance matrix can be calculated exactly in the limit $N \to \infty$

$$C_\infty = \begin{pmatrix} 5/8 & 3/8 \\ 3/8 & 5/8 \end{pmatrix} = \begin{pmatrix} 0.625 & 0.375 \\ 0.375 & 0.625 \end{pmatrix},$$

which is close to our estimate by sampling over the $10,000$ data points. The eigenvectors with corresponding eigenvalues are

$$\vec{n}_1 = \frac{1}{1.41541} \begin{pmatrix} 1.00169 \\ 1 \end{pmatrix}, \qquad \lambda_1 = 0.995282,$$

$$\vec{n}_2 = \frac{1}{1.41302} \begin{pmatrix} -0.998308 \\ 1 \end{pmatrix}, \qquad \lambda_2 = 0.245598.$$

This is a remarkable result. By only studying the sample data, we found a good estimate for $\vec{n}_1$, the first principal component including the variance, i.e., λ_1 is this direction. First (red) and second (green)principal components are also shown in Figure 8.14 alongside the data points.

8.6.2   The Ising phase transition from unsupervised learning

Unsupervised learning is a type of machine learning in which the algorithm is trained on unlabelled data, meaning the data does not come with prede-termined categories or labels. Instead, the algorithm must identify patterns, structures, or relationships in the data without any prior knowledge of what those patterns or relationships might be.

The algorithm is given a dataset and asked to identify any underlying structure or organisation that may exist in the data. Clustering is a common example of unsupervised learning where the algorithm is tasked with grouping similar data points together into clusters. Another example is dimensionality reduc-tion, where the algorithm is tasked with reducing the number of features or variables in the dataset while retaining as much information as possible. This can be useful for visualising complex data or for improving the efficiency of machine learning algorithms. Widespread applications are exploratory data analysis, where the goal is to gain insights into the structure of the data and to identify any patterns or trends that may be present, and anomaly detec-tion. In the latter case, the task is to identify data points that are significantly different from the rest of the dataset.

PCA (see previous subsection) is a powerful tool for unsupervised learning, allowing us to identify and remove noise and redundancy in the data, reduce the number of dimensions in the dataset, and improve the performance of other unsupervised learning algorithms. In this section, we will feed Ising spin configurations to a PCA algorithm and explore whether we can learn some-thing about the Ising phase transition just by analysing the configurations and obtain information about the underpinning probability distribution.

The trusted Ising model (see section 4.4) has served us many times. We just review the basic definitions. Configurations are spins $\sigma = \pm 1$ associated with the sites of a regular square lattice of size $n = L \times L$. We use periodic boundary conditions in this section throughout. If $\langle xy \rangle$ denotes two sites x and y that are nearest neighbours on the lattice, the probabilistic measure is given

$$P = \frac{1}{Z(\beta)} \exp\left\{ \beta \sum_{x,y \in \langle xy \rangle} \sigma_x \sigma_y \right\},$$

where $\beta = 1/T$ is the inverse temperature and Z is the celebrated partition function, which we studied in great detail in section 7.2. We use a Wolff cluster algorithm (see section 6.1.2), adapted to the Ising model, to generate high quality independent spin configurations for several values of β.

In the course of this section, we will "forget" about how these configurations are generated, and will try to "learn" as much as possible about the Ising model. The previous section the configurations have been two-dimensional vectors $\vec{x}_i$, where i labels the sample. In the present case, the vectors $\vec{x}_i \in \mathbb{R}^n$, $n = L \times L$ are given by Ising configurations and only have entries ± 1. Each of these vectors can be visualised in the usual way on a $L \times L$ lattice. For the stack of $i = 1 \ldots N$ samples, we use 30 values of β, i.e.

$$\beta \in \{0.31, 0.32, \ldots, 0.59, 0.6\},$$

and a 1000 configurations for each β, implying $N = 30 \times 1000 = 30,000$. With these definitions, we now can follow the PCA outlined in the previous subsection: we calculate the covariance matrix C (8.41), which is an $n \times n$ matrix ($n = L \times L$). Upon diagonalisation, we find the eigenvalue spectrum, which is shown in Figure 8.15 for the two lattice sizes $L = 10$ and $L = 40$. We find

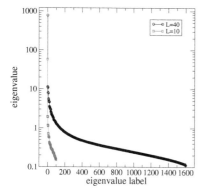
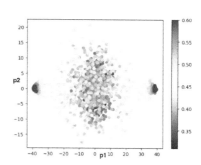

FIGURE 8.15: Eigenvalues of a PCA for the Ising model of siz $L \times L$ (left). Lattice ising configurations projected to the 1st and 2nd principal component ($L = 40$).

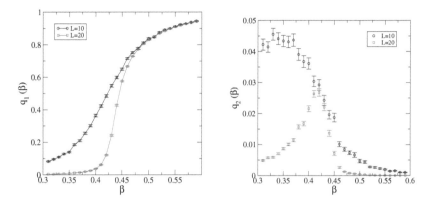

FIGURE 8.16: Averaged 1st and 2nd principal components q_1 and q_2 (see (8.44)) as a function of the inverse temperature β ($L = 40$).

that there is a significant gap between the largest and the remains eigenvalues. This stirs the hope that already the projection to the first principal component would reveal key information about the model. As usual, projection to the kth principal component is defined by:

$$p_i^{(k)} = \vec{x}_i \vec{e}_k \,, \qquad (8.44)$$

where $\vec{e}_k$ is kth eigenvector of the covariance matrix C. Each configuration gives rise to a point in the $p_1 - p_2$ diagram in Figure 8.15, right panel. The colour coding of the point is given by β value with which the configuration was generated. We can clearly see that "hot" configurations, i.e. those with low β, cluster around the origin of the diagram, whereas "cold" configuration (high β) have a double peak structure in p_1.

Note that the ensemble averages of $p_i^{(1)}$ and $p_i^{(2)}$ vanish. A meaningful quantity is the ensemble average of their squares, i.e. $q_{1,2}$:

$$q_k := \frac{1}{|S_\beta|} \sum_{i \in S_\beta} \left(p_i^{(k)} \right)^2 / L^2 \,, \qquad (8.45)$$

where S_β is the set of labels of configurations generated at a specified value β, and $|S_\beta|$ the number of its elements. In both variables, we realise the phase transition of the assigning model at around $\beta \approx 0.44$. The signal is more pronounced if we increase the lattice size from $L = 10$ to $L = 40$.

Bibliography

[1] Kurt Langfeld. Dynamics of epidemic diseases without guaranteed immunity. *Journal of Mathematics in Industry*, 11(1):5, 2021.

[2] Shin ichi Tadaki, Macoto Kikuchi, Minoru Fukui, Akihiro Nakayama, Katsuhiro Nishinari, Akihiro Shibata, Yuki Sugiyama, Taturu Yosida, and Satoshi Yukawa. Phase transition in traffic jam experiment on a circuit. *New Journal of Physics*, 15(10):103034, oct 2013.

[3] Andrey N. Kolmogorov. *Foundations of the Theory of Probability*. Chelsea Pub Co., 2 edition, June 1960.

[4] Makoto Matsumoto and Takuji Nishimura. Mersenne twister: A 623-dimensionally equidistributed uniform pseudo-random number generator. *ACM Transactions on Modeling and Computer Simulation*, 8(1):3–30, jan 1998.

[5] F. James. Ranlux: A fortran implementation of the high-quality pseudorandom number generator of lüscher. *Computer Physics Communications*, 79(1):111–114, 1994.

[6] Martin Luscher. A Portable high quality random number generator for lattice field theory simulations. *Computer Physics Communications*, 79:100–110, 1994.

[7] G. E. P. Box and Mervin E. Müller. A Note on the Generation of Random Normal Deviates. *Annals of Mathematical Statistics (1958), 29(2):610–611*, 1958.

[8] Paula Rooney. Microsoft's CEO: 80-20 Rule Applies To Bugs, Not Just Features. *CRN, The Cannel Co.*, October 2002.

[9] Xavier Gabaix. Power Laws in Economics and Finance. *Annual Review of Economics 1(1):255–294*, 2009.

[10] Benjamin Moll. Inequality and Financial Development: A Power-Law Kuznets Curve. *Princton University Working Paper*, 2012.

[11] Albert Einstein. Investigations on the Theory of the Brownian Movement. *Dover Publications. Retrieved 2013-12-25*, 1956.

[12] L. Hufnagel, D. Brockmann, and T. Geisel. Forecast and control of epidemics in a globalized world. *Proceedings of the National Academy of Sciences*, 101:15124–15129, 2004.

[13] F. Bartumeus, J. Catalan, U. L. Fulco, M. L. Lyra, and G. M. Viswanathan. Optimizing the Encounter Rate in Biological Interactions: Levy versus Brownian Strategies. *Physical Review Letters*, 89:109902, 2002.

[14] Olivier Ibe. *Markov Processes for Stochastic Modeling.* Elsevier; 2 edition (3 Jun. 2013), 2013.

[15] L. D. Landau and E. M. Lifshitz. *Statistical Physics, Part 1 (3rd ed.).* Pergamon Press., 1980.

[16] Hannes Risken. *The Fokker-Planck Equation.* Springer Series in Synergetics, 1996.

[17] Georgio Parisi and Yongshi Wu. Stochastic quantization. *Scientia Sinica*, 24:483, 1981.

[18] Poul H. Damgaard and Helmuth Hffel. Stochastic quantization. *Physics Reports*, 152(5):227–398, 1987.

[19] Steve Brooks, Andrew Gelman, Galin Jones, and Xiao-Li Meng. *Handbook of Markov Chain Monte Carlo.* CRC press, 2011.

[20] David Landau and Kurt Binder. *A Guide to Monte Carlo Simulations in Statistical Physics.* Cambridge University Press, USA, 2005.

[21] Nicholas Metropolis, Arianna W. Rosenbluth, Marshall N. Rosenbluth, Augusta H. Teller, and Edward Teller. Equation of state calculations by fast computing machines. *The Journal of Chemical Physics*, 21(6):1087–1092, 1953.

[22] W. K. Hastings. Monte Carlo sampling methods using Markov chains and their applications. *Biometrika*, 57(1):97–109, 04 1970.

[23] B. Efron. Bootstrap Methods: Another Look at the Jackknife. *The Annals of Statistics*, 7(1):1–26, 1979.

[24] Bradley Efron and Robert J. Tibshirani. *An Introduction to the Bootstrap.* Number 57 in Monographs on Statistics and Applied Probability. Chapman & Hall/CRC, Boca Raton, Florida, USA, 1993.

[25] J. W. Tukey. Bias and Confidence in Not-Quite Large Sample. *The Annals of Mathematical Statistics*, 29(2):614–623, 1958.

[26] B. Efron and C. Stein. The Jackknife Estimate of Variance. *The Annals of Statistics*, 9(3):586–596, 1981.

[27] Bradley Efron. *The jackknife, the bootstrap, and other resampling plans*, volume 38. Siam, 1982.

[28] Lars Onsager. Crystal statistics. i. a two-dimensional model with an order-disorder transition. *Physical Review*, 65:117–149, Feb 1944.

[29] R. Peierls. On ising's model of ferromagnetism. *Mathematical Proceedings of the Cambridge Philosophical Society*, 32(3):477–481, 1936.

[30] K. Huang. *Statistical Mechanics, 2nd Ed.* Wiley India Pvt. Limited, 2008.

[31] P. M. Chaikin and T. C. Lubensky. *Principles of Condensed Matter Physics*. Cambridge University Press, 2000.

[32] M. N. Barber. In *Phase transitions and critical phenomena. Vol.8/* Edited by C. Domb and J. L. Lebowitz. Academic, London, 1983.

[33] Andrea Pelissetto and Ettore Vicari. Critical phenomena and renormalization group theory. *Physics Reports*, 368:549–727, 2002.

[34] F. Y. Wu. The potts model. *Reviews of Modern Physics*, 54:235–268, Jan 1982.

[35] Simon Duane, A. D. Kennedy, Brian J. Pendleton, and Duncan Roweth. Hybrid monte carlo. *Physics Letters B*, 195(2):216–222, 1987.

[36] I. P. Omelyan, I. M. Mryglod, and R. Folk. Optimized verlet-like algorithms for molecular dynamics simulations. *Physical Review E*, 65(5), May 2002.

[37] Tetsuya Takaishi and Philippe de Forcrand. Testing and tuning symplectic integrators for the hybrid monte carlo algorithm in lattice qcd. *Physical Review E*, 73(3): 03670, Mar 2006.

[38] Jesús A. Izaguirre and Scott S. Hampton. Shadow hybrid monte carlo: an efficient propagator in phase space of macromolecules. *Journal of Computational Physics*, 200(2):581–604, 2004.

[39] Ben Calderhead Mark Girolami. Riemann manifold langevin and hamiltonian monte carlo methods. *Statistical Methodology, Series B*, 73(2):123–214, 2011.

[40] Simon Duane, Richard Kenway, Brian J. Pendleton, and Duncan Roweth. Acceleration of Gauge Field Dynamics. *Physics Letters B*, 176:143, 1986.

[41] Tuan Nguyen, Peter Boyle, Norman H. Christ, Yong-Chull Jang, and Chulwoo Jung. Riemannian Manifold Hybrid Monte Carlo in Lattice QCD. *PoS*, LATTICE2021:582, 2022.

[42] Jascha Sohl-Dickstein, Mayur Mudigonda, and Michael Deweese. Hamiltonian monte carlo without detailed balance. *Proceedings of the 31st International Conference on Machine Learning, 2014*, 32, 09 2014.

[43] Robert H. Swendsen and Jian-Sheng Wang. Nonuniversal critical dynamics in monte carlo simulations. *Physical Review Letters*, 58:86–88, Jan 1987.

[44] Ulli Wolff. Comparison Between Cluster Monte Carlo Algorithms in the Ising Model. *Physics Letters*, B228:379–382, 1989.

[45] Shailesh Chandrasekharan. A New computational approach to lattice quantum field theories. *PoS*, LATTICE2008:003, 2008.

[46] Kurt Langfeld. Phase diagram of the quantum O(2)-model in 2+1 dimensions. *Physical Review*, D87(11):114504, 2013.

[47] Christof Gattringer. Flux representation of an effective Polyakov loop model for QCD thermodynamics. *Nuclear Physics*, B850:242–252, 2011.

[48] Christof Gattringer and Carlotta Marchis. Abelian color cycles: a new approach to strong coupling expansion and dual representations for non-abelian lattice gauge theory. *Nuclear Physics*, B916:627–646, 2017.

[49] Michael Creutz. Microcanonical monte carlo simulation. *Physical Review Letters*, 50:1411–1414, May 1983.

[50] Jan Westergren and Sture Nordholm. Melting of palladium clusters – density of states determination by monte carlo simulation. *Chemical Physics*, 290(2):189–209, 2003.

[51] A. M. Ferrenberg and R. H. Swendsen. New Monte Carlo Technique for Studying Phase Transitions. *Physical Review Letters*, 61:2635–2638, 1988.

[52] Bernd A. Berg and Thomas Neuhaus. Multicanonical algorithms for first order phase transitions. *Physical Letters B*, 267:249–253, 1991.

[53] B. A. Berg and T. Neuhaus. Multicanonical ensemble: A New approach to simulate first order phase transitions. *Physical Review Letters*, 68:9–12, 1992.

[54] Fugao Wang and D. P. Landau. Efficient, Multiple-Range Random Walk Algorithm to Calculate the Density of States. *Physical Review Letters*, 86(10):2050, 2001.

[55] R. E. Belardinelli and V. D. Pereyra. Wang-Landau algorithm: A theoretical analysis of the saturation of the error. *The Journal of Chemical Physics*, 127(18):184105, nov 2007.

[56] P. Dayal, S. Trebst, S. Wessel, D. Würtz, M. Troyer, S. Sabhapandit, and S. N. Coppersmith. Performance Limitations of Flat-Histogram Methods. *Physical Review Letters*, 92(9):097201, March 2004.

[57] Thomas Neuhaus and Johannes S. Hager. 2D Crystal Shapes, Droplet Condensation, and Exponential Slowing Down in Simulations of First-Order Phase Transitions. *Journal of Statistical Physics 2003 113:1*, 113(1):47–83, 2003.

[58] Kurt Langfeld, Biagio Lucini, and Antonio Rago. The density of states in gauge theories. *Physical Review Letters*, 109:111601, 2012.

[59] Kurt Langfeld, Biagio Lucini, Roberto Pellegrini, and Antonio Rago. An efficient algorithm for numerical computations of continuous densities of states. *The European Physical Journal C*, 76(6):306, 2016.

[60] Herbert Robbins and Sutton Monro. A Stochastic Approximation Method. *The Annals of Mathematical Statistics*, 22(3):400–407, 1951.

[61] Julius R. Blum. Approximation Methods which Converge with Probability one. *The Annals of Mathematical Statistics*, 25(2):382–386, 1954.

[62] Harold J. Kushner and G. George Yin. *Stochastic Approximation Algorithms and Applications*. Springer, 1997.

[63] K. L. Chung. On a Stochastic Approximation Method. *The Annals of Mathematical Statistics*, 25(3):463–483, 1954.

[64] Christof Gattringer and Pascal Törek. Density of states method for the z3 spin model. *Physics Letters B*, 747:545–550, 2015.

[65] Christof Gattringer, Mario Giuliani, Alexander Lehmann, and Pascal Törek. Density of states techniques for lattice field theories using the functional fit approach (FFA). *POS*, LATTICE2015:194, 2016.

[66] Christof Gattringer, Michael Mandl, and Pascal Törek. New density of states approaches to finite density lattice qcd. *Physical Review D*, 100:114517, Dec 2019.

[67] K. H. Bennemann and J. B. Ketterson, editors. *G. Ahlers, The Pysics of Liquid and Solid Helium*, volume 2, chapter II. Wiley, New York, 1976.

[68] S. Kamal, D. A. Bonn, Nigel Goldenfeld, P. J. Hirschfeld, Ruixing Liang, and W. N. Hardy. Penetration depth measurements of 3d XY critical behavior in $yba_2cu_3o_{6.95}$ crystals. *Physical Review Letters*, 73:1845–1848, Sep 1994.

[69] Peter Grüter, David Ceperley, and Frank Laloë. Critical temperature of bose-einstein condensation of hard-sphere gases. *Physical Review Letters*, 79:3549–3552, Nov 1997.

[70] Michael E. Peskin; Daniel V. Schroeder. *An Introduction to Quantum Field Theory.* Avalon Publishing, 1995.

[71] Ulli Wolff. Collective Monte Carlo Updating for Spin Systems. *Physical Review Letters*, 62:361, 1989.

[72] Aloysius P. Gottlob and Martin Hasenbusch. Critical behaviour of the 3d xy-model: a monte carlo study. *Physica A: Statistical Mechanics and its Applications*, 201(4):593–613, 1993.

[73] Michael Creutz. *Quarks, Gluons and Lattices.* Oxford University Press, 1983.

[74] Heinz J. Rothe. *Lattice Gauge Theories: An Introduction (Fourth Edition)*, volume 43. World Scientific Publishing Company, 2012.

[75] I. Montvay and G. Munster. *Quantum fields on a lattice.* Cambridge Monographs on Mathematical Physics. Cambridge University Press, 3 1997.

[76] Christof Gattringer and Christian B. Lang. *Quantum chromodynamics on the lattice*, volume 788. Springer, Berlin, 2010.

[77] Robert Savit. Duality in Field Theory and Statistical Systems. *Reviews of Modern Physics*, 52:453, 1980.

[78] Chen-Ning Yang and Robert L. Mills. Conservation of Isotopic Spin and Isotopic Gauge Invariance. *Physical Review*, 96:191–195, 1954.

[79] Kenneth G. Wilson. Confinement of Quarks. *Physical Review D*, 10:2445–2459, 1974.

[80] M. Creutz. Monte Carlo Study of Quantized SU(2) Gauge Theory. *Physical Review D*, 21:2308–2315, 1980.

[81] A. D. Kennedy and B. J. Pendleton. Improved Heat Bath Method for Monte Carlo Calculations in Lattice Gauge Theories. *Physics Letters B*, 156:393–399, 1985.

[82] Stephen L. Adler. Over-relaxation method for the monte carlo evaluation of the partition function for multiquadratic actions. *Physical Review D*, 23:2901–2904, Jun 1981.

[83] F R Brown and T J Woch. Overrelaxed heat-bath and metropolis algorithms for accelerating pure gauge monte carlo calculations. *Physical Review Letters, (United States)*, 58:23, 6 1987.

[84] Roberto Petronzio and Ettore Vicari. An overheat bath algorithm for lattice gauge theories. *Physics Letters B*, 254:444–448, 01 1991.

[85] Michael Creutz. Overrelaxation and Monte Carlo Simulation. *Physical Review D*, 36:515, 1987.

[86] Phillipe de Forcrand and Oliver Jahn. Monte Carlo overrelaxation for SU(N) gauge theories. In *3rd International Workshop on Numerical Analysis and Lattice QCD*, pages 67–73, 3 2005.

[87] E. Pietarinen. String Tension in SU(3) Lattice Gauge Theory. *Nuclear Physics B*, 190:349, 1981.

[88] N. Cabibbo and E. Marinari. A New Method for Updating SU(N) Matrices in Computer Simulations of Gauge Theories. *Physics Letters*, B119:387–390, 1982.

[89] David J. Gross and Frank Wilczek. Ultraviolet Behavior of Nonabelian Gauge Theories. *Physical Review Letters*, 30:1343–1346, 1973.

[90] H. David Politzer. Reliable Perturbative Results for Strong Interactions? *Physical Review Letters*, 30:1346–1349, 1973.

[91] Claudio Bonati and Massimo D'Elia. Topological critical slowing down: variations on a toy model. *Physical Review E*, 98(1):013308, 2018.

[92] Christof Gattringer and Kurt Langfeld. Approaches to the sign problem in lattice field theory. *International Journal of Modern Physics A*, 31(22):1643007, 2016.

[93] J. P. Valleau and D. N. Card. Monte carlo estimation of the free energy by multistage sampling. *The Journal of Chemical Physics*, 57(12):5457–5462, 1972.

[94] R. M. Dudley. Sample Functions of the Gaussian Process. *The Annals of Probability*, 1(1):66–103, 1973.

[95] M. B. Marcus and L. A. Shepp. Continuity of gaussian processes. *Transactions of the American Mathematical Society*, 151(2):377–391, 1970.

[96] Karin R. Saoub. *A tour through graph theory*. Textbooks in mathematics. Taylor & Francis, Boca Raton, 2018.

[97] J. E. Besag. Nearest-neighbour systems and the auto-logistic model for binary data. *Journal of the Royal Statistical Society. Series B (Methodological)*, 34(1):75–83, 1972.

[98] David Sherrington and Scott Kirkpatrick. Solvable model of a spin-glass. *Physical Review Letters*, 35:1792–1796, Dec 1975.

[99] S. Kullback and R. A. Leibler. On Information and Sufficiency. *The Annals of Mathematical Statistics*, 22(1):79–86, 1951.

[100] Geoffrey E. Hinton. Training Products of Experts by Minimizing Contrastive Divergence. *Neural Computation*, 14(8):1771–1800, 08 2002.

[101] Leon Bottou. *Stochastic Gradient Descent Tricks*, volume 7700 of *Lecture Notes in Computer Science (LNCS)*, pages 430–445. Springer, neural networks, tricks of the trade, reloaded edition, January 2012.

[102] Robert H. Swendsen and Jian-Sheng Wang. Replica monte carlo simulation of spin-glasses. *Physical Review Letters*, 57:2607–2609, Nov 1986.

[103] E. Marinari and G. Parisi. Simulated tempering: A new monte carlo scheme. *Europhysics Letters*, 19(6):451, jul 1992.

[104] Koji Hukushima and Koji Nemoto. Exchange monte carlo method and application to spin glass simulations. *Journal of the Physical Society of Japan*, 65(6):1604–1608, 1996.

[105] Praveen Agarwal, Mohamed Jleli, and Bessem Samet. *Banach Contraction Principle and Applications*, pages 1–23. Springer Singapore, Singapore, 2018.

Index

For Product Safety Concerns and Information please contact our EU
representative GPSR@taylorandfrancis.com
Taylor & Francis Verlag GmbH, Kaufingerstraße 24, 80331 München, Germany